Materials from Natural Sources

This book comprises interdisciplinary topics including consolidating research activities in all experimental and theoretical aspects of natural advanced materials in the fields of science, engineering, and medicine including structure, synthesis, processing, physico-chemical properties, and applications.

The book covers formulation for natural and organic cosmetics, CO_2 sequestration, drug delivery systems, biosensors, and other related topics. In addition, case studies including some data are presented that demonstrate the growing interest and cost benefits of biomaterials.

The main features of the book are:

1. It discusses the properties and applications of materials from natural sources from an environmental and biomedical engineering perspective.
2. It explores properties of natural materials used in drug delivery, biofuel, adsorbents, waste water treatment, cosmetics, and so forth.
3. It includes studies on recent and emerging biomaterials.
4. It details applications of natural materials in varied areas.
5. It introduces the structure and mechanisms of self-healing material, biomaterial, biofuel, polymers, and adsorbents.

This book is aimed at graduate students and researchers in chemical, environmental, and natural resource engineering.

Emerging Materials and Technologies

Series Editor: Boris I. Kharissov

The *Emerging Materials and Technologies* series is devoted to highlighting publications centered on emerging advanced materials and novel technologies. Attention is paid to those newly discovered or applied materials with potential to solve pressing societal problems and improve quality of life, corresponding to environmental protection, medicine, communications, energy, transportation, advanced manufacturing, and related areas.

The series takes into account that, under present strong demands for energy, material, and cost savings, as well as heavy contamination problems and worldwide pandemic conditions, the area of emerging materials and related scalable technologies is a highly interdisciplinary field, with the need for researchers, professionals, and academics across the spectrum of engineering and technological disciplines. The main objective of this book series is to attract more attention to these materials and technologies and invite conversation among the international R&D community.

Nanofluids
Fundamentals, Applications, and Challenges
Shriram S. Sonawane and Parag P. Thakur

MXenes
From Research to Emerging Applications
Edited by Subhendu Chakroborty

Biodegradable Polymers, Blends and Biocomposites
Trends and Applications
Edited by A. Arun, Kunyu Zhang, Sudhakar Muniyasamy, and Rathinam Raja

Bioinspired Materials and Metamaterials
A New Look at the Materials Science
Edward Bormashenko

Computational Studies
From Molecules to Materials
Edited by Ambrish Kumar Srivastava

2D Semiconductors for Environmental Remediation
Edited by Honey John, Nisha T Padmanabhan, Sona Stanly, and Jith C Janardhanan

Materials from Natural Sources
Structure, Properties, and Applications
Edited by Ramesh Gardas, Neha Patni, and Amita Chaudhary

For more information about this series, please visit: www.routledge.com/Emerging-Materials-and-Technologies/book-series/CRCEMT

Materials from Natural Sources

Structure, Properties, and Applications

Edited by Ramesh Gardas,
Neha Patni, and Amita Chaudhary

CRC Press
Taylor & Francis Group
Boca Raton London New York

CRC Press is an imprint of the
Taylor & Francis Group, an **informa** business

First edition published 2025
by CRC Press
2385 NW Executive Center Drive, Suite 320, Boca Raton FL 33431

and by CRC Press
4 Park Square, Milton Park, Abingdon, Oxon, OX14 4RN

CRC Press is an imprint of Taylor & Francis Group, LLC

ISBN: 9781032538761 (hbk)
ISBN: 9781032636344 (pbk)
ISBN: 9781032636368 (ebk)

DOI: 10.1201/9781032636368

Typeset in Times
by codeMantra

Contents

Preface

Today, the rate of industrial upgrading and material replacement is growing, the scientific and technology revolution is expanding quickly, and new material products are changing every day. Any item or substance that originates from plants, animals, or the Earth is said to be made of natural materials. In the domain of engineering, using natural materials becomes essential. Academicians, entrepreneurs, engineers, and researchers from a variety of professions find these materials to be incredibly intriguing.

Natural resources and their uses in a variety of fields, including wastewater treatment, biosensors, electrochemical processes, adsorbents, and biofuels, among others, comprise a wide range of applications. The purpose of writing this book is to offer a thorough application of materials based on natural resources and review the most recent research and viewpoints from various fields of applications.

The chapter on adsorbent materials describes different natural and synthetic adsorbents, viz. activated carbon, bauxite, alumina, silica gel, biomass, biochar, cellulosic materials, activated carbon nanomaterial, metal organic frameworks, resins, natural and synthetic clays, zeolites, aerogels, and synthetic polymers used for the extraction and removal of a variety of hazardous materials in wastewater produced by many industries, and provides a cost-effective alternative method.

Recent reports on fibre-based materials comprising activated carbon, nanofibres, and modified natural fibres used as adsorbents to remove heavy metals and dyes are discussed in the next chapter. The most commonly used adsorbents for wastewater purification are described, and their preparation method and selective metal or metal or dye remotion are discussed in detail.

Biomaterials are derived from either nature or synthesized material used in the elimination of contaminants passively from treatment plants. The chapter "Biomaterial for Wastewater Treatment" covers the use of biomaterial as biosorbents in treatment technology as it is one of the cheaper, viable, cleaner, and safer technologies to be adopted for removing toxic pollutants.

Various biomaterials containing starch, sucrose, cellulose, and lignocellulose have been explored for the production of bioethanol. Structurally, biomaterials possess polysaccharide networks; it is challenging to break down and exhibit nonavailability of the fermentable sugars to microorganisms. Biomaterials having a complex structure pose a challenge during the conversion of biomaterials into simple sugars, thus making the production of bioethanol commercially non-competitive. The chapter "Biomaterials to Bioethanol: A Practical Approach" presents a practical approach on different strategies and technologies to be developed for the conversion of biomaterials for releasing sugar.

Nowadays biosensors have become a powerful analytical device that converts physical or biological responses into assessable signals. The chapter on biosensors includes its basic introduction, working principle, and its applications. It mainly contains biological detectors, transducers, and signal processors.

Currently, the development of biomaterials with highly controlled functions and characteristics is vital to improve their applications in different areas of research and knowledge. The chapter "Biomaterials in Electrochemical Applications" analyses the main applications of biomaterials in electrochemistry related to the fields of medicine, environment, and energy. Also, the most relevant fundamentals of biosensors and electrochemistry are presented that allow to relate the significant elements in applications.

The chapter on electrochemical sequestration of CO_2 by using natural materials comprises materials, surfaces, provides a comprehensive review on associated tuning of material properties to result in better design of CO_2 sequestration. Further, emphasis was given to the strategies of doping and composite assembly in order to improve the catalytic properties in CO_2 sequestration approaches.

The chapter on materials for drug delivery contains the basic and preliminary aspects of drug delivery systems (DDSs) and different classifications and the need for developing them. The mechanism of drug loading and release together with the need of such delivery systems for different diseases have been described. Additionally, theoretical aspects of the modelling of the drug delivery mechanism have been demonstrated.

Biological surfaces have the ability to retain their liquid repellent behaviour throughout life via biological self-repair and restoration process. Combination of surface chemistry and surface topology has encouraged academicians and scientists to fabricate materials with liquid-repellent and self-cleaning applications. In the chapter "Self-Healing and Self-Cleaning Nature-Inspired Surfaces," the surface wettability of natural surfaces and artificial biomimetic surfaces/materials is highlighted. The underlying mechanism of self-healing and self-cleaning behaviour of various inspired artificial surfaces are also discussed.

A chapter on the understanding of natural materials in cosmetics is also included. Various active ingredients, natural preservatives, and biodegradable and biocompatible surfactants and their formulation for natural and organic cosmetics are discussed in detail.

Moreover, the central issues of sustainability that affect the globe are food loss and food waste. To recreate a sustainable system, there is a need to reorganize the value cycle. The chapter "Role of Upcycled Foods for Global Food Security" addresses the issue by a systematic review of the involvement of various companies to improve on reduction of food waste and then generate gaps that can be addressed for further research especially with regard to the long-term impact of investing in upcycling.

Where and when it is technologically viable, natural materials can replace traditional sources. With that aim authors present this book, which emphasizes fundamentals as a basis for understanding biomaterials and their applications.

We would like to extend our gratitude to all of the authors and collaborators who have shown a keen interest in this multidisciplinary book. This book will definitely aid in greater understanding between disparate scientific disciplines.

Editors

Dr. Ramesh Laxminarayan Gardas is a distinguished researcher and educator in the field of Chemistry. With a PhD in Chemistry from Veer Narmad South Gujarat University, and postdoctoral research experience at prestigious institutions like the University of Coimbra, University of Aveiro (Portugal), and Queen's University Belfast (UK), his expertise lies in ionic liquids' physico-chemical properties and applications. In 2010, he joined the Indian Institute of Technology, Madras, as an Assistant Professor and rose through the ranks to become a full Professor in 2020. His work focuses on 'Chemical Thermodynamics' and 'Phase Equilibria,' primarily exploring eco-friendly solvent systems. Recognized for his contributions, Dr. Gardas serves as Associate Editor of the *Journal of Chemical & Engineering Data,* is a Fellow of the Royal Society of Chemistry, UK, and ranks among the "World's Top 2% Scientists." His impact on the scientific community through research, teaching, and leadership is substantial.

Dr. Neha Patni is working as Assistant Professor (Chemistry) in Nirma University. She has more than 17 years of experience in the field of teaching and research. She obtained her PhD in the area of Electrochemical Applications of Polymers. She has authored two books and ten book chapters published by publication houses of international repute. She has also authored 16 papers published in international referred journals and presented/published more than 50 papers in international conferences/ proceedings. She has been conferred with the prestigious Award of Best Assistant Professor of Chemical Engineering Department for her Teaching Methodology in 2015 and Research Consultancy Awards for publishing quality research paper for the year 2020 from Nirma University. She has also won the Best Research Paper Award six times in various international conferences. The projects guided by her have received various national awards like the Earthian Face Off Challenge, First Position – Technical Essay Competition, Outstanding Young Chemical Engineer Award. She is a life member of several organizations like the Chemical Research Society of India (CRSI), the Indian Society of Technical Education (ISTE), the Association of Chemistry Teachers (ACT), etc. She has been appointed as a reviewer of international referred journals in the area of Polymers, Industrial Chemistry, Energy, and the Environment. She has organised various national and international conferences, workshops, symposiums and webinars. She has also completed five minor research projects under internal research grants. Being actively involved in R & D activities, her areas of expertise are Polymer Science and Technology, Chemical Sensors, Organic Solar Cells, etc.

Dr. Amita Chaudhary is currently working as an Assistant Professor (Chemistry) in the Department of Chemical Engineering, Institute of Technology, Nirma University. She has completed her doctoral studies from the Chemical Engineering Department at Indian Institute of Technology, Delhi. She has published a number of papers in the areas of Ionic Liquid Synthesis, CO_2 Sequestration, Corrosion Science, and Waste Management, etc., in international and national referred journals. She has also presented/published her research work in many international and national conferences/proceedings. She is actively involved in research and is highly passionate about R&D of new technologies. She has edited one book on *Advance Materials and Technologies for Wastewater Treatment* by CRC Press and six book chapters published by publication houses of international repute. She is certified with JRF-NET and GATE in Chemical Sciences. She has won the best paper presentation award at various conferences. She has a published patent on the synthesis of ionic liquid for carbon capture in collaboration with IIT Delhi and FITT, Delhi. She is actively involved in students' Idea Lab projects based on interdisciplinary research and minor funded projects. She is a member of various prestigious organisations like ISTE, IIChE.

Contributors

Madhu Bala
Department of Chemistry, Dr. B.R.
Ambedkar National Institute of
Technology, Jalandhar 144011,
Punjab, India

Dilip Kumar Behara
Department of Chemical Engineering,
JNTUA College of Engineering
Anantapur, Ananthapuramu

L.T. Carvalho
Department of Chemical Engineering,
Engineering School of Lorena,
University of São Paulo, 12602-810
Lorena, SP, Brazil

A. Karthik Chandan
Department of Chemical Engineering,
JNTUA College of Engineering
Anantapur, Ananthapuramu

Amita Chaudhary
Chemical Engineering Department,
Institute of Technology, Nirma
University, Ahmedabad, India

A.I.C. da Silva
Department of Chemistry and
Environmental, State University of
Rio de Janeiro (UERJ), Resende,
Brazil, CEP

Jesús E. Diosa
Grupos de Transiciones de Fase y
Materiales Funcionales (GTFMF),
Departamento de Física, and Centro
de Excelencia en Nuevos Materiales
(CENM), Universidad del Valle,
Ciudad Universitaria Meléndez-A.A.
23360, Cali, Colombia

Ramesh L. Gardas
Department of Chemistry, Indian
Institute of Technology Madras,
Chennai 600 036, India

Shikha Indoria
Department of Chemistry, School of
Chemical Engineering and Physical
Sciences, Lovely Professional
University, Jalandhar 144411, India

Jennyfer Diaz-Angulo
Laboratorio de simulación y
procesos-SIMPROLAB, TV 13 CL
27 BRR Paraiso, Turbaco, Colombia

Keshava Joshi
Department of Chemical Engineering,
SDM College of Engineering and
Technology (V.T.U., Belagavi),
Dharwad 580 002, Karnataka, India

Sandesh K.
NITTE (Deemed to be University),
NMAM Institute of Technology
(NMAMIT), Department of
Biotechnology Engineering, Nitte,
Karnataka 574110, India

Mehul Khimani
Countryside International School,
Nr. Bhesan Railway Crossing, CIS
Barbodhan Road, Surat 394540,
Gujarat, India

C. Uday Kumar
Department of Chemical Engineering,
JNTUA College of Engineering
Anantapur, Ananthapuramu

Nadavala Siva Kumar
Department of Chemical Engineering,
    King Saud University, P.O. Box 800,
    Riyadh 11421, Saudi Arabia

Jose A. Lara-Ramos
Escuela de Ingeniería Química,
    Universidad del Valle, Ciudad
    Universitaria Meléndez-A.A. 23360,
    Cali, Colombia

Fiderman Machuca-Martinez
Escuela de Ingeniería Química,
    Universidad del Valle, Ciudad
    Universitaria Meléndez-A.A. 23360,
    Cali, Colombia

P. Uma Maheshwari
Department of Chemical Engineering,
    JNTUA College of Engineering
    Anantapur, Ananthapuramu

L.S. Maia
Department of Chemistry and
    Environmental, State University of
    Rio de Janeiro (UERJ), Resende,
    Brazil, CEP

S.F. Medeiros
Department of Chemical Engineering,
    Engineering School of Lorena,
    University of São Paulo, 12602-810
    Lorena, SP, Brazil

Neha Mistry
Countryside International School,
    Nr. Bhesan Railway Crossing, CIS
    Barbodhan Road, Surat 394540,
    Gujarat, India

Edgar Mosquera-Vargas
Grupos de Transiciones de Fase y
    Materiales Funcionales (GTFMF),
    Departamento de Física, and Centro
    de Excelencia en Nuevos Materiales
    (CENM), Universidad del Valle,
    Ciudad Universitaria Meléndez-A.A.
    23360, Cali, Colombia

Nabisab Mujawar Mubarak
Department of Petroleum and Chemical
    Engineering, Faculty of Engineering,
    Universiti Teknologi Brunei, Bandar
    Seri Begawan BE1410, Brunei
    Darussalam
Department of Biosciences, Saveetha
    School of Engineering, Saveetha
    Institute of Medical and Technical
    Sciences, Chennai, India

D.R. Mulinari
Department of Mechanic and Energy,
    State University of Rio de Janeiro
    (UERJ), Resende, Brazil, CEP
    27537-000

Parth Naik
Department of Chemistry, UkaTarsadia
    University, Bardoli, Gujarat, India

Lokeshwari Navalgund
Department of Chemical Engineering,
    SDM College of Engineering and
    Technology (V.T.U., Belagavi),
    Dharwad 580 002, Karnataka, India

Ujwal. P
NITTE (Deemed to be University),
    NMAM Institute of Technology
    (NMAMIT), Department of
    Biotechnology Engineering, Nitte,
    Karnataka 574110, India

Paresh Parekh
Department of Chemistry, Veer Narmad
South Gujarat University, Surat
395007, Gujarat, India

Misha Patel
Research and Development Manager,
Beans and Hominy of Teasdale Latin
Foods, USA

Vijay Patel
Navyug Science College, Rander Road,
Surat 395009, Surat, Gujarat, India

Neha Patni
Chemical Engineering Department,
Institute of Technology, Nirma
University, Ahmedabad, India

P.H.F. Pereira
Department of Material and
Technology, State University of
São Paulo, UNESP, 12516-410
Guaratinguetá, SP, Brazil

R. Kiran Kumar Reddy
Smart Materials Research and Device
Technology (SMaRDT) Group,
Department of Pure and Applied
Chemistry, University of Strathclyde,
Thomas Graham Building, Glasgow,
G1 1XL, United Kingdom.

D.S. Rosa
Center for Engineering, Modeling,
and Applied Social Sciences
(CECS), Federal University of ABC
(UFABC), Santo André, Brazil

Kamalika Sen
Department of Chemistry, University
of Calcutta, 92, APC Road, Kolkata
700009, India

Priti Sengupta
Department of Chemistry, University
of Calcutta, 92, APC Road, Kolkata
700009, India

Shobhana Sharma
S.S. Jain Subodh P.G. College, Jaipur,
India

Vinayaka B. Shet
NITTE (Deemed to be University),
NMAM Institute of Technology
(NMAMIT), Department of
Biotechnology Engineering, Nitte,
Karnataka 574110, India

Vickramjeet Singh
Department of Chemistry, Dr. B.R.
Ambedkar National Institute of
Technology, Jalandhar 144011,
Punjab, India

Dharmesh Varade
School of Science, Navrachana
University, Vasna Bhayali Road,
Vadodara 391410, Gujarat, India

Rohit Vekariya
Organic Chemistry Department,
Institute of Science & Technology
for Advanced Studies & Research
(ISTAR), CVM University, Vallabh
Vidyanagar 388 120, Gujarat, India

Yan-Ling Yang
Department of Chemical and Materials
Engineering, Tamkang University,
New Taipei City 25137, Taiwan,
ROC

N.C. Zanini
Center for Engineering, Modeling,
and Applied Social Sciences
(CECS), Federal University of ABC
(UFABC), Santo André, Brazil

1 Introduction to Materials from Natural Sources and Their Applications

Neha Patni and Amita Chaudhary

1.1 INTRODUCTION

Nature is abundant in intelligent materials that can take the place of synthetic and conventional materials in a variety of applications. Natural materials extracted from various natural sources, such as plants or animals, have been the subject of numerous studies. It has been discovered that these natural products have clever and distinctive qualities that frequently outperform those of their synthetic counterparts. Such materials are becoming more and more popular due to their environmental friendliness and sustainability. In general, all materials that are a part of living things are considered to be natural products. Some natural products can also be partially or entirely synthesised chemically. In addition to the food industry, commercially available natural products are now used in pharmaceuticals, cosmetics, wastewater treatment, biosensors, textiles, and fabrics. Recent research has focused on developing a novel class of tissue adhesives that utilise a network of positively charged proteins and polyanionic glycosaminoglycans. It was found in the gel of the snail mucus. Through a number of interactions, the large, pliable sticky matrix may cling to moist tissue. The biomaterial efficiently accelerates the healing of full-thickness skin wounds in both healthy and diabetic individuals and has high haemostatic activity, biocompatibility, and biodegradability. It is highly desirable, but challenging to develop structural materials that are environmentally friendly and have superior mechanical and thermal properties. Considering that it is entirely made of natural raw materials (cellulose nanofibre and mica micro platelets) and outperforms plastics made from petroleum in terms of mechanical and thermal properties, the bioinspired structural material is a high-performance and environmentally friendly alternative structural material to replace plastics. Large-scale applications for natural products include the synthesis of other natural products, pharmaceutical research, the food industry, agricultural science, cosmetics, perfumes, essential oils, and fuel cells and a few of them are discussed in the entire book. This introduction chapter will provide its overview in brief.

1.2 NATURAL PRODUCTS IN DRUG DISCOVERY

Natural product-based drug discovery has the potential to be revitalised in both established and developing areas. The primary source of new drugs for infectious

DOI: 10.1201/9781032636368-1

TABLE 1.1

Mechanism of natural extracts for different fatal diseases

S. no.	Name	Activity	Molecular pathway	References
1	Paclitaxel	Anticancer	Promotes tubulin polymerisation and blocks mitosis	[1]
2	Latanoprost	Antiglaucoma	Stimulates the muscarinic receptors of the ciliary muscle	[2]
3	Epothilone A & epothilone B	Cancer treatment especially taxane-sensitive tumour types (such as breast, lung, and prostate cancers)	By inducing tubulin polymerisation, microtubule stabilisation, cell cycle arrest and apoptosis	[3]
4	Vinblastine	Leukaemia, lymphoma, breast and lung cancer	Inhibits the polymerisation of tubulin and causes arrest of cell division in their metaphase	[4]
5	Digoxin	Cardiac disease	Provides symptomatic relief by the functioning of the Na/K^+ ATPase pump	[5]
6	Vincristine	Anticancer	Blocks DNA-dependent RNA polymerase	[6]
7	Aspirin	Analgesic, anti-pyretic, anti-inflammatory, and anti-thrombotic	Irreversibly binds and inhibits cyclooxygenase enzymes	[7]
8	Reserpine	Antihypertensive	Inhibition of the ATP/Mg^{2+} pump	[8]
9	Curcumin	Anti-inflammatory	Upregulates the activation of peroxisome proliferator-activated receptor-γ	[9]
10	Withanolides	Immunity booster against coronavirus-19 and anti-viral	Modulated proteins and the docking score to boost the immune system	[10]
11	Prostratin	Anti-HIV	By downregulating HIV-1 cellular receptors through the activation of the protein kinase C (PKC) pathway and reducing the HIV-1 latency	[11]

diseases, particularly antibiotics, has long been natural products. Natural products have a favourable record for cancer treatments. Given the current intense interest in methods that could boost response rates to immune checkpoint inhibitors by making "cold" tumours "hot," the ability of some natural products to trigger a selective potent host in an immune reaction against cancer cells represents a significant new opportunity in this field. For example, natural compounds like cardiac glycosides can increase the immunogenicity of cancer cells that are stressed and about to die. These cells can be cured by inducing immunogenic cell death, which is characterised by the release of damage-associated molecular patterns (DAMPs), which may open up new therapeutic options. The details of different natural compounds that help to cure various diseases are tabulated in Table 1.1.

1.3 NATURAL FIBRES IN WASTEWATER TREATMENT

There are numerous methods for treating various agricultural waste types. Since cellulose, hemicellulose, and lignin all have a hydroxyl functional group, they are the primary components of agricultural waste and natural fibre and have the potential to bind a variety of pollutants. However, the process for treating wastewater differs depending on the type of effluent. Based on the type of effluent, the methods can be ionic, chemical, or physical. Heavy metals are the most hazardous pollutants that can be found in effluents from industrial wastewater [12]. Since these heavy metals are very soluble in water, marine organisms can readily consume them. Natural fibres typically require surface modification in order to serve as an adsorbent.

1.4 NATURAL FIBRES IN COSMETICS

The global market for cosmetics has surpassed US$800 billion by 2023 at a growth rate of roughly 7% per year. The cosmetics sector has experienced some of the fastest growth over the previous ten years. East Asian customers are developing a greater understanding of skincare, and thus the concept of healthy skin is beginning to take shape. More and more Korean women and even men are chasing glamour since the "value economy" emerged a few decades ago. As a result, the cosmetics industry is growing, and the East Asian cosmetics market is experiencing growth that has never been seen before. Data indicates that US$60 billion was spent on cosmetics retail in 2021, an increase of 14% from the same period the year before. By 2050, Korean and China are predicted to have the largest consumer cosmetics market in the world, worth roughly US$450 billion [13].

The Greek word "kosmtikos," which means "capable of arrangement, skilled in decoration," is where the English word "cosmetic" originates. Cosmetics are typically applied directly to the exterior surfaces of the human body, which improves appearance, keeps skin healthy, shields from dirt and UV rays, and gets rid of body odour [14]. Shampoos, deodorants, soaps, skin- and haircare items, lipsticks, perfumes, etc., are some examples.

1.5 NATURAL PRODUCTS AS SKIN WHITENING AGENTS

The skin is a vital organ that serves as the first line of defence against external threats. It acts as a protective barrier, shielding us from harmful biotic and abiotic factors that could otherwise harm our bodies. Additionally, the skin plays a crucial role in regulating our body temperature, helping to keep us cool or warm depending on the conditions we are exposed to. However, despite its importance, the skin is also vulnerable to damage caused by environmental exposure. This can result in a range of issues such as ageing, hormone disorders, inflammation, and various skin diseases. In particular, skin pigmentation can occur due to exposure to UV radiation from the sun or other sources, which can cause uneven colouring and dark spots on the skin. To maintain healthy skin and prevent these issues from occurring, it is important to take care of your skin by protecting it from excessive sun exposure and using appropriate skincare products that help nourish and protect your skin from environmental damage. Melanin synthesis and deposit in the epidermis are the reasons for pigmentation. The abundantly found hydroquinone has a phenolic group that aids in lowering the amount of melanin. Thus, it is a common ingredient in cosmetic formulations. Another key ingredient in skin pigmentation treatments is abutin. D-glucose and hydroquinone are the main ingredients. Plants like bearberry contain β-arbutin, which is hydroquinone bound to the β-isomer of D-glucose [15]. One of the most well-known naturally occurring tyrosinase inhibitors is aloesin, a hydroxychromone glucoside. It was separated from the aloe vera plant. Kojic acid is a naturally occurring organic acid that is produced as a by-product by specific fungi species, including Aspergillus and Penicillium [16]. By snatching up copper ions in the active site of tyrosinase, it inhibits the enzyme.

1.6 NATURAL PRODUCTS AS SKIN ANTI-AGEING AGENTS

The skin is composed of three layers, the outer layer of epidermis, the middle layer of dermis, and the underneath or subcutaneous layer. The epidermis gradually thins with age, and the skin's natural capacity to heal itself declines. Additionally, the number of melanocytes declines as we age, resulting in skin that is thinner, paler, and clearer with larger liver spots or age spots. The herb licorice (*Glycyrrhiza glabra*), which is grown all over the world, has long been a popular traditional medicine. Specific substances isolated from licorice include glycyrrhizic acid and glycyrrhetinic acid. The dipotassium salt of glycyrrhizic acid, dipotassium glycyrrhizinate, is a commonly used anti-inflammatory drug [17]. The primary plant pigment delphinidin, which also has antioxidant properties, gives some fruits and flowers their blue colour. According to reports, anthocyanins in their aglycone form effectively prevent UVB-induced skin damage. It has been recommended as an anti-pollution ingredient.

1.7 NATURAL MATERIAL IN SENSORS

Natural materials are vital contenders because of their inherited bioactivity, inherent biocompatibility, decent solution processability, and important flexibility [18]. Any device which can collect the chemical information from the concentration of biomaterials and convert it into signals of importance is termed a biosensor [19, 20]. Biosensors, their general introduction, their working, and their applications are discussed in detail in Figure 1.1.

Biosensors consist of various elements, as shown in Figure 1.1. Sensing elements (or receptors) are used to attach to the specific testing sample of the analyte. An interface offers a working atmosphere for biosensor elements [21]. A transducer converts the physical or chemical information received from the interaction between the sensing elements and the analyte into electrical signals [22]. Moreover, a series of

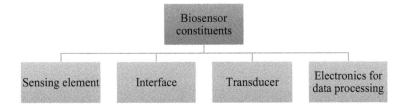

FIGURE 1.1 Constituents of biosensing devices.

electronic equipment including signal amplification, signal processing, and interface circuit are also required for analysing and processing data [23]. Biosensors have wide applications in the fields of medicines, food safety, process control, environmental monitoring, and so on [24].

1.7.1 MATERIALS OF BIOSENSOR INTERFACES

Generally, biosensor interfaces are made up of nanomaterials, polymers, or a metal organic framework (MOF), as illustrated in Figure 1.2. Numerous nanomaterials and structures are mostly utilised in biosensor interfaces like nanospheres [25], nanoporous structures [26], nanotubes [27], and nanowires [28]. Graphene and graphene-based nanomaterials are mostly made of carbon-based materials, due to their high specific surface that offers a display place for biomolecule loading and high conductance, which quickens the transfer of electrons between the biomolecules and the surface [29].

Polymers have two major roles in biosensors. Conducting polymers are utilised for coating or encapsulating materials on the electrode surface, and nonconducting polymers are utilised to immobilise specific receptors on the sensor device [30, 31]. Currently, several polymers have found applications in biosensors like chitosan, agarose, polyethylene glycol, hydrogel, and polyelectrolyte [32].

Use of chitosan in biosensors is important since it is decently biocompatible, adhesive, and able to form an outstanding film that provides strong binding force for proteins and enzymes [33]. Cellulose, on the other hand, comprises glucose-based polymer chains, and is the key constituent of plant cell walls [34]. It has exclusive features such as high transparency, decent dimensional stability, and easy modifiability [35]. It can be further utilised to detect a variety of biomolecules, such as urea, lactic acid, glucose, genes, amino acids, cholesterol, and proteins. Generally, when conducting polymers are used in biosensors, they respond to the different chemicals when exposed and show some physical transformation which can be easily sensed. That is an added advantage [36, 37]. Ammam et al. reported a glutamate microbial sensor based on polypyrrole (PPy), multiwalled carbon nanotubes (MWCNTs), and glutamate oxidase (GluOx) [38].

Metal ions and organic ligands are allied together by strong coordination bonds in the metal organic framework (MOF) [39]. Several kinds of MOFs have been found to be useful to detect DNA, RNA, enzymes, and small molecules when applied in clinical diagnosis such as bioimaging [40].

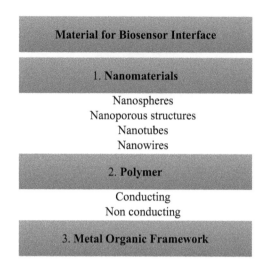

FIGURE 1.2   Various materials used as interface in biosensors.

1.7.2 Structures and Preparation Methods of Stable Biosensor Interfaces

Studies on monolayer membranes have mainly focused on the diverse preparation techniques, as mentioned in Figure 1.3.

The self-assembled monolayer technique (SAM) involves the formation of a monolayer by the chemical adsorption of molecules at the liquid–solid interface. Normally the monolayer is prepared by simply immersing an appropriate solid support into a solution containing molecules which can self-assemble. The Langmuir–Blodgett Technique is another option of self-assembly where the Langmuir monolayer (LM) is formed at the gas–liquid interface. The LM can be relocated to a solid substrate to form a Langmuir–Blodgett (LB) membrane. In this procedure, molecular organisation leads to adsorption [41]. The process of layer-by-layer (LbL) deposition is used to create thin films. The films are created by alternately depositing layers of oppositely charged elements and interspersing them with wash steps. Several methods, including immersion, spin, spray, electromagnetic, and fluid mechanics, can be used to achieve this. Furthermore, in comparison to the two-dimensional structures, three-dimensional structures might deliver better structural stability. Also, for the past 20 years, liquid crystal (LC)-based biosensors have been utilised for unlabelled sensing of biomolecules [19, 42].

1.7.3 Advances and Applications of Nanophotonic Biosensors

The chapter on biomaterials for electrochemical applications incorporates the main applications of biomaterials in electrochemistry related to the fields of medicine, environment, energy. Significant progress has been made over decades in the usage of biosensors, specifically in synthetic biology. Their performance is excellent in detecting the target metabolites, assessing the metabolic processes, and developing rate-limiting enzymes [43, 44]. However, due to their limited dynamic range and

| Monolayer Membrane Techniques | Langmuir–Blodgett (LB) technique Self-assembly monolayer (SAM) technique |

FIGURE 1.3 Techniques of monolayer membrane.

instability, problems with regard to their industrial applications still persist. Such emerging, new and advanced techniques could empower the speedy growth of biosensors with artistic instructions, further inaugurating the evolving field to novel boundaries [45].

1.7.4 NANO-BIOSENSORS AND NANOMATERIALS FOR POLLUTANT DETECTION AND REMEDIATION

Electrochemical-based biosensors have been utilised in the screening of a few poisons, the substance build of interest, and, surprisingly, the recognition of biomolecules connected with illnesses. With the advent of nano-science, the utilisation of nanomaterials in electrochemical biosensors has contributed towards unrivalled conductivity, quick reaction time, and, surprisingly, greater responsiveness [46].

Articles, mixtures, and living beings encounter recognition receptors and catch the analyte based on arrangement. When this happens, the receptor responds by discharging a simple sign due to this response (which can be magnetic, optical, electrochemical, mechanical, or thermal). The readout framework and programming that convert the simple sign into a computerised signal are quantitatively estimated for the portrayal of pollutants [47]. Nanomaterial-based bio-electrochemical sensors offer extra benefits in contrast with the previously existing non-convenient types of location gadgets [46].

The fusing of polyaniline in an electrochemical biosensor has expanded the awareness of biosensors. Because of its superior intercession in the exchange of electrons in either enzymatic or redox responses, it has improved responsiveness and steadiness, and reversibility of redox; moreover, it is easy to combine as per the thickness of polyaniline to be put on the tangible electrode. It has applications in food business, toxicology examination of the climate as well as pathogenic or biochemical examination in the field of medical care [48].

Electrochemical biosensors for clinical diagnostics have been created alignment free because of intermittent adjustment expanding administrator blunder and in some cases even exactness; this is accomplished by a double recurrence strategy where a square wave voltammogram is being produced – this technique for approach enjoys a 20% benefit among the powerful ranges with a restriction of two significant degrees and in this technique undiluted blood can be utilised in a straightforward manner for examination [49]. Execution of nanoparticles (NPs) in biosensors has opened new pathways for recognition and examination and the advancement of novel nanocomposites has been displayed to work on the biosensor in general.

Nanomaterials based nano-biosensors have shown common traits such as lower recognition limits, quicker reaction time, cheap manufacturing costs, and the utilisation of nanomaterials for transportability in contrast with customary identification techniques.

Nano-biosensors and nanofiltration permit the presence of blended contaminations (poisons, synthetic substances, heavy metals) even in minute amounts.

- Execution of nano-biosensors properly avoid utilisation of polluted marine items which could result in illnesses, messes, and so forth.
- Nanoparticles from the nanosensors reaching to the climate brings a higher worry up in the utilisation of nanomaterials in the location and bioremediation of blended marine contamination.
- Marine toxins impact the strength of marine vegetation, causing different sicknesses and deformities, thereby interrupting the marine biological system [45].

1.7.5 ELECTROCHEMICAL SENSORS FOR DETERMINING ANTIOXIDANT ACTIVITY

Novel sensors have been designed for their use in nanobiology and the food sector. For that emphasis has been laid on the rapid and highly sensitive methods where less amount of sample is required along with simple and economic instruments, in order to optimise the usage of research resources [50, 51]. Moving further electrochemical sensors were fabricated for determining antioxidants using several sorts of electrodes, transducers, and receptors. To improvise their performance by aggravating their sensitivity, stability, and selectivity, nanomaterials were also integrated [52]. In order to rapidly prove the antioxidant activity of several organic compounds, electrochemical methods were utilised. For quick comparison of the antioxidant power of numerous compounds like phenols, flavonoids, cinnamic acids, and tannins, etc., the oxidation potentials were measured using cyclic voltammetry (CV), often using a glassy carbon electrode (GCE) [53].

Biosensors stand out for Salmonella recognition as they offer colossal possibilities in the field of food examination. In any case, they are not yet fit to be used as a standard logical device in food enterprises in light of the fact that the innovation is in its early stages. The significant hindrances in improving electrochemical sensors are as follows: (1) accomplishing a low constraint of detection (LOD); (2) invalidating vague adsorption of meddling species; (3) on-location discovery of tests, or transformation of a sensor into a point of-care gadget; (4) safeguarding the sensor's dependability and repeatability in confounded genuine networks [54]. Moreover, when the analyte is a microscopic organism, separation among living and dead cells is a significant test. Electrochemical sensors have shown significant promise as an indicative tool [55].

The development of low-cost detection and analysis techniques for bioanalytes as well as infection biomarkers became an increasingly essential task during and after the COVID-19 pandemic in order to enhance public and individual health. The development of biomaterials and tools to succeed in reaching biosensing goals indicates this difficulty. At the beginning, biological components are in charge of detecting analytes by altering the recognition component's optical or electrochemical properties. Second, the devices must integrate a detector element in order to read and transform one signal into another. Thus, the engineering of biosensors frequently makes use of optical, electrochemical, and spectrofluorometric detector elements [56].

1.8 NATURAL MATERIALS IN THE PRODUCTION OF BIOETHANOL

The chapter on biomaterials to bioethanol (a practical approach) discusses different strategies and technologies under development for the conversion of biomaterials. The continual and rapid depletion of natural resources, a key contributor to global warming, is driving up the need for alternative fuels like biofuels, which are made from lignocellulosic biomass. In order to make biofuels, notably bioethanol, which can be used as a replacement, more affordable and sustainable, forest products, energy crops, agricultural by-products, and municipal solid waste are employed as raw materials. The majority of research has been done on second-generation bioethanol, and the chapter also elaborates on lignocellulosic biomass as having a key role in producing second-generation bioethanol. The biochemical conversion of raw material into bioethanol involves the process of pretreatment followed by hydrolysis and fermentation. Future research aims to manufacture third- and fourth-generation bioethanol from plants and algae, respectively, via genetic engineering. This chapter illustrates and emphasises the crucial function of lignocellulosic material in ethanol production as well as its future significance [57].

The negative side effects and quick depletion of fossil fuels encourage the use of renewable energy sources, such as biofuels [45, 58]. Although first-generation biofuels offer a better sustainability balance in the energy matrix, this balance is still insufficient. The bioethanol sector, for instance, continues to rely on agricultural products like maize and sugarcane, which sequester vast tracts of fertile land that would otherwise be used for food production [59].

However, various wastes that produce lignocellulosic biomass, such as fruit and vegetable waste, can be used to produce second-generation biofuels. The use of these wastes can produce minimal carbon emissions because they are abundant and sustainable [60]. In supermarkets, customers frequently reject fruits and vegetables that have surface flaws, resulting in wastage that costs the US economy US\$34.8 billion a year [61]; 12 million tonnes of fruit are thrown away each year in India [62]. However, the non-edible components of underutilised crops can be used to make biofuels, lowering waste output for a more sustainable and environmentally friendly future [59]. Cellulose (30%–50%), hemicellulose (20%–40%), and polymers that hydrolyse into fermentable mono- and disaccharides make up fruit biomass [63]. The carbohydrates (glucose, fructose, sucrose, and starch) and fibre (pectin) found in lignin (15%–25%) can be utilised for hydrolysis to produce bioethanol [62]. Cellulose is a polysaccharide made up of D-glucose molecules joined linearly by (1,4) glycosidic linkages. This structure gives cellulose its crystalline form. While lignin is a complicated polymer made up of aromatic alcohols, hemicellulose is an amorphous polysaccharide made of sugar molecules with five and six carbons, primarily xylose [64]. Enzymatic hydrolysis can be a viable strategy since it can disassemble complex components like lignin, pectin, cellulose, and hemicellulose to liberate sugars; however, the optimum method for processing fruit biomass for the industrial manufacture of bioethanol has not yet been identified. The fruits banana, apple, mango, and papaya that were unfit for human eating were subjected to enzymatic hydrolysis and fermentation to create bioethanol. The potential of this residue for this use, given its availability, was demonstrated by using banana biomass for the yield of reducing sugars as well as for the production of bioethanol [65].

Cellulose, hemicellulose, and lignin make up the majority of the carbon-neutral and inedible lignocellulose-based biomass. On Earth, lignocellulose is the most accessible and plentiful renewable resource [66]. The intrinsic resistance of lignocellulose and the possible secondary contamination of the by-products and wastes restrict the growth of bioethanol production, even if lignocellulosic biomass has great potential for the production of biofuel and biochemicals [67].

The cell wall's composition (cellulose, hemicellulose, and lignin), level of lignification, and cellulose's crystallinity, which enable lignocellulose to resist chemical and biological deconstruction, are the main causes of the structural recalcitrance [68]. A workable method is needed to lessen the lignocellulosic recalcitrance in order to realise the enormous potential of lignocellulosic biomasses. According to research [69, 70], the combination of phosphoric acid and hydrogen peroxide (PHP) has been found to be an effective way to degrade lignocellulose in benign conditions. Previous studies have demonstrated that PHP pretreatment significantly increases lignocellulose's enzyme accessibility, facilitating fermentation and the formation of bioethanol. As a result, developing a clean and effective PHP process to manufacture bioethanol is essential [71].

Recent years have seen increasing interest in the use of lignocellulosic biomass (LCB) for the manufacture of second-generation (2G) bioethanol [72–74]. This is due to the fact that these biomaterials are broadly accessible, environmentally friendly, and affordable. Furthermore, because they are non-food substrates, even extensive usage does not reopen the "food versus fuel" argument [75]. Reports are starting to come in on the usage of paper mulberry (PM) substrates as feedstocks for the creation of bioethanol. The first successful attempt at 2G ethanol generation from PM wood was made by Wang et al. in 2020. Additionally, a recently created and effective method for disrupting the recalcitrant structure of LCB, PHP, was optimised for pretreatment of PM wood in order to take advantage of the rich cellulose present in its woody portion (59%) for the production of fermentable sugar [76]. How to produce bioethanol from the ideal PHP-pretreated PM wood has not been researched yet [77].

Underutilised biomaterial that is abundantly present in nature has the potential to be converted into bioethanol for the fuel sector's sustainability. Pretreatment is a step that should be chosen based on the structure and composition of the biomaterial. Determining the appropriate pretreatment procedure for the biomaterial depends on releasing the most sugar while avoiding the production of fermentation inhibitors. The choice of microorganism for fermentation should be based on the amount of hexose or pentose sugar released during pretreatment. The basis of the foundation for the bioprocess established in the lab scale reactor is shake flask research. The manufacturing of bioethanol will reach an industrial scale thanks to scale-up experiments.

1.9 NATURAL MATERIALS IN SELF-HEALING AND SELF-CLEANING SURFACES

The chapter on self-healing and self-cleaning nature-inspired surfaces discusses surfaces that can be fabricated by mimicking natural surfaces such as those of plant leaves, insect wings, skin of aquatic animals, etc. Large structural and functional variations are present in naturally occurring materials, and these variations have developed over millions of years as a result of "survival of the fittest" [78–80]. Biological surfaces can adapt to different environmental circumstances over time through mutations and selection. Humans have used natural resources to improve and enjoy their quality of life over the years by observing, learning from, analysing, adopting, and imitating them. The features of biomaterials include their being biodegradable, protective, anti-reflective, self-healing, and self-repairing [80]. As a result, scientists have created a variety of surfaces that are inspired by nature.

Smart surfaces, which draw their inspiration from nature, are those that exhibit stability against environmental wear, tear, and contamination. These surfaces can be created by copying organic ones as those found in plant leaves, insect wings, aquatic animal skin, etc. Through biological self-repair and restoration processes, biological surfaces have the capacity to maintain their liquid-repellent behaviour during the course of life. The self-healing materials automatically repair and restore their properties after being damaged by external stimuli like mechanical or thermal stress. Due to the contamination's super hydrophobicity, self-cleaning surfaces with anti-wetting characteristics can clean the contamination on their own. Superhydrophobic surfaces must have long-lasting resistance to water infiltration, surface contamination, and mechanical abrasion. The inspiration comes from the first observation of a certain function or design, which then inspires the development of a concept or product that has a related function or design [81]. Mimetic is utilised as a later step in the creation of materials or surfaces inspired by nature that can carry out a specific functional task that natural materials normally carry out.

For example, the creation of self-cleaning superhydrophobic surfaces was motivated by the lotus leaf's ability to stay spotless in dirty water. The so-called "lotus effect" is characterised by a very high water contact angle (>150°) and a low contact angle hysteresis (CAH). Moreover, the "rose petal" serves as an inspiration for superhydrophobic surfaces that show high adhesion (large CAH) while the "lotus leaf" serves as an inspiration for superhydrophobic surfaces that reveal low hysteresis (poor droplet adhesion to the surface) [82–84]. In order to manufacture materials with liquid-repellent and self-cleaning uses, academics and scientists have been urged to combine surface chemistry and surface topology. The surface wettability of man-made biomimetic materials and natural surfaces is highlighted in this chapter. The underlying mechanisms of many inspired artificial surfaces' self-healing and self-cleaning behaviour are also covered.

The materials that draw their inspiration from nature are constantly at war with a variety of environmental factors, including those involving harsh conditions that range from a wide temperature range to humid conditions. For technical applications, materials that can withstand environmental wear and tear are highly coveted. Nature serves as an inspiration for process development as well as product design and functionality [81]. The extensive use of materials and surfaces inspired by nature is hampered by their instability against mechanical damage or chemical agents [85–87]. Once more drawing inspiration from nature, it is possible to create surfaces and materials that have the potential to self-heal and maintain the necessary surface wetting qualities. Self-healing and self-cleaning surfaces have been the main topics of discussion in this chapter.

Self-repairing surfaces are able to mend themselves independently after injury with the use of available resources. The self-repairing materials are described using a variety of categories and descriptions [88].The definitions of "self-repairing," "self-healing," and "self-sealing" provided by Speck and colleagues are accurate. The chapter's main topic is self-healing materials, which can be classified into two types: intrinsic self-healing and extrinsic self-healing. The second entails the release of the encapsulated healing agent (rooted inside a vascular system, hollow fibres, or in microcapsules), whereas the former involves healing as an inherent (intrinsic) property of the material [88, 89]. Additional materials can be divided into two categories: autonomous self-healing, which can initiate the healing process on its own (damage can act as a trigger), and non-autonomous self-healing materials, which need an external trigger to induce it [85].

Producing goods for human consumption by mimicking the biological world has both benefits and drawbacks. Soft robotics has recently made use of the development of intelligent, self-healing materials. The capacity to adapt to external circumstances involves not only stability against mechanical wear and tear but also resilience against extremes of temperature and humidity [90]. From biomedicine to soft computing, polymeric materials have been used. The ecological and financial solution is to have a material that demonstrates its robustness against external stimuli (i.e., can self-heal when damaged or self-clean when contaminated). The scientific community has created a wide range of unique chemistries for both extrinsic and intrinsic self-healing techniques during the past few years. As was seen, great work has been put into creating the optimal intrinsic self-healing system that can quickly return to its original qualities at room temperature.

The creation of inexpensive, scalable, and commercially viable nature-inspired materials is still a challenge. The main difficulties in creating and using materials inspired by nature include biocompatibility, mass production, biodegradability, and cost. As a result, finding a trustworthy and long-lasting solution to these issues is essential [91].

1.10 ROLE OF UPCYCLED FOOD

The chapter on the role of upcycled foods for global food security discusses the two vital issues of sustainability that affect the globe. Food loss and food waste are the two main sustainability problems that the entire world is facing. It influences both the altering of biological cycles and climatic change. The value cycle needs to be reorganised in order to construct a sustainable system. The phrase "upcycling" refers to the concept of reintroducing the value cycle, which entails reducing waste creation and promoting efficient resource usage. In essence, the idea of creating sustainable economies has gained popularity in most industries. Therefore, it is crucial to examine this from the perspective of an investor and determine how much upcycling can be done to improve ecosystem sustainability. The chapter on the role of upcycled food addresses the problem by conducting a systematic evaluation of the efforts made by various businesses to reduce food waste. This study identifies gaps that can be filled in future research, particularly regarding the long-term effects of upcycling investments. The chapter also emphasises how upcycling prevents food loss and waste from having unintended negative effects on the environment. In order to increase the global food upcycling process based on the resources available, it is necessary to conduct more studies on how to address the difficulties and gaps. In its simplest form, food upcycling is the process of creating food products from ingredients that ordinarily would not have been consumed by people.

Foods that have been recycled are produced at accredited facilities using waste from legal food companies, and they offer significant economic and environmental benefits [92]. The ingredients used to create the recycled food products are ones that would have otherwise gone to waste. Other examples include repurposing surplus and leftover food to create new ingredients and culinary items. Modifying wasted or leftover food into energy by chemical catalysis or an enzymatic process is an alternative to traditional food upcycling. By adopting the pressure-cooking technique, scientists have developed ways to break down or change food particles and turn them into reusable biofuel. Through this procedure, methane derived from pressure cooking is converted into power and heat.

Food waste is a major issue around the world, and it happens at every stage of the food distribution chain, including procurement, manufacturing, processing, storage, and distribution. Every year, almost one-third of the food that is produced is thrown away and ends up in landfills [93]. Food hunger, financial loss, and other negative environmental effects are the result. Around 1.3 billion tonnes of usable food is lost each year on a global scale. Thus, if we work to reduce food waste, we have a wonderful opportunity to do so and help feed those in need. In addition, food upcycling or food waste management can have a positive financial impact. Food waste is expected to cost the economy US$750 billion annually.

Additionally, the enormous amount of food waste produces about 4.4 billion tonnes of CO_2 equivalent year, which contributes to global warming [94]. Therefore, managing food waste and upcycling food have positive effects on the environment in addition to improving food security and the economy.

When a food product is upcycled, a new product with added value, a sustainable high-quality product, or an existing product is used in a novel way, resources are used as little as possible [95]. The practice of upcycling food has a long history and is founded on the maxim "use what you have." Making the most of what is created while using it to its fullest potential is the idea behind it. In 2020, the phrase "upcycled food" was coined for usage in research and policy. The names "value-added surplus products" and "waste to value food products" were in use prior to the term "upcycled food." Upcycled food is made from nutrients that escape from the inefficiencies in our food supply chain and is wholesome, sustainable, and of the highest quality. According to extensive research, customers would likely pay more for upcycled meals because of their association with the term "recycling" and their advantages for the environment. Upcycled foods can command a greater price than traditional products on the market with the correct messaging and marketing.

It will eventually assist food firms in reducing food waste, maximising the value of their products, and turning profits. If the components or nutrients are seen as garbage instead of resources, consumer acceptance of upcycled foods may suffer. Future research in the area of food waste and food upcycling can be made better by using more real-world study methodologies, a wider variety of cultural contexts, and a stronger emphasis on concepts from consumer behaviour and environmental psychology. Research on how to create the circular (bio) economy that is consumer-focused will be encouraged by the trend towards intervention designs.

1.11 CONCLUSIONS AND PERSPECTIVES

Applications of natural materials look at the many processing phases. Each phase is connected to their characteristics and features, illustrating how the origin of the materials affects the characteristics of the final product through a description of their chemical and structural characteristics. This chapter gave a brief introduction to current research and technical uses of diverse natural materials while taking into account the value-added chain from natural generation to the final product and placing a focus on quality application. However, the materials and the development of alternative and eco-friendly processes for their efficient application and use need to be explored further.

REFERENCES

[1] Z. Chen *et al.*, "Cardiovascular diseases and natural products," *Curr. Rrotein Pept. Sci.*, vol. 20, no. 10, pp. 962–963, 2019, doi: 10.2174/1389203720101990920124756.

[2] S. K. Gupta, R. Agarwal, N. D. Galpalli, S. Srivastava, S. S. Agrawal, and R. Saxena, "Comparative efficacy of pilocarpine, timolol and latanoprost in experimental models of glaucoma," *Methods Find. Exp. Clin. Pharmacol.*, vol. 29, no. 10, pp. 665–671, Dec. 2007, doi: 10.1358/mf.2007.29.10.1147765.

[3] J. M. G. Larkin and S. B. Kaye, "Epothilones in the treatment of cancer," *Expert Opin. Investig. Drugs*, vol. 15, no. 6, pp. 691–702, Jun. 2006, doi: 10.1517/13543784.15.6.691.

[4] A. Tazi *et al.*, "Vinblastine chemotherapy in adult patients with Langerhans cell histiocytosis: a multicenter retrospective study," *Orphanet J. Rare Dis.*, vol. 12, no. 1, p. 95, May 2017, doi: 10.1186/s13023-017-0651-z.

[5] S. Angraal *et al.*, "Digoxin use and associated adverse events among older adults," *Am. J. Med.*, vol. 132, no. 10, pp. 1191–1198, Oct. 2019, doi: 10.1016/j.amjmed.2019.04.022.

[6] S. Lobert, B. Vulevic, and J. J. Correia, "Interaction of vinca alkaloids with tubulin: a comparison of vinblastine, vincristine, and vinorelbine," *Biochemistry*, vol. 35, no. 21, pp. 6806–6814, May 1996, doi: 10.1021/bi953037i.

[7] S. Tanasescu, H. Lévesque, and C. Thuillez, "[Pharmacology of aspirin]," *La Rev. Med. Interne*, vol. 21, Suppl. 1, pp. 18s–26s, Mar. 2000, doi: 10.1016/s0248-8663(00)88721-4.

[8] F. Nozari and N. Hamidizadeh, "The effects of different classes of antihypertensive drugs on patients with COVID-19 and hypertension: a mini-review," *Int. J. Hypertens.*, vol. 2022, p. 5937802, 2022, doi: 10.1155/2022/5937802.

[9] F. Cavaleri, "Presenting a new standard drug model for turmeric and its prized extract, curcumin," *Int. J. Inflam.*, vol. 2018, p. 5023429, 2018, doi: 10.1155/2018/5023429.

[10] P. Khanal *et al.*, "Withanolides from Withania somnifera as an immunity booster and their therapeutic options against COVID-19," *J. Biomol. Struct. Dyn.*, vol. 40, no. 12, pp. 5295–5308, Aug. 2022, doi: 10.1080/07391102.2020.1869588.

[11] G. A. Miana, M. Riaz, S. Shahzad-ul-Hussan, R. Z. Paracha, and U. Z. Paracha, "Prostratin: an overview," *Mini Rev. Med. Chem.*, vol. 15, no. 13, pp. 1122–1130, 2015, doi: 10.2174/1389557515666150511154108.

[12] M. Imran-Shaukat, R. Wahi, and Z. Ngaini, "The application of agricultural wastes for heavy metals adsorption: a meta-analysis of recent studies," *Bioresour. Technol. Reports*, vol. 17, p. 100902, 2022, doi: 10.1016/j.biteb.2021.100902.

[13] J.-K. Liu, "Natural products in cosmetics," *Nat. Products Bioprospect.*, vol. 12, no. 1, p. 40, 2022, doi: 10.1007/s13659-022-00363-y.

[14] M. Jain, S. Shriwas, D. S. Dwivedi, and R. Dubey, "Design, development and characterization of herbal lipstick containing natural ingredients," *Am. J. Life Sci. Res.*, vol. 5, pp. 36–39, May 2017, doi: 10.21859/ajlsr-05021.

[15] X. Zhu, Y. Tian, W. Zhang, T. Zhang, C. Guang, and W. Mu, "Recent progress on biological production of α-arbutin," *Appl. Microbiol. Biotechnol.*, vol. 102, no. 19, pp. 8145–8152, Oct. 2018, doi: 10.1007/s00253-018-9241-9.

[16] S. Chib, V. L. Jamwal, V. Kumar, S. G. Gandhi, and S. Saran, "Fungal production of kojic acid and its industrial applications," *Appl. Microbiol. Biotechnol.*, vol. 107, no. 7–8, pp. 2111–2130, Apr. 2023, doi: 10.1007/s00253-023-12451-1.

[17] T. Lv *et al.*, "Dipotassium glycyrrhizinate relieves leptospira-induced nephritis in vitro and in vivo," *Microb. Pathog.*, vol. 152, p. 104770, 2021, doi: 10.1016/j.micpath.2021.104770.

[18] X. Ma, Z. Jiang, L. Xiang, and F. Zhang, "Natural material inspired organic thin-film transistors for biosensing: properties and applications," *ACS Mater. Lett.*, vol. 4, no. 5, pp. 918–937, 2022, doi: 10.1021/acsmaterialslett.2c00095.

[19] M. Song *et al.*, "Materials and methods of biosensor interfaces with stability," *Front. Mater.*, vol. 7, 2021, doi: 10.3389/fmats.2020.583739.

[20] D. R. Thévenot, K. Toth, R. A. Durst, and G. S. Wilson, "Electrochemical biosensors: recommended definitions and classification," *Biosens. Bioelectron.*, vol. 16, no. 1–2, pp. 121–131, 2001, doi: 10.1016/S0956-5663(01)00115-4.

[21] J. J. Schmidt and C. D. Montemagno, "Bionanomechanical systems," *Ann. Rev. Mater. Res.*, vol. 34. pp. 315–337, 2004, doi: 10.1146/annurev.matsci.34.040203.115827.

[22] A. P. F. Turner, "Biosensors: fundamentals and applications - historic book now open access," *Biosens. Bioelectron.*, vol. 65. p. A1, 2015, doi: 10.1016/j.bios.2014.10.027.

[23] A. Cavalcanti, B. Shirinzadeh, M. Zhang, and L. C. Kretly, "Nanorobot hardware architecture for medical defense," *Sensors*, vol. 8, no. 5, pp. 2932–2958, 2008, doi: 10.3390/s8052932.

[24] A. Kowalczyk, "Trends and perspectives in DNA biosensors as diagnostic devices," *Curr. Opin. Electrochem.*, vol. 23. pp. 36–41, 2020, doi: 10.1016/j.coelec.2020.03.003.

[25] Z. Zhu, F. Gao, J. Lei, H. Dong, and H. Ju, "A competitive strategy coupled with endonuclease-assisted target recycling for DNA detection using silver-nanoparticle-tagged carbon nanospheres as labels," *Chem. - A Eur. J.*, vol. 18, no. 43, pp. 13871–13876, 2012, doi: 10.1002/chem.201201307.

[26] Z. Matharu, A. J. Bandodkar, V. Gupta, and B. D. Malhotra, "Fundamentals and application of ordered molecular assemblies to affinity biosensing," *Chem. Soc. Rev.*, vol. 41, no. 3, pp. 1363–1402, 2012, doi: 10.1039/c1cs15145b.

[27] I. A. M. Frias, C. A. S. Andrade, V. Q. Balbino, and C. P. de Melo, "Use of magnetically disentangled thiolated carbon nanotubes as a label-free impedimetric genosensor for detecting canine Leishmania spp. infection," *Carbon*, vol. 117, pp. 33–40, 2017, doi: 10.1016/j.carbon.2017.02.031.

[28] L. A. Hernández, M. A. del Valle, and F. Armijo, "Electrosynthesis and characterization of nanostructured polyquinone for use in detection and quantification of naturally occurring dsDNA," *Biosens. Bioelectron.*, vol. 79, pp. 280–287, 2016, doi: 10.1016/j.bios.2015.12.041.

[29] X. Fang, J. Liu, J. Wang, H. Zhao, H. Ren, and Z. Li, "Dual signal amplification strategy of Au nanopaticles/ZnO nanorods hybridized reduced graphene nanosheet and multienzyme functionalized Au@ZnO composites for ultrasensitive electrochemical detection of tumor biomarker," *Biosens. Bioelectron.*, vol. 97, pp. 218–225, 2017, doi: 10.1016/j.bios.2017.05.055.

[30] B. Adhikari and S. Majumdar, "Polymers in sensor applications," *Prog. Polym. Sci.*, vol. 29, no. 7, pp. 699–766, 2004, doi: 10.1016/j.progpolymsci.2004.03.002.

[31] S. Cichosz, A. Masek, and M. Zaborski, "Polymer-based sensors: a review," *Polym. Test.*, vol. 67, pp. 342–348, 2018, doi: 10.1016/j.polymertesting.2018.03.024.

[32] E. Sackmann, "Supported membranes: scientific and practical applications," *Science*, vol. 271, no. 5245, pp. 43–48, 1996, doi: 10.1126/science.271.5245.43.

[33] U. Jain, M. Khanuja, S. Gupta, A. Harikumar, and N. Chauhan, "Pd nanoparticles and molybdenum disulfide (MoS_2) integrated sensing platform for the detection of neuromodulator," *Process Biochem.*, vol. 81, pp. 48–56, 2019, doi: 10.1016/j.procbio.2019.03.019.

[34] J. H. Kim, S. Mun, H. U. Ko, G. Y. Yun, and J. Kim, "Disposable chemical sensors and biosensors made on cellulose paper," *Nanotechnology*, vol. 25, no. 9, 2014, doi: 10.1088/0957-4484/25/9/092001.

[35] T. A. Khattab, A. L. Mohamed, and A. G. Hassabo, "Development of durable superhydrophobic cotton fabrics coated with silicone/stearic acid using different cross-linkers," *Mater. Chem. Phys.*, vol. 249, 2020, doi: 10.1016/j.matchemphys.2020.122981.

[36] J. Janata and M. Josowicz, "Conducting polymers in electronic chemical sensors," *Nat. Mater.*, vol. 2, no. 1, pp. 19–24, 2003, doi: 10.1038/nmat768.

[37] J. M. Moon, N. Thapliyal, K. K. Hussain, R. N. Goyal, and Y. B. Shim, "Conducting polymer-based electrochemical biosensors for neurotransmitters: a review," *Biosens. Bioelectron.*, vol. 102, pp. 540–552, 2018, doi: 10.1016/j.bios.2017.11.069.

[38] M. Ammam and J. Fransaer, "Highly sensitive and selective glutamate microbiosensor based on cast polyurethane/AC-electrophoresis deposited multiwalled carbon nanotubes and then glutamate oxidase/electrosynthesized polypyrrole/Pt electrode," *Biosens. Bioelectron.*, vol. 25, no. 7, pp. 1597–1602, 2010, doi: 10.1016/j.bios.2009.11.020.

[39] P. Falcaro, R. Ricco, C. M. Doherty, K. Liang, A. J. Hill, and M. J. Styles, "MOF positioning technology and device fabrication," *Chem. Soc. Rev.*, vol. 43, no. 16, pp. 5513–5560, 2014, doi: 10.1039/c4cs00089g.

[40] K. B. Blodgett, "Films built by depositing successive monomolecular layers on a solid surface," *J. Am. Chem. Soc.*, vol. 57, no. 6, pp. 1007–1022, 1935, doi: 10.1021/ja01309a011.

[41] I. Langmuir, "The mechanism of the surface phenomena of flotation," *Trans. Faraday Soc.*, vol. 15, pp. 62–74, 1920, doi: 10.1039/TF9201500062.

[42] X. Niu, Y. Zhong, R. Chen, F. Wang, and D. Luo, "Highly sensitive and selective liquid crystal optical sensor for detection of ammonia," *Opt. Express*, vol. 25, no. 12, p. 13549, 2017, doi: 10.1364/oe.25.013549.

[43] H. Altug, S. H. Oh, S. A. Maier, and J. Homola, "Advances and applications of nanophotonic biosensors," *Nat. Nanotechnol.*, vol. 17, no. 1, pp. 5–16, Jan. 2022, doi: 10.1038/s41565-021-01045-5.

[44] S. Shi, Y. Xie, G. Wang, and Y. Luo, "Metabolite-based biosensors for natural product discovery and overproduction," *Curr. Opin. Biotechnol.*, vol. 75, 2022, doi: 10.1016/j.copbio.2022.102699.

[45] K. Saravanakumar *et al.*, "Unraveling the hazardous impact of diverse contaminants in the marine environment: detection and remedial approach through nanomaterials and nano-biosensors," *J. Hazard. Mater.*, vol. 433, 2022, doi: 10.1016/j.jhazmat.2022.128720.

[46] A. Hashem, M. A. M. Hossain, A. R. Marlinda, M. Al Mamun, K. Simarani, and M. R. Johan, "Nanomaterials based electrochemical nucleic acid biosensors for environmental monitoring: a review," *Appl. Surf. Sci. Adv.*, vol. 4, p. 100064, Jun. 2021, doi: 10.1016/J.APSADV.2021.100064.

[47] X. Ma, Y. Jiang, F. Jia, Y. Yu, J. Chen, and Z. Wang, "An aptamer-based electrochemical biosensor for the detection of Salmonella," *J. Microbiol. Methods*, vol. 98, no. 1, pp. 94–98, 2014, doi: 10.1016/j.mimet.2014.01.003.

[48] N. Shoaie *et al.*, "Electrochemical sensors and biosensors based on the use of polyaniline and its nanocomposites: a review on recent advances," *Microchim. Acta*, vol. 186, no. 7, Jul. 2019, doi: 10.1007/S00604-019-3588-1.

[49] H. Li, P. Dauphin-Ducharme, G. Ortega, and K. W. Plaxco, "Calibration-free electrochemical biosensors supporting accurate molecular measurements directly in undiluted whole blood," *J. Am. Chem. Soc.*, vol. 139, no. 32, pp. 11207–11213, Aug. 2017, doi: 10.1021/JACS.7B05412/SUPPL_FILE/JA7B05412_SI_001.PDF.

[50] G. Ziyatdinova, S. Omaye, I. G. Munteanu, and C. Apetrei, "A review on electrochemical sensors and biosensors used in assessing antioxidant activity," *Antioxidants*, vol. 11, no. 3, p. 584, Mar. 2022, doi: 10.3390/ANTIOX11030584.

[51] F. Della Pelle and D. Compagnone, "Nanomaterial-based sensing and biosensing of phenolic compounds and related antioxidant capacity in food," *Sensors*, vol. 18, no. 2, Feb. 2018, doi: 10.3390/S18020462.

[52] I. G. Munteanu and C. Apetrei, "Electrochemical determination of chlorogenic acid in nutraceuticals using voltammetric sensors based on screen-printed carbon electrode modified with graphene and gold nanoparticles," *Int. J. Mol. Sci.*, vol. 22, no. 16, Aug. 2021, doi: 10.3390/IJMS22168897.

[53] A. P. Lima, W. T. P. dos Santos, E. Nossol, E. M. Richter, and R. A. A. Munoz, "Critical evaluation of voltammetric techniques for antioxidant capacity and activity: presence of alumina on glassy-carbon electrodes alters the results," *Electrochim. Acta*, vol. 358, Oct. 2020, doi: 10.1016/J.ELECTACTA.2020.136925.

[54] C. Ferrag and K. Kerman, "Grand challenges in nanomaterial-based electrochemical sensors," *Front. Sensors*, vol. 1, 2020, doi: 10.3389/fsens.2020.583822.

[55] S. Mahari and S. Gandhi, "Recent advances in electrochemical biosensors for the detection of salmonellosis: current prospective and challenges," *Biosensors*, vol. 12, no. 6, p. 365, May 2022, doi: 10.3390/BIOS12060365.

[56] G. Bagdži, "Electrochemistry and spectroscopy-based biosensors," *Biosensors*, vol. 13, no. 1, p. 9, Dec. 2022, doi: 10.3390/BIOS13010009.

[57] Z. Anwar, S. Akram, and M. Zafar, "Production of bioethanol from mixed lignocellulosic biomass: future prospects and challenges," pp. 313–326, 2023, doi: 10.1007/978-981-19-6230-1_10.

[58] M. K. Manglam and M. Kar, "Effect of Gd doping on magnetic and MCE properties of M-type barium hexaferrite," *J. Alloys Compd.*, vol. 899, 2022, doi: 10.1016/j.jallcom.2021.163367.

[59] C. K. R. Pocha, S. R. Chia, W. Y. Chia, A. K. Koyande, S. Nomanbhay, and K. W. Chew, "Utilization of agricultural lignocellulosic wastes for biofuels and green diesel production," *Chemosphere*, vol. 290, Mar. 2022, doi: 10.1016/J.CHEMOSPHERE.2021. 133246.

[60] C. Y. Lin and C. Lu, "Development perspectives of promising lignocellulose feedstocks for production of advanced generation biofuels: a review," *Renew. Sustain. Energy Rev.*, vol. 136, Feb. 2021, doi: 10.1016/J.RSER.2020.110445.

[61] S. T. Hingston and T. J. Noseworthy, "On the epidemic of food waste: idealized prototypes and the aversion to misshapen fruits and vegetables," *Food Qual. Prefer.*, vol. 86, Dec. 2020, doi: 10.1016/J.FOODQUAL.2020.103999.

[62] H. Kazemi Shariat Panahi *et al.*, "Bioethanol production from food wastes rich in carbohydrates," *Curr. Opin. Food Sci.*, vol. 43, pp. 71–81, Feb. 2022, doi: 10.1016/J.COFS.2021.11.001.

[63] C. Conesa, L. Seguí, N. Laguarda-Miró, and P. Fito, "Microwaves as a pretreatment for enhancing enzymatic hydrolysis of pineapple industrial waste for bioethanol production," *Food Bioprod. Process.*, vol. 100, pp. 203–213, Oct. 2016, doi: 10.1016/J. FBP.2016.07.001.

[64] J. J. Musci, M. Montaña, E. Rodríguez-Castellón, I. D. Lick, and M. L. Casella, "Selective aqueous-phase hydrogenation of glucose and xylose over ruthenium-based catalysts: influence of the support," *Mol. Catal.*, vol. 495, Nov. 2020, doi: 10.1016/J. MCAT.2020.111150.

[65] D. Paula *et al.*, "Fruit residues as biomass for bioethanol production using enzymatic hydrolysis as pretreatment," Jan. 2023, doi: 10.21203/RS.3.RS-2465028/V1.

[66] V. Ashokkumar *et al.*, "Recent advances in lignocellulosic biomass for biofuels and value-added bioproducts – a critical review," *Bioresour. Technol.*, vol. 344, Jan. 2022, doi: 10.1016/J.BIORTECH.2021.126195.

[67] Y. Wang *et al.*, "Cascading of engineered bioenergy plants and fungi sustainable for low-cost bioethanol and high-value biomaterials under green-like biomass processing," *Renew. Sustain. Energy Rev.*, vol. 137, Mar. 2021, doi: 10.1016/j.rser.2020.110586.

[68] Z. Usmani *et al.*, "Ionic liquid based pretreatment of lignocellulosic biomass for enhanced bioconversion," *Bioresour. Technol.*, vol. 304, p. 123003, May 2020, doi: 10.1016/J.BIORTECH.2020.123003.

[69] J. Qiu *et al.*, "Pretreating wheat straw by phosphoric acid plus hydrogen peroxide for enzymatic saccharification and ethanol production at high solid loading," *Bioresour. Technol.*, vol. 238, pp. 174–181, Aug. 2017, doi: 10.1016/J.BIORTECH.2017.04.040.

[70] Q. Wang *et al.*, "Pretreating lignocellulosic biomass by the concentrated phosphoric acid plus hydrogen peroxide (PHP) for enzymatic hydrolysis: evaluating the pretreatment flexibility on feedstocks and particle sizes," *Bioresour. Technol.*, vol. 166, pp. 420–428, Aug. 2014, doi: 10.1016/J.BIORTECH.2014.05.088.

[71] X. Wan *et al.*, "Conversion of agricultural and forestry biomass into bioethanol, water-soluble polysaccharides, and lignin nanoparticles by an integrated phosphoric acid plus hydrogen peroxide process," *Ind. Crops Prod.*, vol. 191, p. 115969, Jan. 2023, doi: 10.1016/J.INDCROP.2022.115969.

[72] N. Das, P. K. Jena, D. Padhi, M. Kumar Mohanty, and G. Sahoo, "A comprehensive review of characterization, pretreatment and its applications on different lignocellulosic biomass for bioethanol production," *Biomass Convers. Biorefinery*, vol. 13, no. 2, pp. 1503–1527, Jan. 2023, doi: 10.1007/S13399-021-01294-3.

[73] A. Duque, C. Álvarez, P. Doménech, P. Manzanares, and A. D. Moreno, "Advanced bioethanol production: from novel raw materials to integrated biorefineries," *Processes*, vol. 9, no. 2, pp. 1–30, Feb. 2021, doi: 10.3390/PR9020206.

[74] T. Raj *et al.*, "Recent advances in commercial biorefineries for lignocellulosic ethanol production: current status, challenges and future perspectives," *Bioresour. Technol.*, vol. 344, Jan. 2022, doi: 10.1016/j.biortech.2021.126292.

[75] P. Halder, K. Azad, S. Shah, and E. Sarker, "Prospects and technological advancement of cellulosic bioethanol ecofuel production," *Adv. Eco-Fuels a Sustain. Environ.*, pp. 211–236, Jan. 2018, doi: 10.1016/B978-0-08-102728-8.00008-5.

[76] P. C. Ajayo *et al.*, "High yield of fermentable sugar from paper mulberry woods using phosphoric acid plus hydrogen peroxide pretreatment: multifactorial investigation and optimization," *Ind. Crops Prod.*, vol. 180, Jun. 2022, doi: 10.1016/j.indcrop.2022.114771.

[77] P. C. Ajayo *et al.*, "High bioethanol titer and yield from phosphoric acid plus hydrogen peroxide pretreated paper mulberry wood through optimization of simultaneous saccharification and fermentation," *Bioresour. Technol.*, vol. 374, p. 128759, Apr. 2023, doi: 10.1016/J.BIORTECH.2023.128759.

[78] K. Koch and W. Barthlott, "Superhydrophobic and superhydrophilic plant surfaces: an inspiration for biomimetic materials," *Philos. Trans. A. Math. Phys. Eng. Sci.*, vol. 367, no. 1893, pp. 1487–1509, Apr. 2009, doi: 10.1098/RSTA.2009.0022.

[79] J. P. Youngblood and N. R. Sottos, "Bioinspired materials for self-cleaning and self-healing," doi: 10.1557/mrs2008.158.

[80] A. S. H. Makhlouf and R. Rodriguez, "Bioinspired smart coatings and engineering materials for industrial and biomedical applications," *Adv. Smart Coatings Thin Film. Futur. Ind. Biomed. Eng. Appl.*, pp. 407–427, Jan. 2020, doi: 10.1016/B978-0-12-849870-5.00018-5.

[81] N. K. Katiyar, G. Goel, S. Hawi, and S. Goel, "Nature-inspired materials: emerging trends and prospects," *NPG Asia Mater.*, vol. 13, no. 1, pp. 1–16, Jul. 2021, doi: 10.1038/s41427-021-00322-y.

[82] B. Bhushan and Y. C. Jung, "Natural and biomimetic artificial surfaces for superhydrophobicity, self-cleaning, low adhesion, and drag reduction," *Prog. Mater. Sci.*, vol. 56, no. 1, pp. 1–108, Jan. 2011, doi: 10.1016/J.PMATSCI.2010.04.003.

[83] S. Nishimoto and B. Bhushan, "Bioinspired self-cleaning surfaces with superhydrophobicity, superoleophobicity, and superhydrophilicity," *RSC Adv.*, vol. 3, no. 3, pp. 671–690, Dec. 2012, doi: 10.1039/C2RA21260A.

[84] W. Barthlott and C. Neinhuis, "Purity of the sacred lotus, or escape from contamination in biological surfaces," *Planta*, vol. 202, no. 1, pp. 1–8, 1997, doi: 10.1007/S004250050096/METRICS.

[85] O. Speck and T. Speck, "An overview of bioinspired and biomimetic self-repairing materials," *Biomimetics*, vol. 4, no. 1, p. 26, Mar. 2019, doi: 10.3390/BIOMIMETICS4010026.

[86] Z. Ma, H. Li, X. Jing, Y. Liu, and H. Y. Mi, "Recent advancements in self-healing composite elastomers for flexible strain sensors: materials, healing systems, and features," *Sens. Actuators A Phys.*, vol. 329, p. 112800, Oct. 2021, doi: 10.1016/J.SNA.2021.112800.

[87] X. L. Zuo, S. F. Wang, X. X. Le, W. Lu, and T. Chen, "Self-healing polymeric hydrogels: toward multifunctional soft smart materials," *Chinese J. Polym. Sci.*, vol. 39, no. 10, pp. 1262–1280, Oct. 2021, doi: 10.1007/S10118-021-2612-1/METRICS.

[88] D. Döhler, P. Michael, and W. Binder, "Principles of self-healing polymers," *Self-Healing Polym. From Princ. to Appl.*, pp. 5–60, Jun. 2013, doi: 10.1002/9783527670185.CH1.

[89] D. G. Bekas, K. Tsirka, D. Baltzis, and A. S. Paipetis, "Self-healing materials: a review of advances in materials, evaluation, characterization and monitoring techniques," *Compos. Part B Eng.*, vol. 87, pp. 92–119, Feb. 2016, doi: 10.1016/J.COMPOSITESB.2015.09.057.

[90] S. Terryn *et al.*, "A review on self-healing polymers for soft robotics," *Mater. Today*, vol. 47, pp. 187–205, Jul. 2021, doi: 10.1016/J.MATTOD.2021.01.009.

[91] S. Wang and M. W. Urban, "Self-healing polymers," *Nat. Rev. Mater.*, vol. 5, no. 8, pp. 562–583, Jun. 2020, doi: 10.1038/s41578-020-0202-4.

[92] S. Rondeau, S. M. Stricker, C. Kozachenko, and K. Parizeau, "Understanding motivations for volunteering in food insecurity and food upcycling projects," *Soc. Sci.*, vol. 9, no. 3, p. 27, Mar. 2020, doi: 10.3390/SOCSCI9030027.

[93] R. Capone, "Water footprint in the Mediterranean food chain: implications of food consumption patterns and food wastage," *Int. J. Nutr. Food Sci.*, vol. 3, no. 2, p. 26, 2014, doi: 10.11648/j.ijnfs.20140302.13.

[94] M. Chaya, T. Xiang, A. Green, and B. Gu, "Impact of climate change on pests of rice and cassava," *CAB Rev. Perspect. Agric. Vet. Sci. Nutr. Nat. Resour.*, vol. 2021, Jan. 2021, doi: 10.1079/PAVSNNR202116050.

[95] O. Spratt, R. Suri, and J. Deutsch, "Defining upcycled food products," pp. 1–12, 2020, doi: 10.1080/15428052.2020.1790074.

2 Adsorbent Materials

*Neha Mistry, Vijay Patel, Paresh Parekh,
Parth Naik, Nadavala Siva Kumar, Rohit
Vekariya and Mehul Khimani*

2.1 INTRODUCTION

The expeditious expansion of the global population, industrialization, climate changes, and economic development had a major effect in steadily increasing the number of contaminants like heavy metals, colours of synthetic origin, numerous sediments, hazardous chemicals, tremendous radioactive and pharmaceutical [1] products, and various other waste materials from the natural sources in soil, water resources, and other biological frameworks that form the basis for the extraction of raw materials for manufacture and analysis. Even the minute presence of various or some of these basic contaminants can lead to a massive disturbance of various ecosystems including land and aquatic ecosystems, which leads to numerous ill health effects. Due to the constant efforts of various scientists and research methodologies, in recent years, the types of available methodologies have increased to a profound extent. Out of the humongous available techniques for the removal of various hazardous chemicals, adsorption processes have turned out to be one of the best techniques even cost wise [1].

Adsorption broadly refers to the process in which the accumulation of a molecule or chemical or substance takes place at an interfacial level between two different phases (be it a liquid-solid interface or a gas-solid interface). The part that accumulates at the interface is called adsorbate and the surface on which adsorption occurs is called adsorbent. Adsorption differs from absorption, as the latter phenomenon involves the diffusion into the interior of the adsorbing material. The capacity or extent of adsorption of an adsorbent is completely dependent on the surface area. Therefore, porous substances, viz. charcoal, alumina, and silica gel, serve as better adsorbents. Adsorption is a dynamic and reversible process that decreases with temperature rise.

The adsorption phenomenon can broadly be classified into two categories: chemical adsorption and physical adsorption. Chemical adsorption or chemisorptions is usually irreversible because it happens due to the formation of strong chemical bonds on the interface between molecules or ions of the adsorbate surface to that of the adsorbent surface. Physical adsorption occurs via weak van der Waals interaction between particles of the adsorbate and adsorbent and is thus usually reversible. Adsorbents have many benefits over traditional, viz. biodegradability, higher surface area for action, high abundance in nature, higher mechanical strength, easy renewability, cheap cost, and easy manufacturing methods [2].

To understand the overall efficacy and extent of adsorption, a relationship between the amount of gas and liquid physically adsorbed on the given amount of a solid

DOI: 10.1201/9781032636368-2

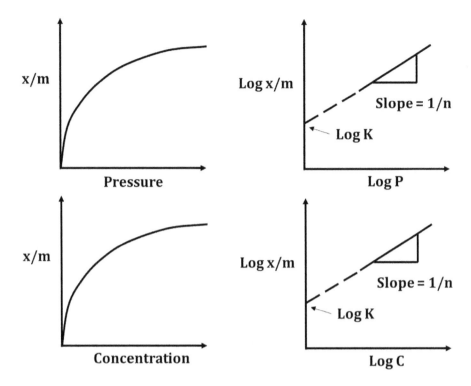

FIGURE 2.1   Adsorption isotherms.

adsorbent at a fixed temperature and equilibrium pressure is plotted to yield an adsorption isotherm. Repeated adsorption and desorption studies on a particular adsorbent will change the shape of isotherms, as described in Figure 2.1, due to a gradual change in the pore structure. Further, adsorption is a reversible process and is exothermic; hence, the concentration of adsorbed gas decreases with an increase in temperature at a constant pressure. There are three commonly used adsorption isotherms, viz. Freundlich, Langmuir, and Brunauer-Emmett-Teller (BET) isotherms [1].

The Freundlich adsorption isotherm gives the quantity of adsorbed gas or liquid substance on one unit of a mass of the solid adsorbent as a function of pressure at a constant temperature. The expression for this isotherm is given as

$$\frac{X}{m} = KP^{\frac{1}{n}} \left(\text{adsorption of gas on the surface of solid} \right) \qquad (2.1)$$

$$\frac{X}{m} = KC^{\frac{1}{n}} \left(\text{adsorption of liquid on the surface of solid} \right) \qquad (2.2)$$

where $n > 1$,

$x =$ gas mass or amount which is adsorbed, $m =$ adsorbent mass, $P =$ pressure of gas, $C =$ concentration of solute, $n =$ constant depending on nature of adsorbent and gas or liquid.

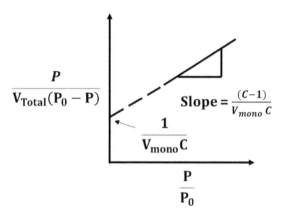

FIGURE 2.2 BET adsorption isotherm representing the formation of multiple adsorption layers.

The Langmuir adsorption isotherm follows certain postulates which state that the gases which are being adsorbed on the surface of a solid adsorbent do not form any layer greater than a molecule in depth. The active number of sites on any adsorbent is fixed in quantity, the adsorption rate is usually proportional to the sites which are unoccupied or empty, and the rate of desorption is inversely proportional to the number of the sites which are filled or occupied [3].

$$\frac{X}{m} = \frac{K_1 P}{1 + K_2 P} \tag{2.3}$$

where K_1, K_2, k = constants, P = pressure.

The BET adsorption phenomenon postulates that the adsorption phenomenon represented in Figure 2.2 involves the formation of many multilayers on the surface rather than a single one.

$$\frac{P}{V_{Total}(P_0 - P)} = \frac{1}{V_{mono}C} + \frac{(C-1)}{V_{mono}C}\left(\frac{P}{P_0}\right) \tag{2.4}$$

where V = volume, P = pressure of adsorbed gas, P_0 = saturated vapour pressure of the adsorbate at temperature T, V_{mono} = volume of gas reduced to standard conditions, C = constant [3].

2.2 SELECTION CRITERIA AND CHARACTERIZATION OF ADSORPTION BEHAVIOUR

2.2.1 FUNCTIONAL GROUPS ON THE SURFACE OF ADSORBENTS

The various types of functional groups which are available on the surface of adsorbents majorly affect the determination and modulation of the interaction of the adsorbent with the adsorbate and other physical properties [3]. If new material is to be used as an adsorbent then it is quite important to determine the various available functional groups on its surface through various analytical techniques such as IR, NMR, etc. The number of hydrogen, carbon, nitrogen, oxygen, and sulphur atoms present in an adsorbent surface can present very useful information about the efficacy and usage of an existing or novel adsorbent's structural composition and efficacy.

1. Surface charge

The adsorption phenomenon is largely affected by the charge present on the surface of the adsorbent (Figure 2.3). Initially, the adsorbent must be at a point of zero charges, which implies the exact pH at which the adsorbent surface is neutral in nature. Determination of the zeta potential or the electro-kinetic potential as briefly depicted in Figure 2.3 at the sheer plane is also required as it affects the concentration of adsorption of ions or particles near the surface material [3].

2. Particle size

The best adsorbents fall in a particular size range. The size range affects the removal rate and capacity as well as their methods and applications of treatments. The first property to look for is the specific surface (BET) area. The adsorption of a gas on an adsorbent surface by complete adsorption allows the formation of a monomolecular level, increasing the BET value to form in a value range slightly below the atmospheric pressure. If nanoparticles are used as substrates, BET is higher, but they are very difficult to separate from the solvent used. Thus, micro-sized particles are preferred. Wherever it is not feasible to determine the unit for physisorption, alternative methods like evaluation of the iodine number, the graph of which is described in Figure 2.4, can prove beneficial for determining the adsorbent surface area [2].

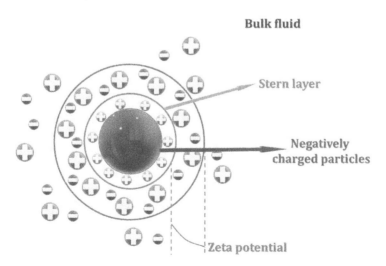

FIGURE 2.3 Surface charge on the surface of adsorbents.

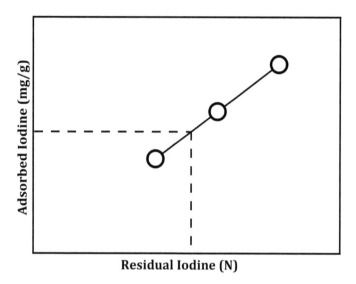

FIGURE 2.4   Schematic diagram of the adsorption of iodine per unit gram of the adsorbent vs. the residual concentration.

Adsorbents are usually available in granular or powdered form. While the size of granular adsorbents falls in the range of a few millimetres, the powdered adsorbents have a particle size range of typically $10–100\,\mu m$. Powdered adsorbents have the advantage of a quicker removal rate and are preferred for suspended mode applications, as observed by Bakkaloglu and co-workers in the pulverized activated carbon for adsorption of carbamazepine from natural sources of water [4]. Granular adsorbents with comparatively larger sizes than the powdered forms are substantially preferred for flow through applications to avoid back pressure. Uniform size distribution is also an important aspect to be considered for maximum adsorption [1].

3. Porosity

It is evident that the higher the number of pores on the material the higher will be its adsorption rate. The advantage of porous materials is that they usually possess higher areas of surface action, which ultimately results in more active sites for action compared to materials that possess lower or no pores at all. Certain materials containing very minute pore surface morphology like mesopores are highly desirable when a high diffusion rate of molecules into the internal surface of the adsorbent is desired [2].

4. Effect of solution pH

The pH of the solution plays a huge role in determining the adsorption rate and kinetics. A high degree of variation in the pH of the solution leads to variation in the extent or the degree of ionization of the molecule of adsorption and effects the surface morphology of the adsorbent. The determination of surface charge (neutral point or zero charges, pH_{pzc}) is a prime indicator of the number of active sites or centres. Zero charges has been the topic of research for various researchers for the determination of surface charges of various adsorbents prepared by them, to get a deeper insight into the adsorbents prepared or extracted. The pH of cation or positive dyes usually should be favoured at a pH greater than pH_{pzc} whereas in negative or anion dyes the pH has to be less than pH_{pzc} [5].

5. Effect of amount of adsorbent

An important parameter is to determine the dosage of the adsorbent for a particular adsorbent under usage in the optimum conditions [6]. In general, as the amount of adsorbent increases, the number of active sites for the removal of the adsorbate particles increase proportionally. With a linear increase in the number of sorption sites at the surface of the adsorbent, an increase in the rate and efficiency of adsorption is observed.

6. Surface area

As the surface area of the adsorbent increases, the number of adsorption sites increases, which increases the amount of adsorption.

7. Effect of contact time

The interference of contact time of the sample to the adsorbent has a huge role to play. The higher the time of reaction or the higher the time of contact of the sample to the adsorbent, the higher is the rate of adsorption. It gives the adsorbate complete time to attain equilibrium with the adsorbent surface. In the preliminary phases, adsorption occurs quickly and then steadily slows down as equilibrium is attained. Metal ions having different concentrations attain equilibrium with the adsorbent at different times and also get affected by the concentration of metal ions, the type of adsorbent used, the starting concentration, and the solution temperature [6].

2.3 NATURE OF ADSORBENTS

Some of the commonly used natural adsorbents are given in Table 2.1.

2.4 BIOMASS

Biomass such as agricultural waste materials like bark, sawdust, etc., offers a promising aspect in the area of adsorption. These agricultural wastes are cost-effective and researchers are widely recognizing the use and effectiveness of these materials as adsorbents worldwide. The wide varieties of substances that come under biomass have various applications as adsorbents. Biomass wastes like the ones obtained from cassava barks and leaves, yam peels, and shells of groundnuts were studied by Thompson and co-workers for the successful removal of Pb(II) ions from wastewater [7]. The maximum adsorption capacities observed with cassava peels, groundnut shell, and yam peel at initial concentrations of 50 mg g^{-1} were 50.1 mg g^{-1}, 46.6 mg g^{-1} and 38.5 mg g^{-1}, respectively. The value of ΔS, R^2, and ΔH for the adsorption with cassava peels are: 67.2 J $mol^{-1}K^{-1}$ 0.945 and 12,762 J mol^{-1}; for that of groundnut shell adsorbent, the, ΔS and R^2, ΔH values are; 130 J $mol^{-1}K^{-1}$ and 0.997, 36,756 J mol^{-1}; while for yam peels adsorbent, the R^2, ΔH, ΔS and values are 0.927, 12,163 J mol^{-1}, 46 J $mol^{-1}K^{-1}$ [7].

TABLE 2.1

Commonly used natural adsorbents

S. no.	Adsorbents	Source	Application
1.	Fuller's earth	Clay obtained from the earth is heated and dried to get a porous structure. Fuller's earth consists essentially of aluminium silicate, silica gel, calcium, and other trace elements.	Decolourizing of clothes and other dyes, dyeing of lubricating oils, adsorption of contaminants like copper and reactive yellow 18 for water purification.
2.	Activated clay especially silk bentonite	Bentonite or other clays are activated by treatment with sulphuric acid and by drying. New age silk fibroin-based bentonite.	Decolorizing of petroleum products, removal of lethal heavy metals from wastewater.
3.	Bauxite	Hydrated alumina, activated by heating at 230°C–815°C.	Removal of phosphorous residue in aqueous solutions especially in seawater.
4.	Alumina	Hydrated aluminium oxide, activated by heating and reduced in size.	Removal of inorganic species like cadmium, lead, arsenic, and fluorides in water.
5.	Activated carbon	Vegetable matter is mixed with calcium chloride, carbonized, and finally the inorganic compounds are leached away. Organic matter is mixed with porous pumice stones, and heated and carbonized to deposit the carbonaceous matter throughout the porous matter. Carbonizing substances like wood, saw dust, coconut shells, fruit pits, coal activated with hot air steam.	Used in the removal of dyes, metal ions, and other organic compounds of low molar mass in water. Finds an extensive use as an adsorbent in the cosmetic industry also.
6.	Silica gel	Obtained from sodium silicate solution after treatment with acid in a hard granular and porous form.	Can adsorb upto 2,200 L of solvent vapours from air and is extensively used as a desiccant to control local humidity.
7.	Cellulosic materials	Naturally occurring material is modified chemically to form micro or nano.	Used in the extraction and discardation of numerous pollutants from wastewaters.

A simple biomass product like sawdust can be converted into various forms like polymerized sawdust by the process of polymerization, which increases the adsorption capacity of the product by improving porosity and specific surface area. Zdenka, Stefan, and co-workers investigated the metal removing capacities of sawdust which was modified in an alkaline solution under a different range of conditions, especially for the extraction and removal of Cu(II) and Zn(II) ions from solutions [8]. The conversion of rice husk to rice husk char to be used as adsorbent has also been studied by Bai and co-workers to determine the kinetics, thermodynamics, and adsorption mechanisms for the extraction of ammonia to be further used as a highly efficient air pollution controller [9].

2.5 BIOCHAR

Anaerobic thermal decomposition of biomass during various processes produces biochar, which is a porous carbonaceous material. The stock which is considered biomass includes organic waste materials produced from forests, wood chips, saw dusts, sludge from sewage, algal deposits, manure, or other solid biorganic wastes. Biochar find extensive usage in the adsorption of toxic heavy metals, pollutants of organic origin, and other nutrients from wastewaters. The physical and chemical properties of biochar depend primarily on the types of feedstock and pyrolysis conditions, i.e., temperature, residence time, reactor type, and heating rate [10]. The new engineered biochar has the advantages of a new larger exposed surface area, an improved capacity of adsorption, and larger functional moieties on the surface for the adsorption of varieties of ions [11].

The release of greenhouse gases in the atmosphere due to the use of biochar-based adsorbents had a tremendous eco-friendly impact, which adds up to the usage efficacy. Moreover, due to the presence of lamellar structure and strong hydrophobicity, biochars can also be used for adsorption of organic high-risk pollutants (HRPs) [12]. Biochar performs adsorption by various mechanisms that have been described in Figure 2.5; it is extensively used in the fields for the improvement of the fertility of the soil, reduction of greenhouse gas emissions, etc. (Figure 2.5) [11].

2.6 CELLULOSIC MATERIALS

In recent times, naturally occurring cellulosic materials are used as cheap adsorbents for the removal of various toxic metals, usually heavy ones, anions of inorganic origin, organics, and micropollutants from the aquatic environment [3]. The modifications to cellulosic materials via chemical processes add up to an increase in the adsorption efficacy of the non-treated variants. The modifications use chemicals of basic or acidic origin or agents of oxidation, compounds of organic origins, etc. [3].

2.7 ACTIVATED CARBON

Activated carbon-based adsorbents are used as adsorbents because of their high chemical and structural stability, their wide range of applications, and lowest density possible. There are two forms of activated carbon available, namely, the commercial activated carbon and activated carbon synthesized from various waste materials.

  i. Commercial activated carbon: various developments and researches have been reported pertaining to the use of various commercial activated carbons as adsorbents for their separations of different dyes from wastewater. Activated carbon adsorbents have been prepared successfully by Chikri and co-workers [13] using various substances like mahogany sawdust, apricot stones [14], red oak [15] to name a few. The adsorption process

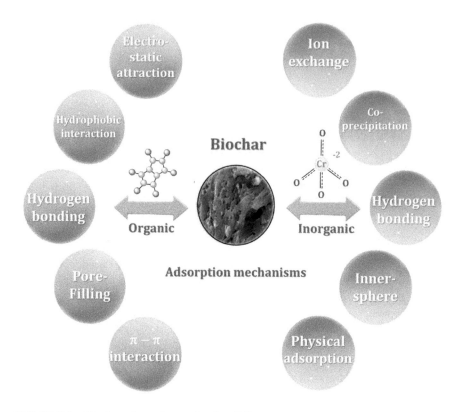

FIGURE 2.5   Biochar adsorption mechanism [12].

of commercially activated carbons is usually very effective and shows an increase in the adsorption of particles from an aqueous solution up to pH ~ 4 and becomes more or less unaltered at higher pH. All adsorption processes are consistent with the Langmuir isotherm adsorption model and the pseudo-second-order kinetics model. Based on the experimental results, they speculated that ion exchange may be the main mechanism for the adsorption of heavy metals for activated carbon [12].

ii. Activated carbons have been synthesized from various waste materials such as orange peels, used tea leaves, etc. [16]. The source or material from which the activated carbon is produced plays a huge role in the efficiency of adsorption. Usually, pseudo-second-order kinetics is observed and all the adsorption phenomena via activated carbon are in accordance with Langmuir and Freundlich isotherms, but, usually, the effect of pH and temperature shows a varied change in the adsorption capacities via differently sourced activated carbon materials. Activated carbon prepared from banana stalks was also used and studied for effective adsorptive removal of 2,4-dichlorophenoxyacetic acid (2,4-D) (196.33 mg g⁻¹) and bentazon (115.07 mg g⁻¹) from aqueous solutions. The process of adsorption kinetics followed both the Freundlich model and the pseudo-second-order kinetics model, and was exothermic in nature with the evolution of heat and quick and rapid. Adsorbents of activated carbon origin were prepared from olive waste to form cakes, which were further investigated to adsorb drugs like ibuprofen, naproxen, diclofenac, and ketoprofen. The results proved that the pK_a affected a lot of the adsorption capacities of these drugs. The adsorption capacity of all the drugs decreased with an increase in the pH [12].

2.8 NANOMATERIALS

Nanomaterials are usually particle sizes ranging from 1 to 100 nm. The effectiveness of these materials based on their efficient adsorption rate and fast-aided mechanisms has garnered huge attention in recent times. Humongous varieties and types of nanoparticles/materials are available now which are either of natural origin or prepared via certain treatments. The basis of the use of nanoparticles as adsorbents pertains to the fact that the materials when reduced to nanometre size usually become quite unstable and readily attract the adsorbate onto their surface. Second, the reduction in size benefits with an increase in the effective adsorption surface area of the particles, resulting in a better and broader perspective for adsorption. Third, the reduction mechanisms employed for the formation of nanoparticles result in the activation of groups containing hydroxyl ions on the surface of these particles, which effectively adsorb the toxic heavy metal ions and organic waste compounds from wastewater. The new age hybrid nanomaterials which are usually two kinds of nanoparticles are highly effective as they offer a wide variety of surfacial functional groups for attraction and adsorption of heavy metals from wastewater.

Nanocomposites of hybrid-based carbon compounds are easily available or are usually synthesized with very simple and cheap mechanisms. They are usually highly user friendly with very low toxic effects and contain large porous sites [17].

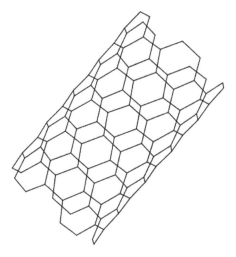

FIGURE 2.6 Carbon nanotubes.

TABLE 2.2
Types and varieties of CNTs

S. no.	CNT	Source	Properties	Application
1.	Single-walled (SWCNTs)	Graphene-based single sheet rolled upon itself available as bundles.	1–2 nm in diameter, length can be varied, bundle structures can form crystal-like construction.	Pollutant adsorption as bare SWCNTs or their modified form from aqueous orgaseous [1], dyes, namely, reactive blue [18].
2.	Double-walled (DWCNTs)	Made by two concentric CNTs in which the outer tube encloses the inner tube.	Greater stiffness and stability provide strength and support.	Adsorption of chromium ions, organic dyes, copper, and zinc ions [18].
3.	Multiwalled (MWCNTs)	Multiple layers of graphene sheets rolled upon each other.	2–50 nm in diameter, lighter in weight.	Adsorption of materials like dyes, namely, alizarin red, H_2S, lead species, zinc, copper [18], antibiotics like sulphonamides, tetracyclines [1], etc.

Graphene-based nanoparticles/nanocomposites are composed of two-dimensional carbon nanoparticles hybridized to form extremely thin sheets of graphene. Their uniqueness lies in their larger specific surface area and adsorption rate. The most effective use is in the adsorption of lead ions. Preparation of graphene sheets and their simultaneous oxidation has resulted in the production of graphene oxide-based nanoparticles for removing dyes from wastewaters [12].

Carbon nanotube (CNT)-based nanohybrid materials are in high demand and are actively used due to the allotropic nature of carbon-based molecules, which has various efficacious advantages including compact size, higher surface area for a specific action, hollow cylindrical structures which implement their usage in numerous industrial processes. The hybrids produced likewise are better in the mechanism of action and removal and the extraction of organic-based contaminants than adsorbents of activated carbon origin. These nanotubes are the members of the Fullerene family and have a wide mechanical strength, and unique electrical as well as thermal properties [12]. Types and varieties of CNTs have been presented in Table 2.2.

Nano metal oxides are nanoparticles mainly made from the metallic oxides of metals like iron oxide, copper oxide, and other metals. Metal oxide nanoparticles provide effective adsorption by making a complex of metal ions with the oxygen atoms of the material. These metal oxide nanoparticles provide a higher range of capacity of adsorption due to their bigger surface area, efficacious distance of diffusion, and a large number of active sites on the surface [12].

2.9 METAL ORGANIC FRAMEWORKS (MOFS)

The usage of MOFs as adsorbents is largely because of the presence of a huge internal area that is hollow, as shown in Figure 2.7. This is because of the presence of

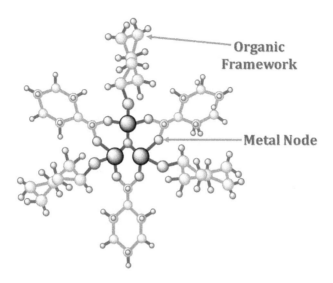

FIGURE 2.7 A metal organic framework depicting the presence of an organic framework and a metal node.

hybrids of organic and inorganic porous materials of crystal origin representing a homogenous effect of positive metal-based ions of the fourth group like titanium, zirconium, etc., which are encircled by linker arms of repetitive cage units or structures [19]. MOFs usually work by forming $\pi-\pi$ bonds and through hydrogen bonding and electrostatic interactions between the adsorbates and the MOF surface. Certain researchers claim that MOFs have an excellent distribution of nanomaterials on their surface [17].

Due to their diverse structures and highly adjustable sizes ranging from 0–3 to 9.8 nm, they fall in the range of sizes of zeolites of microporous origin to mesoporous silicas. As the MOFs are formed due to the interaction of metal nodes and ligands, theoretically a wide variety of organic linkers exist, which makes the formation of a wide range possible for desired functional collaboration by application of various structures and compositions. The disadvantage of MOFs is that they usually possess weak stability in water, lack acidic or alkaline stability, and have low thermal-based stability and strength issues. They usually tend to disintegrate and change their surface morphology, rendering themselves unsuitable for adsorption. The basic problem with the disintegration is due to a number of substitutions reactions with water or hydroxide molecules. Therefore, to make the MOFs stable, it is practical to increase the stability of coordinate bonds [20].

The permanent and high porosity features are what makes them suitable as adsorbents. MOFs largely possess a microporous structure with a pore size ranging from less than 2 nm in diameter, and provide a large area for the adsorption of gases like hydrogen and carbon dioxide. On the one hand, the high micropore volume allows a larger surface area for adsorption; on the other hand, the narrow pores neither allow for the accumulation of large objects and support and holding of functional groups or molecules nor allow quick mass diffusion [19].

MOFs find extensive usage in the removal of organic dyes from wastewaters for the adsorption of organic dyes containing any chromophoric backbone. Syafikah and co-workers synthesized and studied an iron-based MOF of MIL-101 through a facile solvo thermal method for the dye adsorption of methylene blue from wastewater [21].

2.10 RESINS

Resin adsorbents possess a novel organic framework with a 3-D network structure and act as a sponge-like structure with numerous pores; they are heat stable and insoluble in most common solvents. They possess a large surface area. Due to their greater amount of surface area for specific usage and easy methods of regeneration, many of the resin materials are used as adsorbents to adsorb various ions and heavy metals for practical wastewater treatment [12].

There are two kinds of resins, namely, cationic and anionic exchange resins. The resins which are obtained from phenol and formaldehyde are cationic exchange resins whereas those derived from *m*-phenylene diamine are the anionic exchange resins. Cationic exchange resins are generally used for the removal of ions like Ca^{2+} and Mg^{2+} from water, thereby removing the permanent hardness of water. However, cationic resins are unstable and tend to disintegrate in basic solutions [3].

All these issues led to research which contributed to the invention and synthesis of a new resin used for ion-exchange and crosslinked styrene-divinylbenzene resin. These kinds of resins are unique because of their structural ability resulting from various spherical small beads or spheres. To increase the efficacy of the exchange of ions, a sulphonated group was introduced by reaction of the resin with fuming sulphuric acid, which increased the ion-exchanging ability of the resin, especially at higher pH, and proved to be stable at high pH [3].

Ion-exchange resin beads as support groups were synthesized for magnetite (Fe_3O_4) by Sikora and co-workers [22]. They could successfully produce an ion-exchange bead adsorbent which could easily extract heavy metal ions from wastewater and then be separated by a magnetic field. In another approach, Huang and co-workers studied the effects of eight kinds of ion-exchange resins for the removal of bis(hydroxyethyl) terephthalate (BET), a waste colour product produced during glycolysis of polyethylene terephthalate. In this study, resin D201 showed excellent results by adsorbing and removing 95% BET [23].

2.11 CLAYS

Clays possess a layered structure which helps them act as adsorbents. The surface of clays have an identifiable negative charge, which makes them unique in adsorption of positive-charged particles. Depending on the layered structures clays can be broadly divided into the following categories: smectites (montmorillonite, saponite), mica (illite), kaolinite, serpentine, pyrophyllite (talc), vermiculite, and sepiolite [24].

Hybrid clay adsorbents are available today which consist of natural and synthetic clay-based materials. Clays, either of organic or of inorganic origin, are modified by various means to increase the adsorption capability. The most effective and widely used hybridization method used for clay adsorbents is silylation. In the silylation process, silane grafting is done based entirely on the properties of the surface of the clay. The efficiency of adsorption of the silylated clays can be increased manifold by the addition of various functional groups like amino, carboxylic acid, and thiol groups by means of used silanes of organic nature. The incorporated organic functional groups tend to interact efficiently with the adsorbate molecules [24].

They offer various advantages given their low rates, zero to neutral toxic behaviour, high stability in thermal conditions, a greater level of porosity, and a high rate of porous surface. They are also versatile enough and various functional groups can be added on their surface, which renders them useful in various adsorption processes [25]. Therefore, the varieties containing hybridized nanoparticles of clay have been rigorously used in the adsorbent categories for dye removal from various wastewaters. The charge on clay particles which is mostly negative is due to the presence of very fine silicates, which contribute towards an increase in the adsorption capacities of the clay. The adsorption of toxic metals on the surface of clay is because the clay surface is negatively charged. The action of extracting via adsorption is increased multifold in clay because of its enlarged surface area, which ranges upto $800\,m^2$. Within the varieties of clays found around us, montmorillonite clays have the minutest size, thereby increasing the exact surface area for adsorption [3]. Clay has been studied by various researchers for its adsorption efficiency and is used in the extraction of chromium and lead ions [25]. Clays with their common modifications have been described in Table 2.3.

TABLE 2.3
Certain clays with modifications used as adsorbents

Type of clay	Variety	Modifications	Surface charge	Usage in adsorption	References
Kaolinite	Nanoparticles infused with titanium dioxide prepared from kaolin clay	Titanium dioxide nanocomposites	Negative	Cadmium and lead ions	[26]
Bentonite	Absorbent aluminium phyllosilicate	Magnetic bentonite	Negative or neutral	Adsorption of chlorophyll, cations, amines, pesticides, etc.	[27]
Saponite	Aluminosilicates with a nanosized particle belonging to the smectite group of clays	Sap-based hybrids	Negative	Cations of metals range especially copper and nickel ions from wastewater	[28]
Sepiolite	Hydrated magnesium silicate	Calcium oxalate-based composite	Negative	Cations present in aqueous solutions	[29]
Laponite	Synthetic silicate-based clay	Hydrogel	Negative	Methylene blue dye, nickel ions etc.	[30]

2.12 POLYMERIC SYNTHETIC ADSORBENTS

A large variety of polymeric synthetic materials are available to be used readily as adsorbents. The advantage of using polymeric materials for adsorption is that they possess proper porosity, sufficient surface area, a faster rate of adsorption, and higher stability. Polymeric adsorbents are porous, permeable beads of spherical shape like polystyrene-divinylbenzene, which are mostly crosslinked or phenol-formaldehyde polymers which are condensed [21]. Their mechanisms of adsorption and desorption are based on the functional groups and simultaneous structures of the material used.

Commercially available adsorbents which are polymers have huge advantages as they are easily available in various shapes of interest like spheres, sheets, and membranes, etc.; the exact effective surface can be changed easily by numerous variable techniques like brush polymeric arrangement, functionalization, and imprinting of a molecule for the enhancement of the adsorption process. These polymeric adsorbents find wide use in the removal of toxic and/or poisonous metals from polluted water. Rahman and co-workers studied polymer adsorbents grafted in the form of a composite by the usage of diallyl dimethyl ammonium chloride and acrylic acid embedded in a non-woven irradiated polythene sheet. The adsorption maximum was found up to 153.89 mg for copper ions per gram of material and 17.34 mg of chromium per gram of material with an initial concentration of 1,000 ppm and a 24-hour contact time [31]. Table 2.4 provides various adsorbents of polymeric origin used for the extraction of toxic metal ions.

2.13 ZEOLITES

Natural zeolites are hydrated alumino silicate minerals. They offer a primary usage in the elimination of ammonium and heavy metals from wastewater as demonstrated by Guida and co-workers [35]. They are used to remove adsorbates of various sizes, shapes, and polarity and are therefore also termed as molecular sieves. The action of zeolites depends on their ion-exchange capacity [35].

The removal of iron, lead, cadmium, and zinc from mined synthetic water was reviewed by Ugwu and co-workers by studying the effects of various zeolites. The

---

TABLE 2.4

Various adsorbents of polymeric origin used for extraction of toxic metal ions

S.no.	Adsorbent	Adsorbate	References
1	Poly(vinyl pyrrolidinone)	Mercury, lead,	[32]
2	Poly(2)-hydroxy ethyl methacrylate-methacrylolmidophenylalanine)	cadmium	[33]
3	Poly(hydroxy ethyl methacrylate-L-glutamic acid)		[33]
4	Poly(vinyl pyrrolidinone)		[34]
5	Acrylonitrile onto polypropylene fibre having diethylenetriamine group	Mercury	[34]

main area of study was the competitive and collaborative adsorption of a few cationic molecular ions as compared to other cations. The removal of lead ions was successful from acidic as well as neutral solutions; on the other hand, the selective uptake of cadmium and zinc ions was shown to be decreased with increasing concentration and decreasing pH as well [36].

On examination of a zeolite structure, three fundamental components are found to exist: (1) the primary one is a framework with alumino silicates, (2) water containing zeolitic concentrations, and (3) cations which tend to get exchanged. There also occurs a basic formula which caters to all forms of zeolites, which is $M_{x/n}[Al_xSi_yO_{2(x+y)}] \cdot pH_2O$, where M can be any metal ion like sodium, lithium, or cadmium and/or calcium, magnesium, barium, or strontium and n refers to the charge on the cation present [37]. The ion-exchange behaviour of zeolites corresponds to various factors like the structure of the frame, nature, shape and size properties of the ions to be adsorbed, the anion density with respect to the ion's charge ratio, the initial concentration, and the charge of the ions present in the electrolyte solution as studied by Moazeni and co-workers in an Iranian natural zeolite, prepared by placing 200 mL, 1 M NaCl solution in reflux (385 K) for 48 h, which was subsequently dried at 110°C for 24 h [32, 38].

It was concluded that the rate of adsorption in the initial or starting phase is quite rapid due to the presence of a large number of empty sites. With the increase in the initial concentration of Pb^+ ions, the removal efficiency of the zeolite reduced from 94.48% to 14.24% in a contact time of 60 min. It was observed that during the starting phase a greater amount and concentration of the ions which corresponds to about 90% was removed in the initial 15 minutes of the entire 24 hours of contact time provided. For Iranian natural zeolites, it was observed that when the dose range was increased from zero to 50 g L⁻¹, the maximum removal was when the dosage reached about 20 g L⁻¹. High adsorption efficiency was observed at pH = 3. Also, the particle size of the zeolite had a huge role to play, with an increase with decreasing particle size. As adsorption is a phenomenon that involves the surface, smaller-sized sorbents which have a comparatively large surface area can remove lead ions at conditions reaching equilibration. Zeolites are used for removal of surface water, groundwater, and industrial or household wastewater pollutants including inorganic and organic compounds. Some ions adsorbed by zeolites are given in Table 2.5 [38].

TABLE 2.5
Examples of some ions effectively adsorbed on zeolite

Metal ions	Maximum concentration adsorbed (mg g⁻¹)
Pb	65.75
Cu	56.06
Cd	52.12
Ni	34.40
Mn	30.89

2.14 AEROGELS

Aerogels are highly porous light materials. Aerogels are the tailored gels in which liquids are replaced by gas. Aerogels are made up of 99.8% air and are thus quite light in weight. Their interior solid matrix structure space consists mainly of air pockets. Aerogel synthesis includes three primary steps: the gel and sol interactions, maturing, and drying of the final product. The credit for the invention of aerogels goes to Steven Kistler who synthesized the first non-porous ultralight aerogel in 1931. There are certain advantages associated with the usage of aerogels given that they possess high thermal stability, have a large surface area, provide light weight and thus low density, possess high mechanical strength, etc.

There are various types of aerogels which can be prepared by changing the internal liquid matrix structure of the available gels. According to the application and usage of aerogels, they are synthesized in a wide range of shapes and sizes like beads, discs, powders, and monoliths. However, there are three basic classifications of aerogels based on their composition which include inorganics, organics, and even hybrids of organic-inorganic mixtures. Amongst the three categories, inorganic aerogels are the most extensively studied and used and are made from oxides of metals, namely, Si, Al, and Zr, to name a few [17].

The most commonly used inorganic aerogel for adsorption processes is carbon-based aerogels. Carbon aerogels are mostly synthesized by the process of

TABLE 2.6
Various forms of aerogel-based adsorbents and their application and properties

Aerogel	Application	Properties	References
Amine-supported aerogel	Adsorption and removal of carbon dioxide	Low density, high porosity, larger surface area	[40]
Nanocellulose aerogels	Extraction of heavy metals and congo red II dye from wastewater	High porosity and greater surface area	[41]
Silicon oxide and calcium chloride, silicone oxide-lithium bromide inorganic aerogel	Water vapour adsorption	Adsorption properties remain stable and increased	[42]
Activated carbon aerogel	Removal of volatile organic compounds	Excellent regeneration property	[43]
Cellulose nanocrystal aerogels with poly-methyl vinyl ether-co-maleic acid and polyethylene glycol	Removal of methylene blue dye	$116.2\,mg\,g^{-1}$ adsorption of methylene blue dye, can be used for five regeneration cycles	[44]
Graphene oxide-based aerogel	Removal of Pb(II) ions from aqueous solutions	Higher BET surface area applies for effective adsorption	[45]

pyrolysis at temperatures greater than 500°C than organic aerogels. Carbon aerogels provide a wide range of adsorption applications (Table 2.6) ranging from extraction of either organic or inorganic impurities from water for consumption or from industrial wastewater due to their unique properties, which include lower density, 3-D porous structure, active large surface area, etc. [1]. Maryam Hasanpour et al. investigated and reviewed the adsorption of heavy metals from wastewater via aerogels and found it to be governed by various factors like (a) pH, (b) the temperature of the solution, (c) the initial concentration of the heavy metal ions, (d) the composition of the ions, and (e) contact time, etc. [39].

2.15 APPLICATIONS OF ADSORBENTS

1. Extraction of dyes from water/wastewaters

The rapid expansion in globalization and the wide usage of various varieties of dyes of synthetic and semisynthetic origins has led to various environmental issues, which pose a life threat to human health and population, leading to an increase in morbidity and mortality due to significant hazards in the present water systems. The adsorption process is preferred for the removal of these toxic wastes because it has a higher efficiency of dye removal, is quite simple to operate, is cheap and effective, and is easily recyclable or desorption is simple. Various kinds of adsorbents are chosen for the elimination of various dyes from wastewaters. The dyes which pose a threat to the environment and some adsorbents used for their removal are provided in Table 2.7.

Christian and co-workers have demonstrated that nanocomposite α-Fe$_2$O$_3$/ graphene oxide (α-Fe$_2$O$_3$/GO) and polyamidoamine (PAN)-polyamidoamine nanoparticles are the best adsorbents for the removal of dye from aqueous media, with capacity adsorption (q_{max}) of 1,998.14 and 2,000 mg g^{-1}, respectively [57]. Similarly, Ibrahim and co-workers studied the adsorption capacity of walnut husk powder for removal of Azure C dye from its aqueous solution [58].

2. Metal ions and heavy metals

Metal ions pose a serious problem as they act as a pollutant in wastewaters/waters. They cause serious health hazards to humans and domestic stock, leading to varied amounts of damage and livelihood loss. Therefore, optimum elimination of heavy metal ions is important. Panda and co-workers developed a low-cost geopolymer-based adsorbent for the removal of cobalt, nickel, cadmium, and lead ions from an aqueous solution through the adsorption process [59]. Likewise, the removal of heavy metals from wastewaters has also been studied by Zhang and co-workers by using chitosan and its derivatives as adsorbents [60].

The effect of carbon nanotubes containing many walls as adsorbents for removal of heavy metals was researched by Wang and co-workers and it was found that carbon

TABLE 2.7

Environmentally toxic dyes and some adsorbents employed for their removal

Dye	Examples	Adsorbents used	References
	Acid yellow 36, acid orange 7, acid blue 83, acid blue 7	Activated carbon by pomelo peels	[46]
		Modified fly ash	[46]
		Sawdust-based adsorbent	[47]
Acid red 88	Methylene blue, basic red, basic yellow	Modified fly ash	[47]
		Activated carbon from red oak	[17]
		Brazilian low-cost agriculture waste	[48]
Basic blue 26			

Dye	Examples	Adsorbents used	References
	Congo red (CR), black or red direct dye	Walnut-based activated carbon	[49]
		Modified fly ash	[47]
		Graphene oxide	[50]
		Almond shell	[51]
		Spent mushroom waste	[52]

Direct red 28

(Continued)

TABLE 2.7
(Continued)

Dye	Examples	Adsorbents used	References
Mordant black 17	Mordant red, mordant black	Modified fly ash	[47]
		Apricot peels	[17]
		Mesoporous carbon material	[53]
Sulphur green	Sulphur brilliant green, sulphur blue, sulphur black 1	Activated carbon from guava tea wood	[54]
		Areca nut husk	[55]
		Neem leaf powder	[56]
		Activated carbon derived from guava tree wood	[54]

nanotubes containing multiple walls of approximate 60 possess a new magnetic property and furthermore add to the capacity of adsorption, which contributes to the removal efficiency of the adsorption process. These multiwall carbon nanotubes are effective in removing lead ions, which has been proven to be much better than similar adsorbents [61].

Adsorption of Cu by using seed cake made from groundnut powder and cakes produced in a similar manner as those made from coconut powders is best carried out at a pH range of 5 and a temperature range of about 40°C; an initial concentration of approximately 10 mg L^{-1}, a full time of contact of half an hour, and adsorbent dosage of 0.75 g for the groundnut cake and 1 g for the sesame seed cake was investigated by Kumar and co-workers [54]. Simultaneously, magnesium-modified biochars were also investigated and studied by Chen Y and co-workers for the removal of heavy metals like cadmium [10].

2.16 CONCLUSION

The adsorption process involves the accumulation of liquid or gas on the active adsorption surface. The factors that determine the adsorption extent are surface area and charge, the effective groups especially functional groups on the surface, amounts, porosity and particle size of adsorbents, solution pH, adsorbent surface, and time of contact of the adsorbent with the material to be adsorbed. Different natural and commercially available adsorbents find various applications in several industries as they can be employed for the removal of a large number of toxic materials from wastewater. The health hazard caused by the deposition of toxic dyes in wastewater can be minimized by adsorption on suitable and efficient adsorbents that serve as the most suitable option for the treatment of water systems. Its selection over other methods can be justified due to its highly efficient removal rate, ease of operation, cheap and simple method, and adsorbent reusability and recyclable nature. Serious health issues caused by heavy metal ions that are highly toxic and are present in wastewater can be minimized by their adsorption using low-cost polymer-based adsorbents, multiwalled carbon nanotubes, and bio-adsorbents. Adsorption by means of various adsorbents provides a cost-effective alternative method to remove hazardous contaminations from wastewater.

REFERENCES

1. Sabzehmeidani, M.M., S. Mahnaee, M. Ghaedi, H. Heidari, and V.A. Roy, *Carbon based materials: a review of adsorbents for inorganic and organic compounds.* Materials Advances, 2021. **2**(2): pp. 598–627.
2. Ateia, M., D.E. Helbling, and W.R. Dichtel, *Best practices for evaluating new materials as adsorbents for water treatment.* ACS Materials Letters, 2020. **2**(11): pp. 1532–1544.
3. Gupta, A., V. Sharma, K. Sharma, V. Kumar, S. Choudhary, P. Mankotia, B. Kumar, H. Mishra, A. Moulick, and A. Ekielski, *A review of adsorbents for heavy metal decontamination: growing approach to wastewater treatment.* Materials, 2021. **14**(16): p. 4702.
4. Bakkaloglu, S., M. Ersan, T. Karanfil, and O.G. Apul, *Effect of superfine pulverization of powdered activated carbon on adsorption of carbamazepine in natural source waters.* Science of the Total Environment, 2021. **793**: p. 148473.

5. Liu, W., C. Yao, M. Wang, J. Ji, L. Ying, and C. Fu, *Kinetics and thermodynamics characteristics of cationic yellow X-GL adsorption on attapulgite/rice hull-based activated carbon nanocomposites.* Environmental Progress & Sustainable Energy, 2013. **32**(3): pp. 655–662.

6. Yagub, M.T., T.K. Sen, S. Afroze, and H.M. Ang, *Dye and its removal from aqueous solution by adsorption: a review.* Advances in Colloid and Interface Science, 2014. **209**: pp. 172–184.

7. Thompson, C.O., A.O. Ndukwe, and C.O. Asadu, *Application of activated biomass waste as an adsorbent for the removal of lead(II) ion from wastewater.* Emerging Contaminants, 2020. **6**: pp. 259–267.

8. Kovacova, Z., S. Demcak, M. Balintova, C. Pla, and I. Zinicovscaia, *Influence of wooden sawdust treatments on Cu(II) and Zn(II) removal from water.* Materials, 2020. **13**(16): p. 3575.

9. Bai, W., M. Qian, Q. Li, S. Atkinson, B. Tang, Y. Zhu, and J. Wang, *Rice husk-based adsorbents for removing ammonia: kinetics, thermodynamics and adsorption mechanism.* Journal of Environmental Chemical Engineering, 2021. **9**(4): p. 105793.

10. Chen, Y., R. Shan, and X. Sun, *Adsorption of cadmium by magnesium-modified biochar at different pyrolysis temperatures.* BioResources, 2020. **15**(1): pp. 767–786.

11. Xiang, W., X. Zhang, J. Chen, W. Zou, F. He, X. Hu, D.C. Tsang, Y.S. Ok, and B. Gao, *Biochar technology in wastewater treatment: a critical review.* Chemosphere, 2020. **252**: p. 126539.

12. Hu, H., and K. Xu, *Physicochemical technologies for HRPs and risk control.* High-Risk Pollutants in Wastewater. 2020: Elsevier. 169–207.

13. Chikri, R., N. Elhadiri, and M. Benchanaa, *Efficiency of sawdust as low-cost adsorbent for dyes removal.* Journal of Chemistry, 2020. **2020**: 1–17.

14. Abbas, M., *Experimental investigation of activated carbon prepared from apricot stones material (ASM) adsorbent for removal of malachite green (MG) from aqueous solution.* Adsorption Science & Technology, 2020. **38**(1–2): pp. 24–45.

15. Rahman, M.S., *Adsorption of methyl blue onto activated carbon derived from red oak (Quercus rubra) acorns: a 2 6 factorial design and analysis.* Water, Air, & Soil Pollution, 2021. **232**(1): pp. 1–14.

16. Crini, G., *Non-conventional low-cost adsorbents for dye removal: a review.* Bioresource Technology, 2006. **97**(9): pp. 1061–1085.

17. Dutta, S., B. Gupta, S.K. Srivastava, and A.K. Gupta, *Recent advances on the removal of dyes from wastewater using various adsorbents: a critical review.* Materials Advances, 2021. **2**(14): 4497–4531

18. Sadegh, H., G.R. Shahryari, A. Masjedi, Z. Mahmoodi, and M. Kazemi, *A review on carbon nanotubes adsorbents for the removal of pollutants from aqueous solutions.* 2016. **7**(2): 109–120.

19. Jiao, L., J.Y.R. Seow, W.S. Skinner, Z.U. Wang, and H.-L. Jiang, *Metal–organic frameworks: structures and functional applications.* Materials Today, 2019. **27**: pp. 43–68.

20. Zhu, B., G. Huang, Y. He, J. Xie, T. He, J. Wang, and Z. Zong, *Synthesis and characterization of MOFs constructed from 5-(benzimidazole-1-yl) isophthalic acid and highly selective fluorescence detection of Fe(iii) and Cr(vi) in water.* RSC Advances, 2020. **10**(57): pp. 34943–34952.

21. Feng, Y., Y. Wang, Y. Wang, S. Liu, J. Jiang, C. Cao, and J. Yao, *Simple fabrication of easy handling millimeter-sized porous attapulgite/polymer beads for heavy metal removal.* Journal of Colloid and Interface Science, 2017. **502**: pp. 52–58.

22. Sikora, E., V. Hajdu, G. Muránszky, K.K. Katona, I. Kocserha, T. Kanazawa, B. Fiser, B. Viskolcz, and L. Vanyorek, *Application of ion-exchange resin beads to produce magnetic adsorbents.* Chemical Papers, 2021. **75**(3): pp. 1187–1195.

23. Huang, R., Q. Zhang, H. Yao, X. Lu, Q. Zhou, and D. Yan, *Ion-exchange resins for efficient removal of colorants in bis(hydroxyethyl) terephthalate.* ACS Omega, 2021. **6**(18): pp. 12351–12360.

24. Gil, A., L. Santamaría, S. Korili, M. Vicente, L. Barbosa, S. de Souza, L. Marçal, E. de Faria, and K. Ciuffi, *A review of organic-inorganic hybrid clay based adsorbents for contaminants removal: synthesis, perspectives and applications.* Journal of Environmental Chemical Engineering, 2021. **9**(5): p. 105808.

25. Adeyemo, A.A., I.O. Adeoye, and O.S. Bello, *Adsorption of dyes using different types of clay: a review.* Applied Water Science, 2017. **7**(2): pp. 543–568.

26. Awwad, A., M. Amer, and M. Al-aqarbeh, *TiO$_2$-kaolinite nanocomposite prepared from the Jordanian Kaolin clay: adsorption and thermodynamics of Pb(II) and Cd(II) ions in aqueous solution.* Chemistry International, 2020. **6**(4): 168–178.

27. Mockovčiaková, A., and Z. Orolinova, *Adsorption properties of modified bentonite clay.* Cheminé Technologija, 2009. **1**(50): pp. 47–50.

28. Zhou, C.H., Q. Zhou, Q.Q. Wu, S. Petit, X.C. Jiang, S.T. Xia, C.S. Li, and W.H. Yu, *Modification, hybridization and applications of saponite: an overview.* Applied Clay Science, 2019. **168**: pp. 136–154.

29. Xie, H., S. Zhang, J. Liu, J. Hu, and A. Tang, *A novel calcium oxalate/sepiolite composite for highly selective adsorption of Pb(II) from aqueous solutions.* Minerals, 2021. **11**(6): p. 552.

30. Huang, H., Z. Liu, J. Yun, H. Yang, and Z.-l. Xu, *Preparation of Laponite hydrogel in different shapes for selective dye adsorption and filtration separation.* Applied Clay Science, 2021. **201**: p. 105936.

31. Rahman, M.O., N. Rahman, G.F. Ahmed, M.S. Hasan, N.C. Dafader, M.J. Alam, S. Sultana, and F.T. Ahmed, *Synthesis and implication of grafted polymeric adsorbent for heavy metal removal.* SN Applied Sciences, 2020. **2**: pp. 1–10.

32. Wingenfelder, U., C. Hansen, G. Furrer, and R. Schulin, *Removal of heavy metals from mine waters by natural zeolites.* Environmental Science & Technology, 2005. **39**(12): pp. 4606–4613.

33. Denizli, A., B. Garipcan, A. Karabakan, and H. Senöz, *Synthesis and characterization of poly(hydroxyethyl methacrylate-N-methacryloyl-(l)-glutamic acid) copolymer beads for removal of lead ions.* Materials Science and Engineering: C, 2005. **25**(4): pp. 448–454.

34. Ma, N., Y. Yang, S. Chen, and Q. Zhang, *Preparation of amine group-containing chelating fiber for thorough removal of mercury ions.* Journal of Hazardous Materials, 2009. **171**(1–3): pp. 288–293.

35. Guida, S., C. Potter, B. Jefferson, and A. Soares, *Preparation and evaluation of zeolites for ammonium removal from municipal wastewater through ion exchange process.* Scientific Reports, 2020. **10**(1): pp. 1–11.

36. Ugwu, E., A. Othmani, and C. Nnaji, *A review on zeolites as cost-effective adsorbents for removal of heavy metals from aqueous environment.* International Journal of Environmental Science and Technology, 2021. **19**(8) pp. 1–24.

37. Ozaydin, S., G. Kocer, and A. Hepbasli, *Natural zeolites in energy applications.* Energy Sources, Part A, 2006. **28**(15): pp. 1425–1431.

38. Moazeni, M., S. Parastar, M. Mahdavi, and A. Ebrahimi, *Evaluation efficiency of Iranian natural zeolites and synthetic resin to removal of lead ions from aqueous solutions.* Applied Water Science, 2020. **10**(2): pp. 1–9.

39. Hasanpour, M., and M. Hatami, *Application of three dimensional porous aerogels as adsorbent for removal of heavy metal ions from water/wastewater: a review study.* Advances in Colloid and Interface Science, 2020. **284**: p. 102247.

40. Linneen, N.N., R. Pfeffer, and Y. Lin, *Amine distribution and carbon dioxide sorption performance of amine coated silica aerogel sorbents: effect of synthesis methods.* Industrial & Engineering Chemistry Research, 2013. **52**(41): pp. 14671–14679.

41. Shaheed, N., S. Javanshir, M. Esmkhani, M.G. Dekamin, and M.R. Naimi-Jamal, *Synthesis of nanocellulose aerogels and Cu-BTC/nanocellulose aerogel composites for adsorption of organic dyes and heavy metal ions.* Scientific Reports, 2021. **11**(1): pp. 1–11.

42. Yu, Y., Q. Ma, J.-b. Zhang, and G.-b. Liu, *Electrospun SiO₂ aerogel/polyacrylonitrile composited nanofibers with enhanced adsorption performance of volatile organic compounds.* Applied Surface Science, 2020. **512**: p. 145697.

43. Wang, Y., J. Pan, Y. Li, P. Zhang, M. Li, H. Zheng, X. Zhang, H. Li, and Q. Du, *Methylene blue adsorption by activated carbon, nickel alginate/activated carbon aerogel, and nickel alginate/graphene oxide aerogel: a comparison study.* Journal of Materials Research and Technology, 2020. **9**(6): pp. 12443–12460.

44. Liang, L., S. Zhang, G.A. Goenaga, X. Meng, T.A. Zawodzinski, and A.J. Ragauskas, *Chemically cross-linked cellulose nanocrystal aerogels for effective removal of cation dye.* Frontiers in Chemistry, 2020. **8**: p. 570.

45. Gao, C., Z. Dong, X. Hao, Y. Yao, and S. Guo, *Preparation of reduced graphene oxide aerogel and its adsorption for Pb(II).* ACS Omega, 2020. **5**(17): pp. 9903–9911.

46. Liu, Z., and K. Xing, *Removal of acid red 88 using activated carbon produced from pomelo peels by KOH activation: orthogonal experiment, isotherm, and kinetic studies.* Journal of Chemistry, 2021. **2021**: pp. 1–9.

47. Jiang, D., M. Chen, H. Wang, G. Zeng, D. Huang, M. Cheng, Y. Liu, W. Xue, and Z. Wang, *The application of different typological and structural MOFs-based materials for the dyes adsorption.* Coordination Chemistry Reviews, 2019. **380**: pp. 471–483.

48. Caponi, N., G.C. Collazzo, S.L. Jahn, G.L. Dotto, M.A. Mazutti, and E.L. Foletto, *Use of Brazilian kaolin as a potential low-cost adsorbent for the removal of malachite green from colored effluents.* Materials Research, 2017. **20**: pp. 14–22.

49. Li, Z., H. Hanafy, L. Zhang, L. Sellaoui, M.S. Netto, M.L. Oliveira, M.K. Seliem, G.L. Dotto, A. Bonilla-Petriciolet, and Q. Li, *Adsorption of congo red and methylene blue dyes on an ashitaba waste and a walnut shell-based activated carbon from aqueous solutions: experiments, characterization and physical interpretations.* Chemical Engineering Journal, 2020. **388**: p. 124263.

50. Jiang, H., Y. Cao, F. Zeng, Z. Xie, and F. He, *A novel Fe₃O₄/graphene oxide composite prepared by click chemistry for high-efficiency removal of congo red from water.* Journal of Nanomaterials, 2021. **2021**: pp. 1–11 .

51. El Khomri, M., N. El Messaoudi, A. Dbik, S. Bentahar, A. Lacherai, Z.G. Chegini, and M. Iqbal, *Organic dyes adsorption on the almond shell (Prunus dulcis) as agricultural solid waste from aqueous solution in single and binary mixture systems.* 2021. **12**(2): pp. 2022–2040.

52. Tian, X., C. Li, H. Yang, Z. Ye, and H. Xu, *Spent mushroom: a new low-cost adsorbent for removal of congo red from aqueous solutions.* Desalination and Water Treatment, 2011. **27**(1–3): pp. 319–326.

53. Blachnio, M., A. Derylo-Marczewska, S. Winter, and M. Zienkiewicz-Strzalka, *Mesoporous carbons of well-organized structure in the removal of dyes from aqueous solutions.* Molecules, 2021. **26**(8): p. 2159.

54. Mansour, R., A. El Shahawy, A. Attia, and M.S. Beheary, *Brilliant green dye biosorption using activated carbon derived from guava tree wood.* International Journal of Chemical Engineering, 2020. **2020**: pp. 1–12.

55. Baidya, K.S., and U. Kumar, *Adsorption of brilliant green dye from aqueous solution onto chemically modified areca nut husk.* South African Journal of Chemical Engineering, 2021. **35**: pp. 33–43.

56. Vinnilavu, G., and P. Balamurugan, *Colour removal using neem leaf powder.* Indian Journal of Environmental Protection, 2019. **30**(12): pp. 1148–1153.

57. Osagie, C., A. Othmani, S. Ghosh, A. Malloum, Z. KashitarashEsfahani, and S. Ahmadi, *Dyes adsorption from aqueous media through the nanotechnology: a review.* Journal of Materials Research and Technology, 2021. **14**: pp. 2195–2218

58. Ibrahim, H.K., M.A.A.H. Allah, M.A. Al-Da'amy, E.T. Kareem, and A.A. Abdulridha. *Adsorption of basic dye using environmental friendly adsorbent.* IOP Conference Series: Materials Science and Engineering. 2020. **871**: p. 012027. IOP Publishing.

59. Panda, L., S.K. Jena, S.S. Rath, and P.K. Misra, *Heavy metal removal from water by adsorption using a low-cost geopolymer.* Environmental Science and Pollution Research, 2020. **27**(19): pp. 24284–24298.

60. Zhang, Y., Q. Cheng, C. Wang, H. Li, X. Han, Z. Fan, G. Su, D. Pan, and Z. Li, *Research progress of adsorption and removal of heavy metals by chitosan and its derivatives: a review.* Chemosphere, 2021. **279**: p. 130927.

61. Wang, Z., W. Xu, F. Jie, Z. Zhao, K. Zhou, and H. Liu, *The selective adsorption performance and mechanism of multiwall magnetic carbon nanotubes for heavy metals in wastewater.* Scientific Reports, 2021. **11**(1): pp. 1–13.

3 Natural Fiber-Based Adsorbents for Heavy Metals and Dyes Removal

L.S. Maia, A.I.C. da Silva, N.C. Zanini,
L.T. Carvalho, P.H.F. Pereira, S.F. Medeiros,
D.S. Rosa and D.R. Mulinari

3.1 INTRODUCTION

Nowadays, the increasing demand for freshwater, the development of industrialization, and rapid human population growth have resulted in a deficiency in clean water resources. The chemical contamination of water from a wide range of organic and inorganic pollutants has raised the need to develop technologies capable of removing these pollutants in liquid and gaseous waste. Heavy metals and dyes are two examples of the largely widespread pollutants in industrial effluents. Heavy metal contaminants in water are usually hazardous to organisms and can affect the ecosystem. On the other hand, synthetic dyes are also one of the primary sources of water pollution due to the dyeing of textile fibers, papers, leathers, plastics, elastomers, and pharmaceuticals. Thus, the presence of these dyes in natural water bodies reduces the penetration of light, delays photosynthetic activity, inhibits biota growth, and forms bonds between different metal ions that induce micro-toxicity. Decontamination processes involve physical, biological, and chemical techniques. The available technologies are classified into conventional methods, established recovery methods, and emerging methods.

Among the diverse treatment processes developed for water treatment, every process has intrinsic limitations in terms of applicability, cost, and effectiveness. One attractive option is treating wastewater from industry or other activities with an adsorption process to improve its quality for further industrial and agricultural operations. Adsorption is highly effective and the most extensively used method due to its simplicity of performance and relatively low application cost. In addition, adsorption has become a valuable tool in various sectors, with technological and biological importance and practical applications in industry and environmental protection.

This chapter presents the adsorptive removal process of heavy metals and dyes onto natural fiber-based materials comprising activated carbon (AC), nanofibers, modified natural fibers, and the regeneration of adsorbents. Thus, we explore and comprehensively review the multitude of aspects regarding the chemical and physical

DOI: 10.1201/9781032636368-3

nature of natural fiber-based materials and their adsorption ability, the factors affecting the adsorption process, as well as equilibrium, the kinetics model, and mechanism adsorption.

3.2 MATERIALS ADSORBENTS

The adsorption process is a surface phenomenon in which adsorbates transfer onto adsorbents. Over the past decades, adsorption technology has been widely applied for water and wastewater treatment because of its low-cost, efficiency, simplicity, and environmental friendliness [1]. Water quality is crucial for human health, and it can be improved by the adsorption of pollutants such as dyes, heavy metals, phenolic compounds, and pharmaceuticals. Heavy metals are one of the major pollutants in industrial wastewaters, originating from metal complex dyes, pesticides, fertilizers, fixing agents (added to dyes to improve dye adsorption onto the fibers), mordants, pigments, and bleaching agents [2]. Industries must reduce the concentration of these metals below a standard concentration regulated by environmental legislation to correctly discard their waste streams in the environment [3].

There are many ways to remove organic compounds and heavy metals, such as oxidation, reverse osmosis, ion exchange processes, electrodialysis, electrolyte, adsorption, among other methods. However, it is essential to mention that due to the high operating cost of some of these methods, adsorption is one of the best methods given its availability, low cost, and bioavailability of [4]. In the adsorption process, all soluble and insoluble compounds in water can be separated. The accumulation of absorbable materials at the external surface and in the space between surfaces occurs in adsorbent particles.

Adsorption is commonly used for water and wastewater treatment because of its convenience, ease of operation, and simplicity of design [5]. Adsorption separation processes are based on three distinct mechanisms: the steric, equilibrium, and kinetic mechanisms. For the steric mechanism, the pores of the adsorbent material have characteristic dimensions, which allow specific molecules to enter, excluding the others. For the equilibrium mechanisms, there are the abilities of different solids to accommodate different species of adsorbates, which are preferentially adsorbed to other compounds. The kinetic mechanism is based on the different diffusivities of different species in the adsorbent pores [6]. In this context, much research has been devoted to understanding the properties of materials as adsorbents to remove these dangerous compounds [7]. It was already reported that the removal capacity of inorganic, organic, and biological pollutants from wastewater via adsorption could be as high as 99.9% [8].

Adsorption efficiency is related to adsorbents' pore number and size and their surface area [9]. Several types of adsorbents are used to efficiently remove heavy metals from wastewater: both commercial and bioadsorbents [10]. Some types of adsorbents commonly used for the wastewater purification process include activated alumina [11], silica gel [12], limestone [13], AC [14–16], nanofibers [17], modified natural fibers [18], and others. AC is the most commercialized adsorbent globally; however, its high cost and the difficulty of regeneration restrict its use [19]. The use of nano- and biocomposites is also related in the literature. However, the increased cost of these materials and complexities involved in their synthesis leads to numerous research related to the use of low-cost agricultural and bio-based waste materials that can be suitably activated or used to remove inorganic impurities from wastewater. Alternative cheap adsorbents are needed to be developed for the removal of inorganic impurities and metal ion-containing wastewater. There is now a reported utilization of adsorbents made from natural materials, for example, natural clays [20], natural zeolites [21], and polysaccharides such as chitosan and its derivatives [3].

Nowadays, a series of new agricultural and forest waste materials, such as grouts, crumbs, husks, corn cobs, dross and straws, barks, fruit peels, nuts, dates, fruit seeds, and biowaste adsorbents, for example, biochars and biorefinery wastes, have initiated a great interest in the use of such materials as alternate adsorbents. A new term can be frequently found in literature, which is called "green adsorption" and refers to low-cost materials originating from (i) agricultural sources and by-products (fruits, vegetables, foods); (ii) agricultural residues and wastes; and (iii) low-cost sources from which most complex adsorbents are produced (i.e., ACs after pyrolysis of agricultural sources) [22]. These "green adsorbents" are expected to be inferior regarding their adsorption capacity than the conventional adsorbents, but their cost potential makes them competitive. Table 3.1 lists some commonly used natural fiber-based fibers, their preparation methods, and metal- or dye-adsorbed materials.

To improve the adsorption of heavy metals using different alternatives, the authors have investigated fibers based on biological materials. Paramasivam et al. [30] investigated banana pseudostem fiber (BPF) as a biological material to remove the concentration of Pb(II) and observed that pH, agitation speed, and contact time substantially contribute to the removal efficiency. Huang et al. [36] also evaluated the influence of pH values. They noted that at lower pH, protonated amine/imine units in polyaniline/jute fiber (JF) attract negatively charged Cr(VI) anions that oxidize the amine nitrogen atoms producing imine nitrogen and reduced Cr(III) ions. Zhou et al. [24] compared cellulose-based amphoteric adsorbents with amino groups and carboxyl groups. They observed that the removal rate for heavy metal ions (Cr(VI), Cd(II), Cu(II), Zn(II), and Pb(II)) reached 100%. The high densities of amino and carboxyl groups contributed to the excellent performance via specialization and cooperation. Niu et al. [27] evaluated the use of cotton fiber to remove metals such as Cu(II), Cr(III), and Pb(II), while Rahman et al. [29] focused on the use of waste fruit fiber removing a significant amount of 98% of toxic metal ions. Some authors have also studied the use of stable materials capable of regenerating after application, focusing on improving a fiber's adsorption efficiency. For example, Wu et al. [26] applied fiber based on polysaccharides incorporated at TiO_2 nanoparticles to evaluate heavy metal removal, such as Ni(II), Cd(II), Cu(II), Hg(II), Mn(VII), and Cr(VI) ions. The material demonstrated efficient adsorption even after five cycles, around 75% of removal, demonstrating their regeneration capacity and improved application.

Natural fibers are often studied for the removal of dyes and heavy metals from wastewater. Dai et al. [31] studied natural palm leaf fibers (PLSFs) to adsorb reactive dyes and heavy metals by hydroxyl enrichment. The authors obtained adsorption of 83.19 mg g^{-1} for RY3 and adsorption of 189.48 mg g^{-1} for Cr(VI) when applied at lower pH, representing good toxic removal. Some studies for removing methylene blue dye used biochar incorporated with MnO_2, resulting in high removal efficiency of 99.73% [37, 38]. Other studies used metaplexis japonica fiber, achieving 100%

TABLE 3.1
Natural fibers used for heavy metals and dyes removal

Natural fiber-based adsorbents	Preparation method	Adsorbed metals or dyes	References
AC from hemp, sisal, and flax	Impregnation and carbonization	Potential application for wastewater purification	[23]
Cellulose-based amphoteric with a high density of amino and carboxyl groups	Cross-linking reaction followed by chemical modification	Cr(VI), Cd(II), Cu(II), Zn(II), and Pb(II)	[24]
Natural fiber-nanocellulose composite filters derived from flax and agave fibers	Reaction with anionic TEMPO-oxidized cellulose nanofibrils	Cu(II)	[25]
Ternary composite consisting of carboxymethyl chitosan, hemicellulose, and nanosized TiO$_2$ (CHNT)	Incorporation of TiO$_2$ nanoparticles into carboxymethyl chitosan-hemicellulose polysaccharide network	Ni(II), Cd(II), Cu(II), Hg(II), Mn(VII), and Cr(VI)	[26]
Cotton fiber functionalized with tetraethylenepentamine and chitosan (CTPC)	Chemical modification, oxidation, and functionalization with chitosan	Cu(II), Pb(II), and Cr(III) ions	[27]
Cellulose acetate/zeolite fiber	Wet spinning method	Cu(II) and Ni(II)	[28]
Cellulose from waste pandanus fruit fiber and durian rinds	Chemical extraction grafted with poly(acrylonitrile)	Toxic metals from wastewater	[29]
BPF	Fiber extraction with raspador machine	Pb(II)	[30]
Palm leaf sheath fibers (PLSFs)	Carboxymethylation multi-hydroxylation followed by the adsorption of reactive yellow 3 (RY3)	Cr(VI)	[31]
Coir fiber	Layer-by-layer deposition of chitosan and polyacrylamide	Dye from local textile mill effluent treatment	[32]
Bast fibers of flax, ramie, and kenaf	Used as natural bast fibers	Yellow 37 dye	[33]
TiO$_2$-based polymeric hollow fiber	Nonsolvent-induced phase separation (NIPS)	Arsenic	[34]
Dissolved cellulose fiber/microfibrillated cellulose composite	Chemical modification by organic solvent-free approach followed by precipitation	Dyes from aqueous solutions	[35]

(Continued)

TABLE 3.1
(Continued)

Natural fiber-based adsorbents	Preparation method	Adsorbed metals or dyes	References
Polyaniline/JF composite	In situ polymerization by ammonium persulfate (APS) oxidative initiation	Cd(II) and Cr(VI)	[36]
AC fiber derived from the seed hair fibers of metaplexis japonica (MACFs)	Chemical activation, pre-oxidation, and carbonization	Methylene blue	[37]
Lignin-derived porous biochar	Chemical modification with different manganese compounds	Methylene blue	[38]
Juncus effusus (JE) fiber	Chemical modification with an FA resin, mainly constituted by dimethyloldihydroxyethylene urea (DMDHEU)	Disperse red 3B dye	[39]

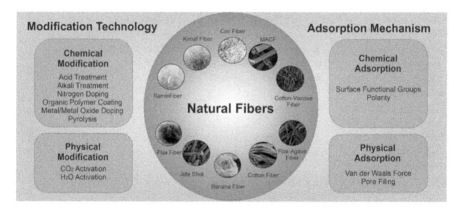

FIGURE 3.1   Physical and chemical mechanism involved in the fiber treatment and natural fibers adsorption process applied to heavy metal and dyes removal adapted [62].

removal under single filtration and good recyclability [37, 38]. Low-cost sorbents have also been evaluated for dyes removal, such as yellow 37 [33].

Figure 3.1 represents the physical and chemical mechanism involved in the fiber treatment and adsorption process, demonstrating some of the possibilities of natural fibers to treat heavy metals and dye compounds.

3.2.1 FACTORS AFFECTING THE ADSORPTION PROCESS

Adsorption phenomena result from combining the forces involved in physical and chemical interactions [6]. Thus, several factors influence the adsorption process, such as (i) effect of pH, (ii) effect of adsorbent dose, (iii) effect of contact time, (iv) effect of temperature, and (v) effect of competing ions [40]. It is a process that depends on several factors, such as the adsorbent nature, the adsorbate, and the operating conditions. Adsorbent characteristics include surface area, pore size, density, functional groups present on the surface, and material hydrophobicity [7]. On the other hand, the adsorbate nature depends on polarity, molecule size, solubility, and acidity or basicity. Operational conditions mainly include temperature, pH, and nature of the solvent [6].

3.2.2 ADSORPTION EQUILIBRIUM

Adsorption balance is generally an essential requirement for information about the analysis of an adsorption separation process. The affinity of an adsorbent on the adsorbate can be quantified by the adsorption equilibrium capacity [41]. The equilibrium isotherms give the adsorption capacity of an adsorbent and the thermodynamic parameters and provide the means to find the interaction between the adsorbent (q_e) and the adsorbate (C_e) [7].

In a solid-liquid system, adsorption removes solutes from the solution and their accumulation at a solid surface. The solute remaining in the solution reaches a dynamic equilibrium with that adsorbed on the solid phase. The amount of adsorbate that an adsorbent can take up as a function of both temperature and concentration of adsorbate, and the process, at a constant temperature, can be described by an adsorption isotherm according to Equation 3.1:

$$q = (C_0 - C_e)Vm \qquad (3.1)$$

where q (mg g^{-1}) is the adsorption capacity; C_0 and C_t (mg L^{-1}) are the initial and final adsorbate concentrations; V is the solution volume (L); m is the adsorbent mass (g) [5].

So we get a graph of q versus C_e. First, however, we need to know how to get the value of C_e and q. The obtained graphics can present several forms, providing important information about the adsorption mechanism, as shown in Figure 3.2.

After equilibrium is reached to obtain the C_e values, the adsorbent is separated from the solution using a membrane filter, filter paper, or centrifugation. The supernatant solution is analyzed to determine the residual concentration adsorbate (C_e). Analytical techniques can determine this (depending on the adsorbate used), such as gas or liquid chromatography, ultraviolet or visible spectrometry, absorption or emission spectrometry, or other suitable means [6].

Several isotherm equations have been proposed with two or more parameters to fit the experimental data on the values of q versus C_e. Among these, we can mention the Langmuir, Freundlich, Redlich-Peterson, BET (Brunauer, Emmett, Teller) isotherm, and Dubinin-Radushkevich equations. However, the most used are the Langmuir, Freundlich, and Temkin equations because they predict a maximum material adsorption capacity (Langmuir model) and the ability to detect the behavior of experimental data [42].

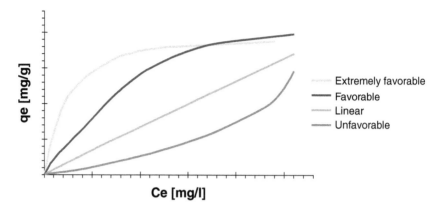

FIGURE 3.2   Illustration of the possible forms of the adsorption isotherms.

TABLE 3.2

Adsorption isotherm models' illustration considering their expressions, parameters, and basic considerations

Adsorption isotherm models	Original forms	Parameters	Basic considerations
Langmuir	$q_e = \dfrac{q_{max}\,k_L\,C_e}{1 + k_L\,C_e}$	q_{max} (mg g^{-1}) and K_L (L mg^{-1}) are the monolayer maximum adsorption capacity and the adsorption constant	Dynamic equilibrium monolayer adsorption between adsorbent and adsorbate
Freundlich	$\log q_e = \log K_f + \dfrac{1}{n}\log C_e$	K_f (mg g^{-1}) and n are the adsorptions constant for Freundlich isotherm and the adsorption intensity	Adsorption on the heterogeneous surface and the content of adsorption enhanced infinitely with a high concentration
Temkin	$q_e = \dfrac{RT}{b}\ln\!\left(k_T C_e\right)$	K_T (L mg^{-1}) is the Temkin isotherm constant, b (J mol^{-1}) is the adsorption energy variation factor, R (J mol^{-1} K^{-1}) is the universal gas constant, and T (K) is the absolute temperature	Heat of adsorption pollutants eliminated directly by coverage owing to adsorbent-adsorbate interactions

Frequently used adsorption isotherm models to describe the adsorbent and the adsorbate are provided in Table 3.2.

3.2.3 ADSORPTION KINETICS MODEL

Kinetics is fundamental to understand the adsorption mechanism and the rate limiting steps. Adsorption kinetics is expressed as the removal rate of the fluid phase

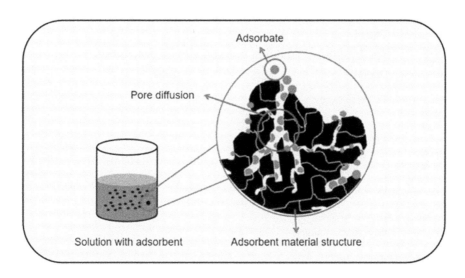

FIGURE 3.3 Illustration of the migration and adsorption kinetics steps.

adsorbate over time, involving the mass transfer of one or more components contained in an external liquid mass into the interior of the adsorbent particle, which should migrate through the macropores to the most interiors of this particle [6]. In principle, adsorption kinetics can be guided by different processes: (i) external mass transfer, which corresponds to the transfer of molecules from the fluid phase to the outer surface of the particle adsorbent through a layer of fluid that surrounds the particle; (ii) pore diffusion, caused by the diffusion of molecules in the fluid into the pores; and (iii) surface diffusion, which corresponds to the diffusion of molecules fully adsorbed along the pore surface. These steps are provided in Figure 3.3.

The adsorption nature depends on evaluating the adsorption kinetics, like the rate and adsorption mechanism, which can occur through physical or chemical phenomena. In a batch adsorption process, the traditional models that are frequently utilized are provided in Table 3.3.

The pseudo-first-order equations are based on five assumptions: adsorption only occurs on localized sites, and the interactions between adsorbates are not considered; the adsorption energy is not dependent on surface coverage; maximum adsorption corresponds to a saturated monolayer of adsorbates on the adsorber surface; the adsorbates' concentrations are constant; and the adsorbate uptake obeys the pseudo-first-order rate. For pseudo-second-order kinetics, the assumptions are almost the same as for the pseudo-first-order model.

Some Elovich model hypotheses include a linear relationship between energy and coverage surface; constant adsorbate concentration and the ion uptake on the adsorbent start small and after some time they are in exponential form. The intra-particle diffusion model is the so-called Weber-Morris intra-particle model. The straight-line form of q_t versus $t_{1/2}$ shows that the adsorption process is controlled only by intra-particle diffusion. The external resistance to mass transfer enveloping the particles is understood to be significant only in the early stages of adsorption. According to the

TABLE 3.3

The traditional equation used on kinetic description

Model	Equation
Pseudo first order	$\dfrac{dq_t}{dt} = K_1\left(q_e + q_t\right)$
Pseudo second order	$\dfrac{dq_t}{dt} = K_2\left(q_e + q_t\right)^2$
Elovich model	$\dfrac{dq_t}{dt} = \alpha \exp\left(-bq_t\right)$
Intra-particle diffusion	$q_t = K_1 t^{\frac{1}{2}} + C$
Decay order	$q_t = q_e + a_1 e^{-\frac{t}{b_1}} + a_1 e^{-\frac{t}{b_2}}$
Shrinking-core	$t = \dfrac{q}{q_e}\tau_1 + \left[1 + 2\left(1 - \dfrac{q}{q_e}\right) - 3\left(1 - \dfrac{q}{q_e}\right)^{2/3}\right]\tau_2$

shrinking-core model, an adsorption process occurs in three sequential steps: diffusion through an external liquid film, diffusion through a saturated shell, and adsorption at the surface of the sorbate-free core. In this model, the two parameters t_1 and t_2 are the characteristic diffusion times for diffusion through the external liquid film and diffusion through the saturated shell. In a continuous adsorption process, the various experimental models for calculating breakthrough curves that can analyze the mass balance of a fixed bed include Bohart-Adams, Thomas, Wolborska, and Yoon-Nelson [7].

3.2.4 ADSORPTION MECHANISM

The adsorption mechanisms can be classified according to their intensity into two types: physical adsorption (physisorption) and chemical adsorption (chemisorption) based on the interaction strength between the adsorbate and the substrate [41–43]. In physical adsorption, the binding of the adsorbate to the surface of the adsorbent involves a relatively weak interaction that can be attributed to hydrogen bonding, van der Waals forces, electrostatic forces, and hydrophobic interactions, which are similar to molecular cohesion forces [44, 45]. On the other hand, chemical adsorption involves the strong interaction of H electrons and metal electrons in such a way that it leads to splitting of the molecule, formation of covalent bonds with metal electrons, and perturbations in the metal electron system, as shown in Figure 3.4 [46]. This essentially results in a new chemical bond and, therefore, it is much stronger than in the case of physisorption [6].

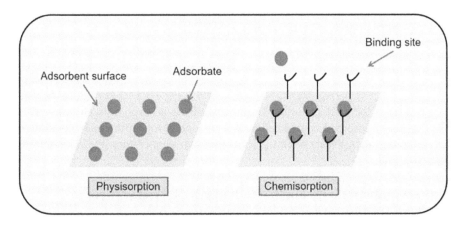

FIGURE 3.4   Schematic representation of the physisorption and chemosorption process.

TABLE 3.4
Difference between chemisorption and physisorption [38]

Properties	Physical adsorption	Chemical adsorption
Adsorption force	Van der Waals force, hydrogen bond interaction	Chelating formation, electrostatic attraction, and ion exchange
Adsorption heat	$<40\,mJ\,mol^{-1}$	$80–400\,kJ\,mol^{-1}$
Adsorption rate	Fast	Sometimes slow
Adsorption temperature	Low	High
Adsorption layer	Monolayer or multilayer	Monolayer
Adsorption-desorption process	Reversible	Irreversible

The concepts of chemisorption and physisorption are distinct, but the two adsorption mechanisms are not entirely independent. Physisorption is an exothermic process, and its adsorption enthalpy is low, nearly 20 to $40\,kJ\,mol^{-1}$ [5]. In addition, physisorption depends on the surface area of the adsorbent. As surfaces area increases, the extent of adsorption also increases. Chemisorption has high specificity; that is, it is peculiar, and it takes place only if there is a chemical bonding between the adsorbent and the adsorbate. Chemisorption is irreversible, and it favors high pressure. Due to chemical bonding, the enthalpy of adsorption of chemisorption is high, nearly 80 to $240\,kJ\,mol^{-1}$. Chemisorption depends on the surface area. As the surface area of the adsorbent increases, chemisorption also increases [47]. The main differences between chemical and physical adsorption are listed in Table 3.4.

Adsorption rates are not good criteria for distinguishing adsorption types (chemical and physical). Chemical adsorption can be fast if the activation energy is high or low, and it can be slow if the activation energy is high. Physical adsorption is generally fast, but it can be slow if it involves the occupation of a porous medium [6].

Moreover, isosteric heat is considered one of the fundamental quantities of adsorption studies, representing the proportion of the change in the infinitesimal of the adsorbent enthalpy to the infinitesimal change in the amount adsorbed. Isosteric heat is calculated from the Van't Hoff thermodynamics Equation 3.2,

$$\frac{\Delta H^\circ}{R_{gT_2}} = -\left(\frac{\delta \ln P}{\partial T}\right) C_\bullet \tag{3.2}$$

where ΔH° is the differential molar enthalpy of the adsorption in (J mol^{-1}), C_μ is the maximum adsorbed concentration, δ is the coefficient of thermal expansion of the saturation concentration. If C_μ decreases with temperature, the isosteric heat increases [42].

3.3 ADSORPTION BY NATURAL FIBER-BASED MATERIALS FOR HEAVY METALS AND DYES REMOVAL

It is estimated that by 2025, half of the world's population will live in water-stressed areas [48]. Safe drinking water and sanitation are one of the 17 Sustainable Development Goals set by the World Health Organization (WHO) to be guaranteed by 2030 for the world's population [49]. Without access to adequate sanitation measures associated with lack of regulation or non-enforcement of industrial waste, humans are exposed to different types of contamination in water, such as harmful microorganisms [50], emerging organic contaminants [51], dyes [52], and heavy metals [9]. Rapid industrialization and urbanization resulting from the globalized world of the last decades have had positive consequences in advancing various technologies and facilities. Nevertheless, water pollution from heavy metals and dye contamination has negatively accompanied these processes, generating worldwide concern of ecological and health damage due to their toxic characteristics like carcinogenicity, persistence (non-biodegradability), and bioaccumulative effect in the food chain [53, 54]. The toxicity of heavy metals is associated with their ability to modify proteins that lose their natural functions, causing improper functioning or cell death. In this case, heavy metals interact with proteins forming complexes involving carboxylic acid, amine, and diol functional groups [55].

Some heavy metals, such as Ni, in specific small concentrations can even participate in the metabolism of animals and plants [56]. Thus, we can classify such metals as potentially toxic metals. Organizations such as the US Environmental Protection Agency stipulate maximum concentration limits (MCLs) representing the highest contaminant level allowed in drinking water [57]. According to Afroze and Sen [53, 57], the heavy metals that pose the most risk to humans and the environment are arsenic (As) with MCL of 0.010 ppm, mercury (Hg) with MCL of 0.002 ppm, and cadmium (Cd) with MCL of 0.005 ppm. Higher concentrations over long periods can cause cancer, skin lesions, cardiovascular disease, neurotoxicity, and diabetes from As exposure [55], kidney damage and hypertension from Cd exposure [55], and kidney damage from Hg exposure [57]. Other heavy metals, or potentially toxic elements [54], can be cited, such as antimony (Sb) [58], barium (Ba) [59], chromium (Cr) [60], cobalt (Co) [61], lead (Pb) [59], molybdenum (Mo) [62], nickel (Ni) [63], selenium (Se) [64], thallium (Tl) [65], tin (Sn) [66], tungsten

(W) [67], uranium (U) [68], vanadium (V) [69], copper (Cu) [70], zinc (Zn) [71], and manganese (Mn) [72].

Although heavy metals can be discharged into waterways naturally (action of time/erosion in natural rocks and ore deposits), growing and uncontrolled anthropogenic action (industrial, mining, and agricultural activities) is a primary ecological/health concern [73]. Specifically, industrial sources of different heavy metals polluting watercourses can be exemplified. Arsenic pollution sources can come from mining activities, the combustion of fossil fuels, petroleum refining, sewage sludge [74]. For Cd, the pollution sources are the production of batteries, fertilizers, paint, and stabilizers. Hg pollution usually comes from refineries and coal-fired power plants [53]. These are just a few examples; obviously, there are other polluting sources for the other heavy metals mentioned previously. Thus, it is very alarming that most of the end products surrounding human life may have generated some heavy metals in their production process, mistakenly disposed of in high concentrations in water bodies, potentially reaching drinking water sources and posing a health hazard to flora and fauna [49].

Unfortunately, heavy metals are not the only pollutant in water bodies generated by industries. The textile industry plays an essential role in the world economy, where 90% of the dyes produced worldwide are destined for fabric dyeing [75]. The high demand for textile fabric can produce toxic chemical contaminants primarily due to colored aromatic and ionizing dyes. Dyeing processes usually generate large volumes of coloring agents and other chemicals, which, if improperly disposed of, can also generate health/environmental nuisances [53]. Contamination of water bodies by dyes can affect aquatic organisms and cause mutations, skin irritations, and allergies in humans that come into contact with such harmful substances [48]. Dyes can be classified as non-biodegradable compounds in nature due to their inert and complex recalcitrant molecular structures. Their inherent color makes them easily identifiable and visible at a low concentration of 1 ppm in water. This characteristic can modify the water quality, harming biological processes such as photosynthesis and aquatic life through eutrophication that interferes with the light penetration through the water [75, 76]. Furthermore, trace metals like Cr, As, Cu, and Zn can be found in wastewater from textile industries [75].

As mentioned, heavy metals and dyes present a great danger to the natural environment and humans if left untreated in water bodies [53]. Therefore, it is necessary to develop ways to deal with those contaminants. According to Kunjuzwa et al. [77], conventional techniques applied in water treatment systems such as coagulation, flocculation, sedimentation, filtration by layers of sand and gravel, and disinfection have been used for centuries due to their efficiency for contaminants commonly observed in water bodies. Nevertheless, these techniques have negative features such as the centralization of the process, the high cost of implementation, maintenance, and configuration. Other methods can be discussed to retain heavy metals of wastewaters, such as chemical precipitation [78], coagulation-flocculation [79], ion exchange [80], electrochemical (EC) treatment [81], hydrogels [82], aerogels [83], membranes [84], and adsorption [85]. Among all the methods mentioned, the simplicity of design, ease of operation [53], heavy metals recovery, high selectivity, less sludge volume produced [86], and high efficiency make adsorption an exciting and successful alternative for retaining heavy metals and dyes [53]. One way to deal with adsorption using abundant and inexpensive sources is the use of natural-source materials [53], such as bioadsorbents, and natural adsorbents such as zeolites [87], clay [88], chitosan [3], red mud [89], and green sorbents (agricultural waste materials and natural fibers in general) [90].

Natural fibers are classified into three groups: plant, animal, and mineral fibers [91]. Proteins are mostly the principal components of animal fibers, while cellulose is in plant fibers [92]. Furthermore, botanical types are the most common classification for natural fibers. They are divided into five categories based on this approach: bast, leaf, seed, core, grass, and other types such as roots and woods [93–95]. The bast fibers extracted from the inner bark of trees include jute, flax, cannabis, ramie, and kenaf, leaf fibers extracted from the leaf, such as banana, sisal, abaca, agave, and pineapple, while seeds fibers extracted from dried seeds of the plant include cotton, mango, grape, coir, and kapok, grass and reeds such as wheat, maize, core fibers such as kenaf, hemp, and jute and rice, and other types such as roots and wood [96, 97]. For instance, both bast and core fibers have jute, flax, hemp, and kenaf, while agave, coconut, and oil palm have fruit and stem fibers. Animal fibers include hair, wool, silk, and collagen, while mineral fibers include asbestos, erionite and other zeolites, and mineral talc [91]. These fibers are biodegradable, inexpensive, of low density, non-toxic, available worldwide, energy-efficient, and eco-friendly [97]. Cellulose, hemicelluloses, and lignin are the main components of natural fiber, with trace contents of organic extractives and inorganic minerals [98]. Generally, the contents of cellulose, hemicelluloses, and lignin account for over 90 wt% in most lignocellulosic biomass [99]. Vegetal fibers are mainly formed by a complex structure of cellulose (composed of crystalline, disorganized, organized, and aggregated parts) [100], which are primarily present in amorphous and non-cellulosic regions [101].

According to Rojas et al. [102], the use of natural fibers for wastewater contamination treatments is gradually increasing. Natural fibers have a high surface area due to complex organic geometry, favoring adsorption processes with high sorption efficiency and capacity, and are also low-cost. According to Gallardo-Rodríguez, Rios-Rivera and Bennevitz [103], while sorption processes using synthetic polymers can reach up to US\$1,000 m^{-3}, the use of natural materials can cost ten times less. Vegetable fibers stand out among natural fibers due to their porous structure, good mechanical properties [36], and lignocellulosic composition, which can interact with heavy metal ions through ion exchange providing high adsorption power [102]. Moreover, when these natural fibers are agricultural by-products, their use is even more advantageous since they are abundant non-conventional cost-effective adsorbents for removing metal ions and dyes. Furthermore, the destination of agro-industrial residues of natural fibers for research and solutions to retain heavy metals/dyes is a sustainable and intelligent route [53]. Although natural fibers can be used untreated and/or in microscopic size for heavy metals and dyes sorption in aqueous media [36, 53], it is worth mentioning that the use of AC from natural fibers [104], nanosized natural fibers (nanofibers) [48], and modified natural fibers [18] can be intelligent alternatives to improve adsorption processes.

3.4 ACTIVATED CARBON

The adsorption process is a widely applied technique for wastewater decontamination. Among the options for adsorbent materials, AC has attracted much attention in the scientific community.

AC is an exceedingly porous carbonaceous material due to its large surface area ranging from 500 to 3,000 m^2g^{-1}. This excellent feature ensures that AC has a high adsorption capacity [7]. In addition, it presents good charge-holding capacity, high mechanical strength, and thermal and chemical stability. Also, it has low reactivity with acids and bases, and it has a hydrophobic surface and can be relatively easily regenerated [105–107]. All these characteristics contribute toward environmental protection, resulting in their increased production and utilization [108].

Commercial AC can be found in different physical forms, such as granular, powdery, fibrous [7]. Among them, powdered activated carbon (PAC) and granulated activated carbon (GAC) are prominent [109]. PAC has a large specific surface area, which contributes to excellent adsorption efficiency. However, this material requires a longer contact time with the substrate, and the separating process (filtration, sedimentation, and centrifugation) can be challenging compared to GAC, restricting its use [110]. GAC has good mechanical strength, high durability, good fluidity, and ease of recovery by filtration [110]. It can be used for small and large-scale adsorption [109].

AC has been widely used in various vital fields due to its fascinating physicochemical characteristics. It can be used not just for water and wastewater treatment generated by industries and municipals [111, 112], but also for the removal of contaminants such as dyes [113–115] and heavy metals [116–118], gas storage [119], energy storage [120], elimination of taste and odor [121],and recently in electronic devices [122].

An essential parameter for the use of AC is porosity. This factor corresponds to the pore volume that can contain the fluid/molecule, and the pore's total volume present on the material [123], and can vary according to the preparation process [124].

Considering the presence of the porous complex network in AC's structure, their pores can be classified according to their diameter: micropores (2–10 nm), mesopores (10–100 nm), and macropores (>100 nm), as shown in Figure 3.5 [113].

Based on a study developed by De Gisi et al. [5], in the presence of macropores and mesopores, AC has densities (defined as pore space volume for a gram of carbon) greater than 1 cm^3g^{-1} and between 0.85 and 1.0 cm^3g^{-1}, respectively. In the presence of micropores, AC has a density of 0.85 cm^3g^{-1}.

Another crucial parameter that must be evaluated is pore volume. This factor is associated with the total volume of empty (or pores) present in the adsorbent material. According to Henning and Von Kienle [124], the pore volume of AC is greater than 0.2 mL g^{-1}.

3.4.1 AC FROM NATURAL FIBER-BASED RESIDUES

Throughout history, industries have produced millions of tons of AC from tons of coal, wood, lignite, which generates several harmful wastes for the environment.

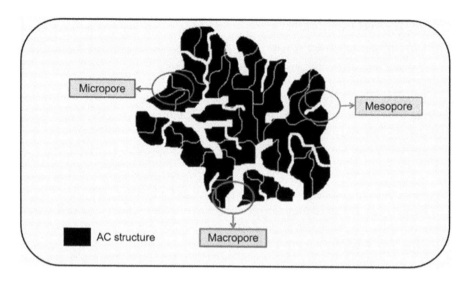

FIGURE 3.5 Representation and classification of pores present in AC.

Therefore, in recent years, reducing waste disposal has become one of the main strategies of industries for developing ecologically correct products. In this context, an increase in the use of natural fiber-based residue for obtaining AC has been observed [125, 126].

According to the literature, the natural fiber-based residue has attracted attention as the only energy source with many ecological advantages such as low cost, abundant availability, biodegradable resource, renewable, and high lignocellulosic composition [114, 127]. Besides, these biomasses have a higher percentage of carbon and lower ash content [113, 115]. Considered as a "green" adsorbent [128], several natural fiber-based residues can be used for the production of AC, such as nutshell [129], corn cobs [130], açaí [127], banana peels [113], sugarcane [131], palm [114], European trees [132], peanut shells [128], moringa oil [133], cucumis [134], pineapple crown [115], and others.

3.4.2 METHODOLOGY FOR THE PRODUCTION OF AC FROM NATURAL FIBER-BASED RESIDUES

When designing the conversion of lignocellulosic feedstock to AC, some parameters must be evaluated, such as activation process and reaction temperature. These points are essential since their different preparation conditions determine the AC performance. Currently, ACs are mainly obtained by activation and pyrolysis processes at high temperatures [113, 135].

The activation process is an important step to improve the properties of the adsorbent materials' pores and can be carried out by three methods: chemical, physical, or a combination of both methods [108].

Chemical activation consists of using acid and basic chemical agents such as $ZnCl_2$ [136], H_3PO_4 [134, 137], H_2SO_4 [138], NaOH [113–115], KOH [139], K_2CO_3

[107], and others. Chemical reagents are responsible for the degradation or dehydration of organic components in natural fibers to prevent depolymerization of the hydrocarbon on the surface [135]. When natural fiber is submitted for acid activation, an increase in functional groups (such as carboxylic, phenolic, and carbonyl) occurs on the surface of the AC. Alkaline treatment tends to form -OH groups, ensuring quick ion exchange [140].

It is noteworthy that the main advantages of this method are that it can be performed in a single step at lower temperatures, higher yields, and shorter activation times [113]. Excessive use of chemical reagents can cause severe environmental impacts [141]. On the other hand, this process can become economical and environmentally friendly if recovery and recycling are carried out during the activation method [135].

Physical activation uses appropriate gas such as carbon dioxide (CO_2), steam, or both as activating agents. According to the literature, CO_2 has shown remarkable performance by providing a slow reaction rate and low reactivity at high temperatures. This method requires a longer activation time to ensure more mesopores and macropores. Nonetheless, this factor can cause a reduction in pore volume and surface area [135, 142]. It can also be argued that this process generates a lower environmental impact as it does not generate chemical residues. However, it is a technique that operates at high temperatures for long periods and requires greater energy use, which can generate severe environmental impacts depending on the energy source [141].

In both activation methods conducted, pyrolysis is a process of irreversible thermal degradation of any lignocellulosic matter through chemical and physical changes at high temperatures (300–700°C) under oxygen-free conditions [141]. The solid waste generated is AC, and its yield can vary from 13% to 90% wt according to the adopted temperature [135].

As mentioned in the previous section, pyrolysis temperature is an essential parameter when converting the activated natural fiber into AC. This factor can influence the quantity and quality of the obtained adsorbent material [135]. Some studies report that the rapid degradation process of lignocellulosic material at high temperatures into volatile compounds results in low yield. However, a higher percentage of fixed carbon is obtained under the same conditions, resulting in AC with a better porous structure [135, 143].

3.4.3 AC Application from Natural Fiber-Based Residues for Heavy Metals and Dyes Removal

AC presents an excellent adsorption performance for heavy metals and dyes. Heavy metals are widely found in products and production chains such as photographic constituents, petrochemical products, fertilizer production, battery manufacturing [144]. Dyes are extensively used in industries such as textile, printing, cosmetics, ink, and others [113].

The excessive use of these contaminants is responsible for infecting food, water, air, and land, which causes severe environmental problems and affects many life forms [145].

Table 3.5 shows relevant studies about the precursor, the methodology adopted to prepare AC, and their application.

TABLE 3.5

Some relevant studies about the use of AC and their application

Precursor	Activation process	Surface area $(m^2 g^{-1})$	Application	References
JF	Chemical activation ($ZnCl_2$)	1,028.614	Lithium ion batteries	[146]
Spent coffee ground	Chemical (KOH) and physical (steam) activation	981.12 (physical) 2,109.08 (chemical)	Supercapacitor applications	[122]
Coconut shell	Chemical ($Ce(NO_3)_3 \cdot 6H_2O$ and $Co(NO_3)_2 \cdot 6H_2O$) and physical activation	1,080 (5% CeO_2/AC) 1,005 (5% CeO_2/AC-P)	Removal of elemental mercury	[147]
Palm fibers (sheath and stem)	Chemical activation (NaOH)	1,102 (sheath) 1,172 (stem)	Removal of methylene blue dye	[114]
Dates stems	Chemical activation ($ZnCl_2$) and impregnation using different metals (Ca, Co, Mg, and Al)	1,342 561 (impregnation using Al metal)	Fluoride removal	[148]
Polyacrylonitrile	Chemical activation (N,N-dimethylformamide)	118.7	Removal of uranium	[149]
Balanites aegyptiaca	Chemical (H_2SO_4) and physical activation	670.35	Removal of hexavalent chromium	[117]
Honeydew peels	Chemical (HNO_3) and physical activation	20.12	Removal of Cr^{3+} and Zn^{2+}	[118]
Mangosteen shell	Chemical activation (H_2SO_4)	209	Ni(II) removal	[150]
Chicken feathers (CFs) and eggshell (ES)	Chemical (HCl) and physical activation	668	Cd^{2+}, Cu^{2+}, Pb^{2+}, Ni^{2+}, and Zn^{2+} removal	[151]
Sugar	Chemical activation (H_2SO_4)	3.88 (SC) 739.08 (La/Co-SC)	Adsorption of methyl orange	[152]
Algal dead	Chemical activation (H_2SO_4, H_3PO_4, HNO_3, HCl, CH_3COOH, SDS, and NaOH) and physical activation	–	Adsorption of Congo red dye	[153]
Pineapple crown	Chemical (NaOH) and physical activation	–	Removal of methylene blue dye	[115]
Oak wood	Chemical (ZnO/Fe_3O_4) and physical activation	48	Methyl violet 2B removal	[154]

(Continued)

TABLE 3.5
(Continued)

Precursor	Activation process	Surface area $(m^2 g^{-1})$	Application	References
Oil palm	Chemical (KOH) and physical activation	–	Malachite green removal	[155]
Solanum melongena	Chemical (H_3PO_4) and physical activation	211.38	Acid yellow 17 dye removal	[156]
Rotten carrots	Chemical ($ZnCl_2$ and HCl) and physical activation	1,365.5 (AC-600) 1,509.7 (AC-700) 1,405.8 (AC-800)	Degradation of rhodamine B dye	[157]
Dragon fruit peels	Chemical (KOH) and physical activation	756.3	Methylene blue dye adsorption	[137]
Ashitaba	Chemical ($ZnCl_2$) and physical activation	1,626.96 (WSAC) 1,228.53 (AWAC)	Congo red and methylene blue removal	[129]
Banana peel	Chemical (NaOH) and physical activation	–	Removal of methylene blue dye	[113]
Peanut shell	Chemical ($FeCl_3 \cdot 6H_2O$) and physical activation	722.34	Malachite green	[158]
Palm oil mill	Chemical ($ZnCl_2$) and physical activation	1,058	Dye removal	[159]
Sunflower pith	Chemical (NaOH and KOH) and physical activation	2,690 (NaOH) 2,090 (KOH)	Methylene blue dye adsorption	[160]
Peanut shell	Chemical (H_3PO_4) and physical activation	1,138	Dye adsorption	[161]
Ziziphus lotus	Chemical (H_3PO_4) and physical activation	–	Cationic dyes: BY28 and BR46	[162]
Banana trunk	Chemical (H_3PO_4) and physical activation	1,173.16	Methylene blue	[163]
Tea leaf	Chemical (H_3PO_4) and physical activation	1,313.4 (MTLACs) 1,169.3 (MTLAC-SA)	Rhodamine B, methylene blue, crystal violet, brilliant green, and orange G	[164]

TABLE 3.5
(Continued)

Precursor	Activation process	Surface area $(m^2 g^{-1})$	Application	References
Seed hull	Physical activation	614	Malachite green dye	[165]
Banana peel	Microwave pyrolysis (KOH/NaOH)	1,130 (BPK)	Malachite green dye	[166]
Pineapple crown leaf	Chemical (KOH) and physical activation	71.75 (AC-1) 128.94 (AC-2) 383.17 (AC-3)	Methyl violet dye removal	[167]
Rice	Chemical (H_2SO_4) and physical activation	–	Adsorption crystal violet	[164]
Coconut husk	Chemical (KOH) and physical activation	1,448	Fluoride	[168]

3.5 NANOFIBERS

Nanoscience and nanotechnology are multi-disciplinary and emerging fields that can influence many branches of science involving materials. Nanotechnology is used to understand and control the phenomena of matter at the nanoscale. Generally, nanotechnology deals with nanomaterials by both top-down (deconstruction from larger or bulk parts to form smaller, in this case, nanoscale parts) and bottom-up (construction from atoms or molecules) approaches [169]. Materials in bulk dimensions have measurements above 100 nm. When this dimension is reduced, the material ceases to be bulk to become a nanometric material. Nanomaterials are configured as nanoscale structures, where at least one of their dimensions is in the range of 1 to 100 nm [170]. This dimensional range allows nanomaterials to be two-dimensional (2D) or three-dimensional (3D), or even having only one or no dimensions (1D and 0D, respectively) [171]. Thus, nanostructures can vary in morphology and present themselves as quantum dots [172], nanowires [173], nanocrystals [174], nanorods [174], nanosheets [175], nanotubes [176], and nanofibers [177]. Compared to other shapes, nanofibers can be advantageous for their usual high surface area and high porosity, enabling advanced applications [170], for example, industrial processes [178], building materials [179], drug delivery [180], diagnostics [181], and for environmental applications like wastewater treatments [17]. There are several classifications to define techniques to obtain or synthesize nanofibers. Pira et al. [182] define the methods for obtaining nanofibers in chemical (acid hydrolysis [183], (2,2,6,6-tetramethylpiperidin-1-yl)oxyl (TEMPO)-mediated oxidation [184] and enzymatic reactions [185]) and mechanical processes (high-pressure homogenization [186], cryocrushing

[187], microfluidization [188], high-intensity ultrasonic treatments [77], and ball milling [189]). Kenry and Lim [170] classify nanofiber synthesis techniques as current and emerging; some of the techniques mentioned in their work (among others) and different applications of nanofibers in heavy metal or dyes removal found in the literature are provided in Table 3.6.

Among the techniques mentioned in Table 3.6, ball milling is an emerging technology that is more ecological since it avoids chemical solvents in the grinding process, increasing the sustainable appeal of obtaining nanostructures by this technique [182]. Another factor that enhances the eco-friendly aspect of nanofibers is the use of bio-based nanofibers. According to Gandavadi et al. [48], the use of bio-based nanofibers (such as nanocellulose, nanochitosan, and protein-based) can be advantageous because they have efficiency similar to synthetic materials and a remarkable tendency to adsorb heavy metals and dyes. Their interaction in contaminated water can justify this fact by size exclusion, covalent/non-covalent interaction, and solution diffusion processes, favored by their superior surface area compared to microfibers. In their study, the authors also pointed out that modification of these nanofibers may be required for further interaction with heavy metals and dye pollutants.

3.6 SURFACE MODIFICATION OF NATURAL FIBERS

The adsorption of heavy metals and dye on solid supports, either modified or non-modified, has been widely studied in recent years. Thus, natural fibers have gained considerable interest for use as adsorbent materials because of their attractive properties, including biodegradability, low density, non-toxicity, and low cost. In addition, these fibers are mainly composed of cellulose and lignin, presenting great potential for removing metallic ions while also being economically attractive [18]. However, some disadvantages of natural fibers, such as biocompatibility and hydrophilic properties, can be overcome by several surface modifications and chemical methods of treatment.

Cellulose is the most abundant non-toxic and organic compound extracted from natural fibers with high content and is susceptible to modifications [18]. Thus, residues obtained from natural fibers have been an interesting material for cellulose separation through chemical treatments [18]. Chemical modifications activate or introduce new functional groups, making the fibers interact effectively with other chemical groups and improving the compatibility between fibers and the insert group, such as inorganic molecules like phosphate sulfate [199].

Overall, natural fiber modification alters chemical, physical, thermal, and mechanical properties. It depends on various factors like method preparation, time, temperature, and chemical concentration.

Several chemicals, with mechanical and biological modifications (including their combinations), have been reported to extract cellulose from lignocellulosic biomass [200]. Some essential chemical treatments that have been able to modify the properties of fibers are alkali treatment (NaOH, KOH) [98, 201], acylation [202], sodium bicarbonate treatment [203], sodium chlorite treatment [204], silane treatment [205], acetylation [206], permanganate treatment [207], benzoylation [208], sodium chlorite treatment [204], and oxygen peroxide (H_2O_2) bleaching [209].

TABLE 3.6

Status of different techniques to obtain nanofibers, their description, and examples of nanofiber application in heavy metal or dyes removal

Technique status	Nanofiber fabrication/ synthesis technique	Technique description	Application in heavy metal or dyes removal
Current	Electrospinning/ electrospun	Electrohydrodynamic process, where the liquid droplet is electrified to generate a jet, and consequent stretching and elongation to generate a nanofiber [190]	[191]
	Self-assembly	Organization of individual macromolecules into ordered and stable nanoscale structures [170]	[192]
	Sonochemical method	Application of powerful ultrasound radiation in molecules causing a chemical to form nanofibers [193]	[77]
	Template-based synthesis	The template is a central structure that forms a network, and its removal creates a filled cavity with nanofeatures [194]	–
	Polymerization	As an example, polyaniline (PANI) nanofibers were fabricated via an interfacial polymerization method [195]	[196]
Emerging	CO_2 laser supersonic drawing	Fibers are melted and then passed through supersonic airflow to achieve the supersonic drawing of nanofibers [170]	–
	Solution blow spinning	An airbrush with a compressed gas source ejects concentrated polymer solution in a target, forming nanofibers [77]	[197]
	Plasma-induced synthesis	Bombardment of radicals on the electrode surface, followed by atomic vapor deposition, causing an expansion of plasma and condensation of solution medium. Finally, there is an in situ reaction of oxygen and growth of nanofibers [170]	[86]
	Centrifugal jet spinning	Materials' viscoelasticity and mass transfer are similar to electrospinning but with centrifugal force [170]	[198]
	Electrohydrodynamic direct writing	Electrical/mechanical forces project an ink for a large-scale direct writing of solution-based materials to form nanofibers [170]	–
	Ball milling	Mechanical technique to grind materials into nanometric particles and blend materials. Through the friction of balls with the material, both contained in a closed container that rotates, providing the formation of nanofibers [182]	[189]

Some physical methods are also used to modify the surface of natural fibers; stretching, calendaring, electric discharge, and the production of hybrid yarns do not change the chemical composition of the fibers. Among these methods, electric discharge (corona, cold plasma) is highlighted. Corona treatment is one of the common techniques for surface oxidation activation of natural fibers, altering the surface energy of the natural fibers and causing an increase in the number of aldehyde groups [210].

One interesting application is the chemical modification of cellulose fibers with metal oxide particles (Cell/M_xO_y). The experimental methodology of the fiber coating process depends on the form in which the cellulose is obtained, as fiber or a membrane. Cellulose treatment with a precursor reagent can be made in an aqueous or non-aqueous solvent as fiber forms. Figure 3.6 shows the schematic representation of a Lewis acid with cellulose.

The reaction is carried out under anhydrous conditions in the first step as a donor-acceptor bond is primarily formed between $ZrOCl_2 \cdot 8H_2O$ and the oxygen of the C^1-O-C^5 and C^1-O-C^4 bonds. The attack on the standard oxide bond provides a ring-opening resulting in a molecule rupture to form microfibers. Ammonium solution is required to promote the chemical link of metallic oxide with fiber surfaces [18].

Another chemical method most used to modify natural fibers is mercerization, also referred to as alkaline treatment, which removes surface impurities and the occurrence of fibrillation, obtaining a fibrous material with greater surface area and a smaller diameter. This method involves modifying fibers' structure and chemical composition using an aqueous solution of sodium hydroxide (NaOH), as shown in Figure 3.7. The concentration, temperature, and time of action of the solution depend

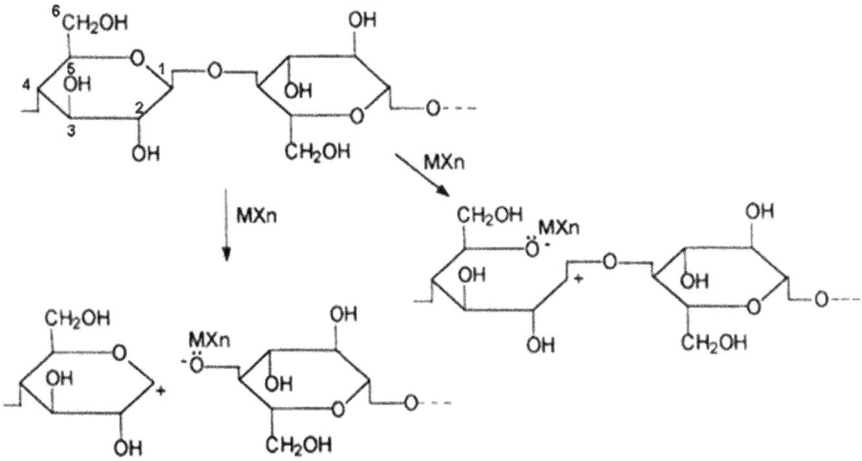

FIGURE 3.6 Reaction of the Lewis acid with cellulose and formation of the donor-acceptor bond.

$$\text{Fiber-OH} \ + \ \text{NaOH} \longrightarrow \text{Fiber-O}^-\text{-Na}^+ \ + \ H_2O$$

FIGURE 3.7 Schematic representation of alkali treatment in cellulosic fibers.

on the characteristics of the fibers and their application. The type of alkali and its concentration influence the degree of swelling [211].

However, these techniques cause secondary pollutions due to the chemical reagents employed to prepare these adsorbent materials.

Among more environmentally friendly surface modification methods, reactive cold plasmas are interesting but not yet fully explored. Several species, such as electrons, ions, and free radicals, are created in the physical plasma. These species collide with the surface of the material surrounded by plasma-breaking chemical bonds, leading to the formation of new ones and inserting functional groups on the surface, thus promoting surface modifications. Moreover, surfaces are modified under plasma treatments at low treatment time, low cost, and extremely low waste production compared to the conventional chemical process for surface modification [212]. Additionally, plasma treatment can introduce various functional groups on the natural fiber surface, forming strong covalent bonds. Also, surface etching by plasma treatment may improve the surface roughness and result in a better interface with the matrices through mechanical interlocking.

Table 3.7 shows various types of application methods of natural fibers modified for heavy metal and dyes removal.

3.7 ADSORBENTS REGENERATION

As discussed in the previous sections, an adsorbent material must have a high adsorption efficiency. This factor is influenced by the availability of a good surface area that provides free binding sites for contaminants' sorption [252]. In a given degree of use, these adsorbents reach a saturation point losing their adsorption capacity and are incorrectly discarded, which can cause severe environmental damage [114]. For this reason, the circular economy is becoming relevant for economical and environmentally sustainable development [122].

Thus, regeneration is proposed to reaffirm the importance of environmental protection and follow the required regulations [253]. This process reduces the accumulation of contaminant-laden materials in the environment [156]. Also, it recovers the original porous structure with little or no damage to the material [254, 255]. Literature reports some regeneration techniques for adsorbent materials, including thermal (TH), chemical (CH), wet air oxidation (WAO), EC, bio regeneration (BR), steam (ST), electro-Fenton (EF) regeneration, and microwave (MW) irradiation [252, 256], which are selected according to the contaminant's nature, the operating cost, and the conditions of the processes [114].

The WAO process involves adsorbent material oxidation by injecting O_2 in an aqueous medium and the application of heat (150–250°C) and pressure (10–50 bar). This method is based on converting toxic substances into species with less environmental impact [257]. Meanwhile, the ST technique consists of heating the system using steam. In this case, the adsorbent material can only be regenerated when the medium is heated enough, ensuring fast and great removal of the contaminants [258]. Like ST, in the MW method, the adsorbent must be exposed to microwave irradiation under different operating conditions (power level and irradiation times). The energy present in this system is converted into heat, causing desorption [258, 259].

TABLE 3.7

Some application methods of natural fibers modified for heavy metal and dyes removal

Type of natural fibers	Functional groups/modifying agent	Modification/adsorbent	Adsorption	References
Orange bagasse	Methyl ester part of orange pectic acid	Saponified orange juice residue	Fluoride	[213]
Corn stalk	2,2,6,6-Tetrametilpiperidina-1-oxila (TEMPO)	Oxidized cellulose TEMPO-oxidized cellulose (TOCS)	Cd(II)	[214]
Cotton fibers	ZnO and Ag NPs	ZnCt (cotton impregnated zinc oxide NPs) AgCt (cotton impregnated silver NPs)	Hg(II), Ni(II), Cr(III), Co(II), Pb(II)	[215]
Pineapple leaf	Ethylenediaminetetraacetic acid (EDTA), carboxymethyl (CM)	Cell-EDTA and cell-CM	Cd(II) Pb(II)	[216]
Coconut coir	H_3PO_4	Modified agricultural waste, prepared by chemical impregnation of coconut coir by H_3PO_4	Pyridine	[217]
Orange peel	Alkali saponification	Saponification orange peel	Cd(II)	[218]
Sugarcane bagasse	Mercerization	Mercerized sugarcane bagasse	Cd(II), Cu(II) Cd(II), Pb(II)	[219]
Palm waste	Cryogel modified with GO/Fe_3O_4	$CNFs/GO/Fe_3O_4$ cryogel	Pb(II), Hg(II) Cr(VI)	[220]
Corchorus olitorius (jute)	H_2SO_4 HCl	α-Cellulose/H_2SO_4 α-Cellulose/HCl	Cr(III), Pb(II), Cd(II)	[221]
Rice bran	Chemical treatment	Cellulose/hemicellulose and lignin	Cd(II)	[222]
Coconut husk	Modified with tannin	Cellulose fiber/tannin	Pb(II), Cd(II), Cu(II)	[223]
Coffee husks	Untreated	–	Cu(III), Cd(II), Zn(II), Cr(VI)	[224]
Maize cob (MC)	H_2O, NaOH, and HNO_3 solutions	MC_H_2O, MC_NaOH, and MC_HNO_3	Pb(II), Ni(II)	[225]
Untreated watermelon rind (WMR)	Modified with xanthate (X) and citric acid (CA)	X-WMR and CA-WMR	As(III) and As(V)	[226]
Commercial cellulose	ZrO_2	Cell/ZrO_2	Ni(II)	[227]
Tucumã cake, *in natura* (TCN)	Chemical and thermal treatment	Tucumã cake thermally treated (TCT)	Cationic and anionic dyes	[228]
Banana steam	Chemical treatment	AC	Methylene blue	[229]

(Continued)

TABLE 3.7
(Continued)

Type of natural fibers	Functional groups/modifying agent	Modification/adsorbent	Adsorption	References
Jute	Chemical treatment	Alkali steam treated lignocellulosic (AL)	Methylene blue	[230]
Sugarcane bagasse	Alkali treatment	Biosorbent	Methylene blue	[231]
Sugarcane bagasse	Untreated	Biosorbent	Methylene blue	[232]
Pineapple (*Ananas comosus*) plant stem	Hot water treatment	Biosorbent	Congo red	[233]
Lemongrass leaf	Chemical treatment	Biosorbent	Crystal violet	[234]
Banana peel power	Pretreatment	Biosorbent	Reactive black 5 and Congo red	[235]
Garlic peel	Hot water treatment	Adsorbent	Direct red 12B	[236]
Mushroom waste	Untreated	Adsorbent	Direct red 5B, direct black 22, direct and black 71	[237]
Pumpkin peels	Pretreatment	AC	Methylene blue	[238]
Grape peel	Treated by microwave and conventional hydrothermal	Biosorbent	Methylene blue	[239]
Pomelo peel	Chemical precipitation method (ZrO_2)	Zirconium oxide-modified pomelo peel biochar (ZrBC)	Sulfate (SO_4^{2-})	[240]
Sugarcane bagasse	Chemical precipitation method (ZrO_2)	Cell/$ZrO_2 \cdot nH_2O$	Sulfate (SO_4^{2-})	[241]
Coffee husk	Chemical treatment	Biochar	Cd(II), Pb(II)	[242]
Tangerine peel	Hydrothermal precipitation method	Nanocomposites (T-Fe_3O_4)	Pb(II)	[243]
Cellulose derived from wood pulp	Hydrolysis-precipitation method	TiO_2/cellulose (CE-Ti) nanocomposites	Phosphate (PO_4^{2-})	[244]
Peanut husk	Chemical treatment	Peanut husk powder (PHP)	Pb(II), Mn(II), Cd(II), Ni(II)	[245]
Banana peel	Untreated	—	Cd(II)	[246]
Litchi peels	Chemical treatment	Litchi peels coated with Fe_3O_4	Pb(II)	[247]
Pomegranate peel	Chemical treatment	Biosorbent-bimetallic hydroxide composite (Z-La/Peel)	Phosphate (PO_4^{2-})	[248]
Corn stalks (CS)	Chemical treatment	Modified cellulose from CS (MCCS)	Phosphate (PO_4^{2-}), nitrate (NO_3^{2-})	[249]
Banana peel	Untreated	Banana (*Musa sapientum*) peel	Pb(II)	[250]
Pistachio shell	Chemical treatment	Biochar	Pb(II)	

Another regeneration technique is EC, which involves converting contaminants into CO_2 and water through EC anode oxidation. This happens when two electrodes are applied to the adsorbent material favoring oxidative degradation [258, 260]. EF regeneration was developed based on the goals of electro-desorption and electro-oxidation. This technique consists of advanced oxidation of organic compounds caused by the in situ production of hydroxyls through a cathode [261, 262]. In contrast, the BR technique is associated with the metabolism of microorganisms. This process occurs in three stages: consumption of organic matter (food source) by microbes, induction of hydrolysis due to released exoenzymes, and microbial degradation causing slow desorption [263].

Table 3.8 provides examples of regeneration techniques for AC. However, the main methods are thermal and chemical regeneration.

3.7.1 THERMAL REGENERATION

Thermal regeneration is a fast, simple, low-cost, efficient process [12, 290] and is characterized by complex physical and chemical processes. The adsorbent material

TABLE 3.8
Some examples of regeneration techniques for different adsorbents of AC

Different adsorbents of natural fiber-based residues	Application	Regeneration techniques	Efficiency (%)	References
Cypress sawdust (AC)	H_2S removal	Thermal regeneration	80.08 (four cycles)	[264]
Cotton (AC)	Adsorption of nonpolar volatile organic compounds	Electro-thermal regeneration	100 (one cycle)	[265]
Lotus (modified fiber)	Methyl orange	Thermal regeneration	(Four cycles)	[266]
Jamun seed (AC)	Adsorption of fluoride	Chemical regeneration	96 (– cycles)	[267]
Soya (modified fiber)	Methylene blue	Thermal-chemical regeneration	80 (three cycles)	[268]
Bamboo (AC)	Perfluorooctane sulfonate, and perfluorooctanoate removal	Thermal-chemical regeneration	71 (two cycles)	[269]
Pine cone (AC)	Anionic dye	Chemical regeneration	61.43 (four cycles)	[270]
Coconut husk (modified fiber)	Heavy metal ions	Chemical regeneration	76 (fifth cycle)	[223]
Date pits (AC)	Amino acid l-tryptophan	Thermal regeneration	59.9 (four cycles)	[271]

(Continued)

TABLE 3.8
(Continued)

Different adsorbents of natural fiber-based residues	Application	Regeneration techniques	Efficiency (%)	References
Durian shells (modified fiber)	Ammonia nitrogen	Chemical regeneration	73 (five cycles)	[272]
Sycamore sawdust (nanofibers)	Aniomic dye	Chemical regeneration	90 (29 cycles)	[273]
Manihot esculenta (AC)	Fluoride	Chemical regeneration	91.3 (one cycle)	[274]
Luffa (modified)	Uranium(VI)	Thermal-chemical regeneration	90 (eight cycles)	[275]
Rice husk (AC)	TMA and H$_2$S gas	Thermal regeneration	–	[276]
Cornulaca monacantha (AC)	Congo red dye	Chemical regeneration	75.75 (one cycle)	[277]
Reed (AC)	Methylene blue	Chemical regeneration	85.15 (one cycle)	[278]
Vitis vinifera (AC)	Methylene blue	Chemical regeneration	99 (– cycles)	[279]
Sugar (AC)	Rhodamine B dye	Thermal regeneration	98 (five cycles)	[280]
Peanut shells (AC)	Malachite green	Thermal-chemical regeneration	30.85 (four cycles)	[281]
Almond peanut (modified fiber)	Methylene blue	Chemical regeneration	90 (four cycles)	[282]
Coconut husk (AC)	Fluoride	Chemical regeneration	98.8 (three cycles)	[168]
Vatica rassak wood (AC)	Pb(II) adsorption	Chemical regeneration	10.72 (five cycles)	[283]
Coconut shell (AC)	Salt adsorption	EC regeneration	– (100 cycles)	[284]
Bagasse and natural attapulgite (modified fiber)	Reactive brilliant ed X-3B	Chemical regeneration	83.7 (1-5-AB)	[285]
Vinasse (AC)	Methylene blue	Chemical regeneration	99 (five cycles)	[286]
Ceiba speciosa (AC)	Phenol effluent	Chemical regeneration	89.02 (five cycles)	[287]
Apple (modified fiber)	Congo red and malachite green	Thermal-chemical regeneration	61 (Congo red) 46 (malachite green)	[288]
Acalypha indica (AC)	Dyes	Chemical regeneration	86 (five cycles)	[289]

is submitted at high temperatures (~850°C), and, during the heating process, volatile compounds adsorbed are released, and less volatile compounds are decomposed. From the environmental point of view, this method is unsuitable since the technique uses ovens that release large amounts of CO_2, contaminants in a partially oxidized state, and particulate matter. Besides, it results in a lower carbon yield, generating ash content and affecting the adsorption capacity [258, 291, 292].

When studying the thermal regeneration at 800°C of lotus biomass-derived biochar, Hou [266] reported an increase in methyl orange dye removal efficiency by biochar, and after five adsorption-desorption cycles a slight decrease in surface area was observed.

3.7.2 CHEMICAL REGENERATION

Chemical regeneration is the desorption or decomposition of adsorbates using chemical reagents which react on the adsorbents' surface and their pores [252, 291]. This technique occurs through a chemical reaction or a change in pH, which can modify the adsorbent-analyte interaction (electrostatic, hydrophobic, π–π interaction, hydrogen bonding) and consequently increase the desorption of the adsorbate [293]. One of the advantages of this method is having a recovery using distillation. However, this process involves a high cost of reagents, the risk of pollution, and incomplete regeneration [252, 292].

Lyu [273] studied the use of KOH (0.1 mol L^{-1}) and H_2SO_4 (0.1 mol L^{-1}) as the desorption reagent and the active reagent for the regeneration of polyaniline/cellulose nanofiber aerogel. Acid red G and methylene blue removal efficiency of around 97.05% and 72.39%, respectively, were obtained after ten consecutive adsorption-desorption cycles.

Xiao et al. [275] proposed chemical regeneration of modified luffa fiber for removal of uranium(VI) from uranium mine water by HCl solution (0.50 mol L^{-1}). The authors observed that with this adsorbent a slight reduction of adsorption capacity and desorption efficiency were obtained. After eight cycles of regeneration, the desorption efficiency was higher than 90%.

Guo [281] investigated the recovery of Fe-Mg bimetallic magnetic AC from peanut shells by chemical regeneration (deionized water, HCl (0.1 mol L^{-1})), NaOH (0.1 mol L^{-1}), and CH_3CH_2OH (95%)) and thermal regeneration (at 298K). The author observed that after four adsorption-desorption cycles, the removal efficiency of malachite green dye was 30.85%. In contrast, by way of thermal regeneration, this efficiency was 100%.

3.7.3 FACTORS AFFECTING REGENERATION

The regeneration process is influenced by factors such as pH, temperature, and time. The pH is an essential factor that can modify adsorbents' and adsorbates' physical and chemical properties, resulting in a loss or gain in adsorption capacity. Another parameter that plays a vital role in regeneration processes is temperature. This factor must be taken into careful consideration for good results. In this respect, regeneration experiments with dry, hot, and humid air have been done successfully [291]. The time required for the regeneration of adsorbents also greatly affects the efficiency process. Excellent time administration is essential to prevent the material from getting damaged or to prevent loss in its adsorption efficiency. All these factors mentioned above are essential for effective regeneration and directly influence the number of adsorption-desorption cycles of the material [256, 291].

It is also worth mentioning that regeneration depends on the nature of the adsorption process (physical or chemical). In physical adsorption, the material needs to be subjected to heating, decreasing the pressure for desorption. In chemical adsorption, the rupture of ionic or covalent bonds between adsorbent and adsorbate is realized through a higher energy supply than the absorptive force [255].

3.7.4 Regeneration Challenges

Regeneration is the most efficient technique for the recovery of adsorbents. However, studies have reported some challenges from an environmental point of view that need to be solved [291]. It is important to highlight that each regeneration method has advantages and disadvantages or is sometimes the correct choice. Suppose the procedure generates more waste and requires more operational work than the adsorption process, and development of the adsorbent is more straightforward than recovery. In that case, other methods must be adopted, such as incineration or disposal in landfills [256]. According to Dotto and McKay [256], developing alternative and eco-friendly regeneration techniques is one of the main challenges in adsorbent recovery. For the authors, minimizing costs and waste generation, maximization of cycles, and adsorbate recovery are essential for more and more efficiencies.

3.8 CONCLUSIONS AND PERSPECTIVES

This chapter summarized recent reports in which fiber-based materials comprising AC, nanofibers and modified natural fibers were used as adsorbents to remove heavy metals and dyes. The adsorption process is commonly used for water and wastewater treatment because of its convenience, ease of operation, and simplicity of design. Once more research was devoted to understanding the properties of materials as adsorbents for the removal of these dangerous compounds, the most commonly used adsorbents for wastewater purification were described and their preparation method and selective metal or metal or dye remotion. The multitude of aspects regarding the chemical and physical nature of natural fiber-based materials and their adsorption ability, the factors affecting the adsorption process, as well as equilibrium, kinetics model, and mechanism adsorption, could be gathered in a single work, making it easier for the reader to access. AC is mainly applied as an adsorbent for pollutants removal from wastewater. A major concern about the use of adsorbents is the regeneration of the materials after use. The loss of adsorption efficiency after regeneration may limit its use.

Furthermore, regeneration of adsorbents, mainly carbon, after exhausting remains a challenge. In this context, alternative AC materials should be available, inexpensive, and chemically regenerable with efficient quantitative recovery. An interesting perspective concerning the limitation imposed by regeneration consists of commonly used adsorbent materials by green adsorbent materials. The "green adsorption" term refers to the use of low-cost materials from agricultural sources and by-products, agricultural residues and wastes, and low-cost sources. However, this subject and the development of alternative and eco-friendly regeneration techniques need to be explored further.

REFERENCES

[1] J. Wang and X. Guo, *Adsorption isotherm models: Classification, physical meaning, application and solving method*, Chemosphere 258 (2020), p. 127279.

[2] K. Rao, M. Mohapatra, S. Anand and P. Venkateswarlu, *Review on cadmium removal from aqueous solutions*, Int. J. Eng. Sci. Technol. 2 (2011), pp. 81–103.

[3] U. Upadhyay, I. Sreedhar, S.A. Singh, C.M. Patel and K.L. Anitha, *Recent advances in heavy metal removal by chitosan based adsorbents*, Carbohydr. Polym. 251 (2021), p. 117000.

[4] S. Danehpash, P. Farshchi, E. Roayaei, J. Ghoddousi and A.H. Hassani, *Investigation the advantages of some biocompatible adsorbent materials in removal of environmental pollutants in aqueous solutions*, Ukr. J. Ecol. 8 (2018), pp. 199–202.

[5] S. De Gisi, G. Lofrano, M. Grassi and M. Notarnicola, *Characteristics and adsorption capacities of low-cost sorbents for wastewater treatment: A review*, Sustain. Mater. Technol. 9 (2016), pp. 10–40.

[6] R.F. Nascimento, D.Q. Melo, A.C.A. Lima, A.L. Barros, C.B. Vidal and G.S.C. Raulino, *2- Equilíbrio de Adsorção*, in Adsorção: Aspectos teóricos e aplicações ambientais, R.F. do Nascimento, A.C.A. de Lima, C.B. Vidal, D. de Quadros Melo, G.S.C. Raulino, eds., 2014, pp. 25–50.

[7] M.M. Sabzehmeidani, S. Mahnaee, M. Ghaedi, H. Heidari and V.A.L. Roy, *Carbon based materials: A review of adsorbents for inorganic and organic compounds*, Mater. Adv. 2 (2021), pp. 598–627.

[8] I. Ali, M. Asim and T.A. Khan, *Low cost adsorbents for the removal of organic pollutants from wastewater*, J. Environ. Manage. 113 (2012), pp. 170–183.

[9] V. Kumar, R.D. Parihar, A. Sharma, P. Bakshi, G.P. Singh Sidhu, A.S. Bali et al., *Global evaluation of heavy metal content in surface water bodies: A meta-analysis using heavy metal pollution indices and multivariate statistical analyses*, Chemosphere 236 (2019), p. 124364.

[10] Renu, M. Agarwal and K. Singh, *Heavy metal removal from wastewater using various adsorbents: A review*, J. Water Reuse Desalin. 7 (2017), pp. 387–419.

[11] E. Im, H.J. Seo, D.I. Kim, D.C. Hyun and G.D. Moon, *Bimodally-porous alumina with tunable mesopore and macropore for efficient organic adsorbents*, Chem. Eng. J. 416 (2021), p. 129147.

[12] Z. Sun, Y. Liu and W. Hong, *Facile synthesis of porous hydrated magnesium silicate adsorbent from ordinary silica gel*, Mater. Lett. 272 (2020), p. 127886.

[13] J. Vijayeeswarri, M. Geethapriyai and V. Ramamurthy, *Community level defluoridation of groundwater with limestone derived adsorbent*, Process Saf. Environ. Prot. 127 (2019), pp. 9–15.

[14] G. Zhang, B. Lei, S. Chen, H. Xie and G. Zhou, *Activated carbon adsorbents with micro-mesoporous structure derived from waste biomass by stepwise activation for toluene removal from air*, J. Environ. Chem. Eng. 9 (2021), p. 105387.

[15] W. Yang, K. Cheng, R. Tang, S. Wu, H. Wang, Z. Han et al., *Ion exchange resin derived magnetic activated carbon as recyclable and regenerable adsorbent for removal of mercury from flue gases*, J. Energy Inst. 97 (2021), pp. 225–232.

[16] K. Kuśmierek, A. Świątkowski, T. Kotkowski, R. Cherbański and E. Molga, *Adsorption on activated carbons from end-of-life tyre pyrolysis for environmental applications. Part I. Preparation of adsorbent and adsorption from gas phase*, J. Anal. Appl. Pyrolysis 157 (2021), 105205.

[17] J.-C. Kim, S.-I. Oh, W. Kang, H.-Y. Yoo, J. Lee and D.-W. Kim, *Superior anodic oxidation in tailored Sb-doped SnO₂/RuO₂ composite nanofibers for electrochemical water treatment*, J. Catal. 374 (2019), pp. 118–126.

[18] W.A. Paixão, L.S. Martins, N.C. Zanini and D.R. Mulinari, *Modification and characterization of cellulose fibers from palm coated by ZrO particles for sorption of dichromate ions*, J. Inorg. Organomet. Polym. Mater. 30 (2020), pp. 2591–2597.

[19] M.F. Mendes and S.S. Freitas, *Estudo de materiais vegetais como adsorventes para a remoção dos agrotóxicos trifluralina, clorpirifós e α-endossulfam de solução aquosa*, in Estudos Interdisciplinares em Ciências Exatas e da Terra, A.F. Neves, M.H. de Paula, P.H.R. dos Anjos, A.G. da Silva, eds., 2016, pp. 57–70.

[20] A. Gil, L. Santamaría, S.A. Korili, M.A. Vicente, L.V. Barbosa, S.D. de Souza et al., *A review of organic-inorganic hybrid clay based adsorbents for contaminants removal: Synthesis, perspectives and applications*, J. Environ. Chem. Eng. 9 (2021), p. 105808.

[21] T.H. Cheng, R. Sankaran, P.L. Show, C.W. Ooi, B.L. Liu, W.S. Chai et al., *Removal of protein wastes by cylinder-shaped NaY zeolite adsorbents decorated with heavy metal wastes*, Int. J. Biol. Macromol. 185 (2021), pp. 761–772.

[22] G.Z. Kyzas and M. Kostoglou, *Green adsorbents for wastewaters: A critical review*, Materials 7 (2014), pp. 333–364.

[23] M. Dizbay-Onat, U.K. Vaidya, J.A.G. Balanay and C.T. Lungu, *Preparation and characterization of flax, hemp and sisal fiber-derived mesoporous activated carbon adsorbents*, Adsorpt. Sci. Technol. 36 (2018), pp. 441–457.

[24] H. Zhou, H. Zhu, F. Xue, H. He and S. Wang, *Cellulose-based amphoteric adsorbent for the complete removal of low-level heavy metal ions via a specialization and cooperation mechanism*, Chem. Eng. J. 385 (2020), p. 123879.

[25] A. Mautner, Y. Kwaw, K. Weiland, M. Mvubu, A. Botha, M.J. John et al., *Natural fibre-nanocellulose composite filters for the removal of heavy metal ions from water*, Ind. Crops Prod. 133 (2019), pp. 325–332.

[26] S. Wu, J. Kan, X. Dai, X. Shen, K. Zhang and M. Zhu, *Ternary carboxymethyl chitosan-hemicellulose-nanosized TiO₂ composite as effective adsorbent for removal of heavy metal contaminants from water*, Fibers Polym. 18 (2017), pp. 22–32.

[27] Y. Niu, W. Hu, M. Guo, Y. Wang, J. Jia and Z. Hu, *Preparation of cotton-based fibrous adsorbents for the removal of heavy metal ions*, Carbohydr. Polym. 225 (2019), p. 115218.

[28] F. Ji, C. Li, B. Tang, J. Xu, G. Lu and P. Liu, *Preparation of cellulose acetate/zeolite composite fiber and its adsorption behavior for heavy metal ions in aqueous solution*, Chem. Eng. J. 209 (2012), pp. 325–333.

[29] M.L. Rahman, Z.J. Wong, M.S. Sarjadi, S. Soloi, S.E. Arshad, K. Bidin et al., *Heavy metals removal from electroplating wastewater by waste fiber-based poly(amidoxime) ligand*, Water 13 (2021), pp. 1–20.

[30] S.K. Paramasivam, D. Raja Panneerselvam, D. Panneerselvam, K.N. Shiva and U. Subbaraya, *Influence of operating environments on adsorptive removal of lead (Pb(II)) using banana pseudostem fiber: isotherms and kinetic study*, J. Nat. Fibers 19 (2022), pp. 4485–4495.

[31] W. Dai, J. Zhang, Y. Xiao, W. Luo and Z. Yang, *Dual function of modified palm leaf sheath fibers in adsorbing reactive yellow 3 and Cr(VI) from dyeing wastewater*, J. Polym. Environ. 29 (2021), pp. 3854–3866.

[32] M.L. Mathew, A. Gopalakrishnan, C.T. Aravindakumar and U.K. Aravind, *Low – Cost multilayered green fiber for the treatment of textile industry waste water*, J. Hazard. Mater. 365 (2019), pp. 297–305.

[33] G.Z. Kyzas, E. Christodoulou and D.N. Bikiaris, *Basic dye removal with sorption onto low-cost natural textile fibers*, Processes 6 (2018), p. 166.

[34] S.H. Lee and S.S. Kim, *A new hollow-fiber adsorbent material for removing arsenic from groundwater*, J. Chem. 2019 (2019), p. 2715093.

[35] Y. Li, H. Xiao, Y. Pan and L. Wang, *Novel composite adsorbent consisting of dissolved cellulose fiber/microfibrillated cellulose for dye removal from aqueous solution*, ACS Sustain. Chem. Eng. 6 (2018), pp. 6994–7002.

[36] Q. Huang, D. Hu, M. Chen, C. Bao and X. Jin, *Sequential removal of aniline and heavy metal ions by jute fiber biosorbents: A practical design of modifying adsorbent with reactive adsorbate*, J. Mol. Liq. 285 (2019), pp. 288–298.

[37] D. Wang, Z. Wang, X. Zheng and M. Tian, *Activated carbon fiber derived from the seed hair fibers of Metaplexis japonica: Novel efficient adsorbent for methylene blue*, Ind. Crops Prod. 148 (2020), p. 112319.

[38] X.J. Liu, M.F. Li and S.K. Singh, *Manganese-modified lignin biochar as adsorbent for removal of methylene blue*, J. Mater. Res. Technol. 12 (2021), pp. 1434–1445.

[39] J. Hu, W. Liu, L. Xia, G. Yu, H. Huang, H. Guo et al., *Preparation of a cellulose-based adsorbent and its removal of disperse Red 3B dye*, Cellulose 28 (2021), pp. 7909–7924.

[40] S. Iftekhar, D.L. Ramasamy, V. Srivastava, M.B. Asif and M. Sillanpää, *Understanding the factors affecting the adsorption of lanthanum using different adsorbents: A critical review*, Chemosphere 204 (2018), pp. 413–430.

[41] S.-L.L. Lingling Liu, X.-B. Luo and L. Ding, *4-Application of nanotechnology in the removal of heavy metal from water*, in Micro and Nano Technologies, X. Luo, F. Deng, eds., 2019, pp. 83–147.

[42] M.A. Al-Ghouti and D.A. Da'ana, *Guidelines for the use and interpretation of adsorption isotherm models: A review*, J. Hazard. Mater. 393 (2020), p. 122383.

[43] R.A. Sims, S.L. Harmer and J.S. Quinton, *The role of physisorption and chemisorption in the oscillatory adsorption of organosilanes on aluminium oxide*, Polymers 11 (2019), p. 410. doi: 10.3390/polym11030410

[44] P. Zhang, Y.P. Chen, W. Wang, Y. Shen and J.S. Guo, *Surface plasmon resonance for water pollutant detection and water process analysis*, TrAC – Trends Anal. Chem. 85 (2016), pp. 153–165.

[45] A. Kaushal, *Adsorption phenomenon and its application in removal of lead from waste water: A review*, Int. J. Hydrol. 1 (2017), pp. 38–47.

[46] A.A. Pisarev, *1-Hydrogen adsorption on the surface of metals*, in Gaseous Hydrogen Embrittlement of Materials in Energy Technologies, R.P. Gangloff, B.P. Somerday, eds., Elsevier B.V., 2012, pp. 3–26.

[47] A.J. Emerson, A. Chahine, S.R. Batten and D.R. Turner, *Synthetic approaches for the incorporation of free amine functionalities in porous coordination polymers for enhanced CO_2 sorption*, Coord. Chem. Rev. 365 (2018), pp. 1–22.

[48] D. Gandavadi, S. Sundarrajan and S. Ramakrishna, *Bio-based nanofibers involved in wastewater treatment*, Macromol. Mater. Eng. 304 (2019), pp. 1–15.

[49] M. Radfard, M. Yunesian, R. Nabizadeh, H. Biglari, S. Nazmara, M. Hadi et al., *Drinking water quality and arsenic health risk assessment in Sistan and Baluchestan, Southeastern Province, Iran*, Hum. Ecol. Risk Assess. 25 (2019), pp. 949–965.

[50] A.E. Onyango, M.W. Okoth, C.N. Kunyanga and B.O. Aliwa, *Microbiological quality and contamination level of water sources in Isiolo county in Kenya*, J. Environ. Public Health 2018 (2018), p. 2139867. doi: 10.1155/2018/2139867.

[51] S. Rojas and P. Horcajada, *Metal-organic frameworks for the removal of emerging organic contaminants in water*, Chem. Rev. 120 (2020), pp. 8378–8415.

[52] A. Tkaczyk, K. Mitrowska and A. Posyniak, *Synthetic organic dyes as contaminants of the aquatic environment and their implications for ecosystems: A review*, Sci. Total Environ. 717 (2020), p. 137222.

[53] S. Afroze and T.K. Sen, *A review on heavy metal ions and dye adsorption from water by agricultural solid waste adsorbents*, Water. Air. Soil Pollut. 229 (2018), p. 225. https://doi.org/10.1007/s11270-018-3869-z

[54] S. Nawar, S. Cipullo, R.K. Douglas, F. Coulon and A.M. Mouazen, *The applicability of spectroscopy methods for estimating potentially toxic elements in soils: State-of-the-art and future trends*, Appl. Spectrosc. Rev. 55 (2020), pp. 525–557.

[55] M. Lahkar and K.G. Bhattacharyya, *Heavy metal contamination of groundwater in Guwahati city, Assam, India*, Int. Res. J. Eng. Technol. 06 (2019), pp. 1520–1525.

[56] F.F. Manzini, M. de F. Lima and E.P. Ciscato, *Study on the effect of excess lead, cadmium and nickel on plant and human organism*, XV Fórum Ambient. 1 (2019), pp. 252–266.

[57] U.S. EPA, *National Primary Drinking Water Guidelines*, Epa 816-F-09-004/2009.

[58] J.C. Fei, X.B. Min, Z.X. Wang, Z. hua Pang, Y.J. Liang and Y. Ke, *Health and ecological risk assessment of heavy metals pollution in an antimony mining region: A case study from South China*, Environ. Sci. Pollut. Res. 24 (2017), pp. 27573–27586.

[59] F. Pragst, K. Stieglitz, H. Runge, K.D. Runow, D. Quig, R. Osborne et al., *High concentrations of lead and barium in hair of the rural population caused by water pollution in the Thar Jath oilfields in South Sudan*, Forensic Sci. Int. 274 (2017), pp. 99–106.

[60] X. Su, A. Kushima, C. Halliday, J. Zhou, J. Li and T.A. Hatton, *Electrochemically-mediated selective capture of heavy metal chromium and arsenic oxyanions from water*, Nat. Commun. 9 (2018), p. 4701. https://doi.org/10.1038/s41467-018-07159-0

[61] F. Chen, Z.I. Khan, A. Zafar, J. Ma, M. Nadeem, K. Ahmad et al., *Evaluation of toxicity potential of cobalt in wheat irrigated with wastewater: Health risk implications for public*, Environ. Sci. Pollut. Res. 28 (2021), pp. 21119–21131.

[62] Z. Zhang, K. Chen, Q. Zhao, M. Huang and X. Ouyang, *Comparative adsorption of heavy metal ions in wastewater on monolayer molybdenum disulfide*, Green Energy Environ. 6 (2021), pp. 751–758.

[63] D. Jiang, Y. Yang, C. Huang, M. Huang, J. Chen and T. Rao, *Removal of the heavy metal ion nickel(II) via an adsorption method using flower globular magnesium hydroxide*, J. Hazard. Mater. 373 (2019), pp. 131–140.

[64] Y. Zhu, W. Huang, Z. Liu and D. Wang, *Application of supramolecular nano-material adsorbent in the treatment of heavy metal pollution in acid selenium-rich soils in south China*, Integr. Ferroelectr. 217 (2021), pp. 69–81.

[65] H. Li, J. Xiong, T. Xiao, J. Long, Q. Wang, K. Li et al., *Biochar derived from watermelon rinds as regenerable adsorbent for efficient removal of thallium(I) from wastewater*, Process Saf. Environ. Prot. 127 (2019), pp. 257–266.

[66] A.M. Adam, H.A. Saad, A.A. Atta, M. Alsawat, M.S. Hegab, T.A. Altalhi et al., *An environmentally friendly method for removing Hg(II), Pb(II), Cd(II) and Sn(II) heavy metals from wastewater using novel metal – Carbon-based composites*, Crystals 11 (2021), pp. 1–12.

[67] L. Zhang, X. Jiao, S. Wu, X. Song and R. Yao, *Study on optimization of tungsten ore flotation wastewater treatment by response surface method (Rsm)*, Minerals 11 (2021), pp. 1–14.

[68] X. Yi, Z. Xu, Y. Liu, X. Guo, M. Ou and X. Xu, *Highly efficient removal of uranium(VI) from wastewater by polyacrylic acid hydrogels*, RSC Adv. 7 (2017), pp. 6278–6287.

[69] D. Fang, X. Liao, X. Zhang, A. Teng and X. Xue, *A novel resource utilization of the calcium-based semi-dry flue gas desulfurization ash: As a reductant to remove chromium and vanadium from vanadium industrial wastewater*, J. Hazard. Mater. 342 (2018), pp. 436–445.

[70] S.A. Al-Saydeh, M.H. El-Naas and S.J. Zaidi, *Copper removal from industrial wastewater: A comprehensive review*, J. Ind. Eng. Chem. 56 (2017), pp. 35–44.

[71] E.I. Ugwu and J.C. Agunwamba, *A review on the applicability of activated carbon derived from plant biomass in adsorption of chromium, copper, and zinc from industrial wastewater*, Environ. Monit. Assess. 192 (2020), p. 240. DOI: 10.1007/s10661-020-8162-0

[72] M.D. Yahya, A.S. Aliyu, K.S. Obayomi, A.G. Olugbenga and U.B. Abdullahi, *Column adsorption study for the removal of chromium and manganese ions from electroplating wastewater using cashew nutshell adsorbent*, Cogent Eng. 7 (2020), p. 1748470.

[73] U. ur Rehman, S. Khan and S. Muhammad, *Associations of potentially toxic elements (PTEs) in drinking water and human biomarkers: A case study from five districts of Pakistan*, Environ. Sci. Pollut. Res. 25 (2018), pp. 27912–27923.

[74] J. Wang, T. Zhang, M. Li, Y. Yang, P. Lu, P. Ning et al., *Arsenic removal from water/wastewater using layered double hydroxide derived adsorbents, a critical review*, RSC Adv. 8 (2018), pp. 22694–22709.

[75] S. Mani, R.N. Bharagava and P. Chowdhary, *Textile wastewater dyes: Toxicity profile and treatment approaches*, in Emerging and Eco-Friendly Approaches for Waste Management, R.N. Bharagava, P. Chowdhary, eds., 2018, pp. 1–435.

[76] M.R. Gadekar and M.M. Ahammed, *Modelling dye removal by adsorption onto water treatment residuals using combined response surface methodology-artificial neural network approach*, J. Environ. Manage. 231 (2019), pp. 241–248.

[77] N. Kunjuzwa, L.N. Nthunya, E.N. Nxumalo and S.D. Mhlanga, *The use of nanomaterials in the synthesis of nanofiber membranes and their application in water treatment*, in Advanced Nanomaterials for Membrane Synthesis and Its Applications, W-J. Lau, A.F. Ismail, A. Isloor, A. Al-Ahmed, eds., Elsevier Inc., 2018, pp. 101–125.

[78] Q. Chen, Y. Yao, X. Li, J. Lu, J. Zhou and Z. Huang, *Comparison of heavy metal removals from aqueous solutions by chemical precipitation and characteristics of precipitates*, J. Water Process Eng. 26 (2018), pp. 289–300.

[79] S.W. Puasa, *Plant-based Tacca leontopetaloides Biopolymer Flocculant (TBPF) produced high removal of heavy metal ions at low dosage nurul*, Processes 9 (2021), pp. 1–14.

[80] Y. Ibrahim, V. Naddeo, F. Banat and S.W. Hasan, *Preparation of novel polyvinylidene fluoride (PVDF)-Tin(IV) oxide (SnO$_2$) ion exchange mixed matrix membranes for the removal of heavy metals from aqueous solutions*, Sep. Purif. Technol. 250 (2020), p. 117250.

[81] V. Ya, N. Martin, Y.H. Chou, Y.M. Chen, K.H. Choo, S.S. Chen et al., *Electrochemical treatment for simultaneous removal of heavy metals and organics from surface finishing wastewater using sacrificial iron anode*, J. Taiwan Inst. Chem. Eng. 83 (2018), pp. 107–114.

[82] C.B. Godiya, S.M. Sayed, Y. Xiao and X. Lu, *Highly porous egg white/polyethyleneimine hydrogel for rapid removal of heavy metal ions and catalysis in wastewater*, React. Funct. Polym. 149 (2020), p. 104509.

[83] C.A. García-González, T. Budtova, L. Durães, C. Erkey, P. Del Gaudio, P. Gurikov et al., *An opinion paper on aerogels for biomedical and environmental applications*, Molecules 24 (2019), pp. 1–15.

[84] A.M. Nasir, P.S. Goh, M.S. Abdullah, B.C. Ng and A.F. Ismail, *Adsorptive nanocomposite membranes for heavy metal remediation: Recent progresses and challenges*, Chemosphere 232 (2019), pp. 96–112.

[85] A. Zhang, X. Li, J. Xing and G. Xu, *Adsorption of potentially toxic elements in water by modified biochar: A review*, J. Environ. Chem. Eng. 8 (2020), p. 104196.

[86] D. Cao, Z. Cao, G. Wang, X. Dong, Y. Dong, Y. Ye et al., *Plasma induced graft co-polymerized electrospun polyethylene terephalate membranes for removal of Cu^{2+} from aqueous solution*, Chem. Phys. 536 (2020), p. 110832.

[87] M. Hong, L. Yu, L. Wang, J. Zhang, Z. Chen, L. Dong et al., *Heavy metal adsorption with zeolites: The role of hierarchical pore architecture*, Chem. Eng. J. 359 (2019), pp. 363–372.

[88] B.O. Otunola and O.O. Ololade, *A review on the application of clay minerals as heavy metal adsorbents for remediation purposes*, Environ. Technol. Innov. 18 (2020), p. 100692.

[89] H. Mi, L. Yi, Q. Wu, J. Xia and B. Zhang, *Preparation and optimization of a low-cost adsorbent for heavy metal ions from red mud using fraction factorial design and Box-Behnken response methodology*, Colloids Surfaces A Physicochem. Eng. Asp. 627 (2021), p. 127198.

[90] R. Chakraborty, A. Asthana, A.K. Singh, B. Jain and A.B.H. Susan, *Adsorption of heavy metal ions by various low-cost adsorbents: A review*, Int. J. Environ. Anal. Chem. 102 (2022), pp. 342–379.

[91] A. Karimah, M.R. Ridho, S.S. Munawar, D.S. Adi, Ismadi, R. Damayanti et al., *A review on natural fibers for development of eco-friendly bio-composite: characteristics, and utilizations*, J. Mater. Res. Technol. 13 (2021), pp. 2442–2458.

[92] B. Aaliya, K.V. Sunooj and M. Lackner, *Biopolymer composites: A review*, Int. J. Biobased Plast. 3 (2021), pp. 40–84.

[93] R.M. Rowell, *The use of biomass to produce bio-based composites and building materials*, in Advances in Biorefineries, K. Waldron, ed., Elsevier, 2014, pp. 803–818.

[94] C. Kuranchie, A. Yaya and Y.D. Bensah, *The effect of natural fibre reinforcement on polyurethane composite foams – A review*, Sci. African 11 (2021), p. e00722.

[95] T. P. Tumolva, J. K. R. Vitoa, J. C. R. Ragasaa and R. M. C. Dela Cruza, *Mechanical properties of natural fiber reinforced polymer (NFRP) composites using anahaw (Saribus rotundifolius) fibers treated with sodium alginate*, J. Chem. Eng. Mater. Sci. 8 (2017), pp. 66–73.

[96] K. Joseph, R.D. Tolêdo Filho, B. James, S. Thomas and L.H. de Carvalho, *A review on sisal fiber reinforced polymer composites*, Rev. Bras. Eng. Agrícola e Ambient. 3 (1999), pp. 367–379.

[97] A. Gupta, R. Sharma and R. Katarne, *Experimental investigation on influence of damping response of composite material by natural fibers - A review*, Mater. Today Proc. 47 (2021), pp. 3035–3042.

[98] P.H.F. Pereira, H.L. Ornaghi, V. Arantes and M.O.H. Cioffi, *Effect of chemical treatment of pineapple crown fiber in the production, chemical composition, crystalline structure, thermal stability and thermal degradation kinetic properties of cellulosic materials*, Carbohydr. Res. 499 (2021), p. 108227.

[99] G. Kappler, D.M. Souza, C.A.M. Moraes, R.C.E. Modolo, F.A. Brehm, P.R. Wander et al., *Conversion of lignocellulosic biomass through pyrolysis to promote a sustainable value chain for Brazilian agribusiness*, in Lignocellulosic Biorefining Technologies, A.P. Ingle, A.K. Chandel, S.S. da Silva, eds., Wiley, 2020, pp. 265–283.

[100] B. Duchemin, *Size, shape, orientation and crystallinity of cellulose Iβ by X-ray powder diffraction using a free spreadsheet program*, Cellulose 24 (2017), pp. 2727–2741.

[101] A. Jabbar, J. Militký, J. Wiener, B.M. Kale, U. Ali and S. Rwawiire, *Nanocellulose coated woven jute/green epoxy composites: Characterization of mechanical and dynamic mechanical behavior*, Compos. Struct. 161 (2017), pp. 340–349.

[102] C.M.M. Rojas, M.F. Rivera Velásquez, A. Tavolaro, A. Molinari and C. Fallico, *Use of vegetable fibers for PRB to remove heavy metals from contaminated aquifers-comparisons among cabuya fibers, broom fibers and ZVI*, Int. J. Environ. Res. Public Health 14 (2017), pp. 4–10.

[103] J.J. Gallardo-Rodríguez, A.C. Rios-Rivera and M.R. Von Bennevitz, *Living biomass supported on a natural-fiber biofilter for lead removal*, J. Environ. Manage. 231 (2019), pp. 825–832.

[104] S. Shrestha, G. Son, S.H. Lee and T.G. Lee, *Isotherm and thermodynamic studies of Zn(II) adsorption on lignite and coconut shell-based activated carbon fiber*, Chemosphere 92 (2013), pp. 1053–1061.

[105] Y. Li, Y. Liu, W. Yang, L. Liu and J. Pan, *Adsorption of elemental mercury in flue gas using biomass porous carbons modified by microwave/hydrogen peroxide*, Fuel 291 (2021), p. 120152.

[106] M. Om Prakash, G. Raghavendra, S. Ojha and M. Panchal, *Characterization of porous activated carbon prepared from arhar stalks by single step chemical activation method*, Mater. Today Proc. 39 (2020), pp. 1476–1481.

[107] A.N. Prusov, S.M. Prusova, M.V. Radugin and A.V. Bazanov, *Synthesis and characterization of flax shive activated carbon*, Fullerenes Nanotub. Carbon Nanostructures 29 (2020), pp. 232–243.

[108] E.M. Mistar, T. Alfatah and M.D. Supardan, *Synthesis and characterization of activated carbon from Bambusa vulgaris striata using two-step KOH activation*, J. Mater. Res. Technol. 9 (2020), pp. 6278–6286.

[109] M.C. Silva, L. Spessato, T.L. Silva, G.K.P. Lopes, H.G. Zanella, J.T.C. Yokoyama et al., *H₃PO₄–activated carbon fibers of high surface area from banana tree pseudo-stem fibers: Adsorption studies of methylene blue dye in batch and fixed bed systems*, J. Mol. Liq. 324 (2021), p. 114771.

[110] M.U. Dao, H.S. Le, H.Y. Hoang, V.A. Tran, V.D. Doan, T.T.N. Le et al., *Natural core-shell structure activated carbon beads derived from Litsea glutinosa seeds for removal of methylene blue: Facile preparation, characterization, and adsorption properties*, Environ. Res. 198 (2021), p. 110481.

[111] T.E. Oladimeji, B.O. Odunoye, F.B. Elehinafe, O.R. Obanla and O.A. Odunlami, *Production of activated carbon from sawdust and its efficiency in the treatment of sewage water*, Heliyon 7 (2021), p. e05960.

[112] A. Melliti, V. Srivastava, J. Kheriji, M. Sillanpää and B. Hamrouni, *Date Palm Fiber as a novel precursor for porous activated carbon: Optimization, characterization and its application as Tylosin antibiotic scavenger from aqueous solution*, Surf. Interfaces. 24 (2021), p. 101047.

[113] L.S. Maia, L.D. Duizit, F.R. Pinhatio and D.R. Mulinari, *Valuation of banana peel waste for producing activated carbon via NaOH and pyrolysis for methylene blue removal*, Carbon Lett. 31 (2021), pp. 749–762.

[114] L.S. Maia, A.I.C. da Silva, E.S. Carneiro, F.M. Monticelli, F.R. Pinhati and D.R. Mulinari, *Activated carbon from palm fibres used as an adsorbent for methylene blue removal*, J. Polym. Environ. 29 (2021), pp. 1162–1175.

[115] A.I.C. da Silva, G. Paranha, L.S. Maia and D.R. Mulinari, *Development of activated carbon from pineapple crown wastes and its potential use for removal of methylene blue*, J. Nat. Fibers 19 (2022), pp. 5211–5226.

[116] S. Wu, P. Yan, W. Yang, J. Zhou, H. Wang, L. Che et al., *ZnCl₂ enabled synthesis of activated carbons from ion-exchange resin for efficient removal of Cu²⁺ ions from water via capacitive deionization*, Chemosphere 264 (2021), p. 128557.

[117] U. Yunusa, A. Ibrahim, Y. Abdullahi and T. Abdullahi, *Hexavalent chromium removal from simulated wastewater using biomass-based activated carbon: Kinetics, mechanism, thermodynamics and regeneration studies*, Algerian Journal of Engineering, 04 (2021), pp. 30–44.

[118] Z.M. Yunus, N. Othman, A. Al-Gheethi, R. Hamdan and N.N. Ruslan, *Adsorption of heavy metals from mining effluents using honeydew peels activated carbon; isotherm, kinetic and column studies*, J. Dispers. Sci. Technol. 42 (2021), pp. 715–729.

[119] Y. Chiang, P. Chiang, C.Huang, *Effects of pore structure and temperature on VOC adsorption on activated carbon*, Carbon 39 (2001), pp. 523–534.

[120] A.K. Thakur, R. Sathyamurthy, R. Velraj, I. Lynch, R. Saidur, A.K. Pandey et al., *Sea-water desalination using a desalting unit integrated with a parabolic trough collector and activated carbon pellets as energy storage medium*, Desalination 516 (2021), p. 115217.

[121] O.E. Ohimor, O.P. Ikechukwu, D.V. Limited and O.D. Temisa, *Deodorization of hydrogen sulphide contaminated water by biosorption on deodorization of hydrogen sulphide contaminated water by biosorption on coconut fibre activated carbon*, J. Eng. 11 (2021), pp. 5–12.

[122] A. Adan-Mas, L. Alcaraz, P. Arévalo-Cid, F.A. López-Gómez and F. Montemor, *Coffee-derived activated carbon from second biowaste for supercapacitor applications*, Waste Manag. 120 (2021), pp. 280–289.

[123] N. Issaadi, A. Aït-Mokhtar, R.B. Hamami and Ameur, *8-Effect of variability of porous media properties on drying kinetics: Application to cement-based materials*, in Advances in Multi-Physics and Multi-Scale Couplings in Geo-Environmental Mechanics, F. Nicot, O. Millet, eds., Elsevier B.V., 2018, pp. 243–289.

[124] K.-D. Henning and H. von Kienle, *Chapter 9-Activated carbon*, in Industrial Carbon and Graphite Materials, Volume I: Raw Materials, Production and Applications, I, D.D.L. Chung, ed., 2021.

[125] R. Zakaria, N.A. Jamalluddin and M.Z. Abu Bakar, *Effect of impregnation ratio and activation temperature on the yield and adsorption performance of mangrove based activated carbon for methylene blue removal*, Results Mater. 10 (2021), p. 100183.

[126] R.A. Canales-Flores and F. Prieto-García, *Taguchi optimization for production of activated carbon from phosphoric acid impregnated agricultural waste by microwave heating for the removal of methylene blue*, Diam. Relat. Mater. 109 (2020), p. 108027.

[127] L.S. Queiroz, L.K.C. de Souza, K.T.C. Thomaz, E.T. Leite Lima, G.N. da Rocha Filho, L.A.S. do Nascimento et al., *Activated carbon obtained from Amazonian biomass tailings (acai seed): Modification, characterization, and use for removal of metal ions from water*, J. Environ. Manage. 270 (2020), p. 110868.

[128] A.E. Orduz, C. Acebal and G. Zanini, *Activated carbon from peanut shells: 2,4-D desorption kinetics study for application as a green material for analytical purposes*, J. Environ. Chem. Eng. 9 (2021), p. 104601.

[129] Z. Li, H. Hanafy, L. Zhang, L. Sellaoui, M. Schadeck Netto, M.L.S. Oliveira et al., *Adsorption of Congo red and methylene blue dyes on an ashitaba waste and a walnut shell-based activated carbon from aqueous solutions: Experiments, characterization and physical interpretations*, Chem. Eng. J. 388 (2020), p. 124263.

[130] Z. Liu, Y. Sun, X. Xu, X. Meng, J. Qu, Z. Wang et al., *Preparation, characterization and application of activated carbon from corn cob by KOH activation for removal of Hg(II) from aqueous solution*, Bioresour. Technol. 306 (2020), p. 123154.

[131] M.E. Peñafiel, J.M. Matesanz, E. Vanegas, D. Bermejo, R. Mosteo and M.P. Ormad, *Comparative adsorption of ciprofloxacin on sugarcane bagasse from Ecuador and on commercial powdered activated carbon*, Sci. Total Environ. 750 (2021), p. 141498.

[132] A. Jain, M. Ghosh, M. Krajewski, S. Kurungot and M. Michalska, *Biomass-derived activated carbon material from native European deciduous trees as an inexpensive and sustainable energy material for supercapacitor application*, J. Energy Storage 34 (2021), p. 102178.

[133] F. Bahador, R. Foroutan, H. Esmaeili and B. Ramavandi, *Enhancement of the chromium removal behavior of Moringa oleifera activated carbon by chitosan and iron oxide nanoparticles from water*, Carbohydr. Polym. 251 (2021), p. 117085.

[134] A. El Kassimi, Y. Achour, M. El Himri, M.R. Laamari and M. El Haddad, *Process optimization of high surface area activated carbon prepared from cucumis melo by H_3PO_4 activation for the removal of cationic and anionic dyes using full factorial design*, Biointerface Res. Appl. Chem. 11 (2021), pp. 12662–12679.

[135] D. Duan, D. Chen, L. Huang, Y. Zhang, Y. Zhang, Q. Wang et al., *Activated carbon from lignocellulosic biomass as catalyst: A review of the applications in fast pyrolysis process*, J. Anal. Appl. Pyrolysis 158 (2021), p. 105246.

[136] M.O. Bello, N. Abdus-Salam, F.A. Adekola and U. Pal, *Isotherm and kinetic studies of adsorption of methylene blue using activated carbon from ackee apple pods*, Chem. Data Collect. 31 (2021), p. 100607.

[137] A.H. Jawad, M. Bardhan, M.A. Islam, M.A. Islam, S.S.A. Syed-Hassan, S.N. Surip et al., *Insights into the modeling, characterization and adsorption performance of mesoporous activated carbon from corn cob residue via microwave-assisted H$_3$PO$_4$ activation*, Surf. Interfaces. 21 (2020), p. 100688.

[138] O.D.C. Salcedo, D.P. Vargas, L. Giraldo and J.C. Moreno-Piraján, *Study of mercury [Hg(II)] adsorption from aqueous solution on functionalized activated carbon*, ACS Omega 6 (2021), pp. 11849–11856.

[139] R. Shahrokhi-Shahraki, C. Benally, M.G. El-Din and J. Park, *High efficiency removal of heavy metals using tire-derived activated carbon vs commercial activated carbon: Insights into the adsorption mechanisms*, Chemosphere 264 (2021), p. 128455.

[140] N.U.M. Nizam, M.M. Hanafiah, E. Mahmoudi, A.A. Halim and A.W. Mohammad, *The removal of anionic and cationic dyes from an aqueous solution using biomass-based activated carbon*, Sci. Rep. 11 (2021), pp. 1–17.

[141] N. Abuelnoor, A. AlHajaj, M. Khaleel, L.F. Vega and M.R.M. Abu-Zahra, *Activated carbons from biomass-based sources for CO$_2$ capture applications*, Chemosphere 282 (2021), p. 131111.

[142] S. Rovani, A.G. Rodrigues, L.F. Medeiros, R. Cataluña, É.C. Lima and A.N. Fernandes, *Synthesis and characterisation of activated carbon from agroindustrial waste - Preliminary study of 17β-estradiol removal from aqueous solution*, J. Environ. Chem. Eng. 4 (2016), pp. 2128–2137.

[143] M.J. Pinto Brito, M.P. Flores Santos, E.C. De Souza Júnior, L.S. Santos, R.C. Ferreira Bonomo, R. Da Costa Ilhéu Fontan et al., *Development of activated carbon from pupunha palm heart sheaths: Effect of synthesis conditions and its application in lipase immobilization*, J. Environ. Chem. Eng. 8 (2020), p. 104391.

[144] F.S.A. Khan, N.M. Mubarak, Y.H. Tan, M. Khalid, R.R. Karri, R. Walvekar et al., *A comprehensive review on magnetic carbon nanotubes and carbon nanotube-based buckypaper for removal of heavy metals and dyes*, J. Hazard. Mater. 413 (2021), p. 125375.

[145] K. Lellala, *Sulfur embedded on in-situ carbon nanodisc decorated on graphene sheets for efficient photocatalytic activity and capacitive deionization method for heavy metal removal*, J. Mater. Res. Technol. 13 (2021), pp. 1555–1566.

[146] Y. Dou, X. Liu, X. Wang, K. Yu and C. Liang, *Jute fiber based micro-mesoporous carbon: A biomass derived anode material with high-performance for lithium-ion batteries*, Mater. Sci. Eng. B Solid-State Mater. Adv. Technol. 265 (2021), p. 115015.

[147] W. Yu, L. Zhang, H. Xu, H. Wang, X. Peng, S. Wu et al., *Highly dispersed transition metal oxide-supported activated carbon prepared by plasma for removal of elemental mercury*, Fuel 286 (2021), pp. 1–10.

[148] S. Bakhta, Z. Sadaoui, U. Lassi, H. Romar, R. Kupila and J. Vieillard, *Performances of metals modified activated carbons for fluoride removal from aqueous solutions*, Chem. Phys. Lett. 754 (2020), p. 137705.

[149] C.K. Aslani and O. Amik, *Active carbon/PAN composite adsorbent for uranium removal: Modeling adsorption isotherm data, thermodynamic and kinetic studies*, Appl. Radiat. Isot. 168 (2021), p. 109474.

[150] D. Anitha, A. Ramadevi and R. Seetharaman, *Biosorptive removal of Nickel(II) from aqueous solution by Mangosteen shell activated carbon*, Mater. Today Proc. 45 (2021), pp. 718–722.

[151] A. Rahmani-Sani, P. Singh, P. Raizada, E. Claudio Lima, I. Anastopoulos, D.A. Giannakoudakis et al., *Use of chicken feather and eggshell to synthesize a novel magnetized activated carbon for sorption of heavy metal ions*, Bioresour. Technol. 297 (2020), p. 122452.

[152] X. Liu, Y. Wu, H. Ye, J. Chen, W. Xiao, W. Zhou et al., *Modification of sugar-based carbon using lanthanum and cobalt bimetal species for effective adsorption of methyl orange*, Environ. Technol. Innov. 23 (2021), p. 101769.

[153] M. Maqbool, S. Sadaf, H.N. Bhatti, S. Rehmat, A. Kausar, S.A. Alissa et al., *Sodium alginate and polypyrrole composites with algal dead biomass for the adsorption of Congo red dye: Kinetics, thermodynamics and desorption studies*, Surf. Interfaces. 25 (2021), p. 101183.

[154] R. Foroutan, R. Mohammadi, A. Ahmadi, G. Bikhabar, F. Babaei and B. Ramavandi, *Impact of ZnO and Fe_3O_4 magnetic nanoscale on the methyl violet 2B removal efficiency of the activated carbon oak wood*, Chemosphere 286 (2022), p. 131632.

[155] Y.L. Pang, W.S. Teh, S. Lim, A.Z. Abdullah, H.C. Ong and C.-H. Wu, *Enhancement of adsorption-photocatalysis of malachite green using oil palm biomass-derived activated carbon/titanium dioxide composite*, Curr. Anal. Chem. 17 (2020), pp. 603–617.

[156] M.C. Kannaujiya, A.K. Prajapati, T. Mandal, A.K. Das and M.K. Mondal, *Extensive analyses of mass transfer, kinetics, and toxicity for hazardous acid yellow 17 dye removal using activated carbon prepared from waste biomass of Solanum melongena*, Biomass Conv. Bioref.13 (2023), pp. 99–117. https://doi.org/10.1007/s13399-020-01160-8

[157] S.A. Hira, M. Yusuf, D. Annas, H.S. Hui and K.H. Park, *Biomass-derived activated carbon as a catalyst for the effective degradation of Rhodamine B dye*, Processes 8 (2020), p. 926. https://doi.org/10.3390/pr8080926

[158] F. Guo, L. Xiaolei, J. Xiaochen, Z. Xingmin, G. Chenglong and R. Zhonghao, *Characteristics and toxic dye adsorption of magnetic activated carbon prepared from biomass waste by modified one-step synthesis*, Colloids Surfaces A Physicochem. Eng. Asp. 555 (2018), pp. 43–54.

[159] J.R. García, U. Sedran, M.A.A. Zaini and Z.A. Zakaria, *Preparation, characterization, and dye removal study of activated carbon prepared from palm kernel shell*, Environ. Sci. Pollut. Res. 25 (2018), pp. 5076–5085.

[160] M. Baysal, K. Bilge, B. Yılmaz, M. Papila and Y. Yürüm, *Preparation of high surface area activated carbon from waste-biomass of sunflower piths: Kinetics and equilibrium studies on the dye removal*, J. Environ. Chem. Eng. 6 (2018), pp. 1702–1713.

[161] H. Wu, R. Chen, H. Du, J. Zhang, L. Shi, Y. Qin et al., *Synthesis of activated carbon from peanut shell as dye adsorbents for wastewater treatment*, Adsorpt. Sci. Technol. 37 (2019), pp. 34–48.

[162] N. Boudechiche, M. Fares, S. Ouyahia, H. Yazid, M. Trari and Z. Sadaoui, *Comparative study on removal of two basic dyes in aqueous medium by adsorption using activated carbon from Ziziphus lotus stones*, Microchem. J. 146 (2019), pp. 1010–1018.

[163] M. Danish, T. Ahmad, S. Majeed, M. Ahmad, L. Ziyang, Z. Pin et al., *Use of banana trunk waste as activated carbon in scavenging methylene blue dye: Kinetic, thermodynamic, and isotherm studies*, Bioresour. Technol. Rep. 3 (2018), pp. 127–137.

[164] M. Goswami and P. Phukan, *Enhanced adsorption of cationic dyes using sulfonic acid modified activated carbon*, J. Environ. Chem. Eng. 5 (2017), pp. 3508–3517.

[165] M. Mohammad, S. Maitra and B.K. Dutta, *Comparison of activated carbon and physic seed hull for the removal of malachite green dye from aqueous solution*, Water. Air. Soil Pollut. 229 (2018), p. 45. https://doi.org/10.1007/s11270-018-3686-4

[166] R.K. Liew, E. Azwar, P.N.Y. Yek, X.Y. Lim, C.K. Cheng, J.H. Ng et al., *Microwave pyrolysis with KOH/NaOH mixture activation: A new approach to produce micro-mesoporous activated carbon for textile dye adsorption*, Bioresour. Technol. 266 (2018), pp. 1–10.

[167] W. Astuti, T. Sulistyaningsih, E. Kusumastuti, G. Yanny and R. Sari, *Bioresource technology thermal conversion of pineapple crown leaf waste to magnetized activated carbon for dye removal*, Bioresour. Technol. 287 (2019), p. 121426.

[168] M. Talat, S. Mohan, V. Dixit, D.K. Singh, S.H. Hasan and O.N. Srivastava, *Effective removal of fluoride from water by coconut husk activated carbon in fixed bed column: Experimental and breakthrough curves analysis*, Groundwater Sustain. Dev. 7 (2018), pp. 48–55.

[169] S. Chakraborty, B.W. Jo and Y.S. Yoon, *Development of nano cement concrete by top-down and bottom-up nanotechnology concept*, in Smart Nanoconcretes and Cement-Based Materials, M.S. Liew, P. Nguyen-Tri, T.A. Nguyen, S. Kakooei, eds., 2020, pp. 183–213.

[170] Kenry and C.T. Lim, *Nanofiber technology: Current status and emerging developments*, Prog. Polym. Sci. 70 (2017), pp. 1–17. https://doi.org/10.1016/j.progpolymsci.2017.03.002

[171] T.A. Saleh, *Nanomaterials: Classification, properties, and environmental toxicities*, Environ. Technol. Innov. 20 (2020), p. 101067.

[172] S. Campuzano, P. Yáñez-Sedeño and J.M. Pingarrón, *Carbon dots and graphene quantum dots in electrochemical biosensing*, Nanomaterials 9 (2019), pp. 1–18.

[173] K. Yu, X. Pan, G. Zhang, X. Liao, X. Zhou, M. Yan et al., *Nanowires in energy storage devices: Structures, synthesis, and applications*, Adv. Energy Mater. 8 (2018), pp. 1–19.

[174] S. Kang, G.M. Biesold, H. Lee, D. Bukharina, Z. Lin and V.V. Tsukruk, *Dynamic chiro-optics of bio-inorganic nanomaterials via seamless co-assembly of semiconducting nanorods and polysaccharide nanocrystals*, Adv. Funct. Mater. 2104596 (2021), p. 2104596.

[175] Z. Wang and B. Mi, *Environmental applications of 2D molybdenum disulfide (MoS₂) nanosheets*, Environ. Sci. Technol. 51 (2017), pp. 8229–8244.

[176] G. Vilardi, T. Mpouras, D. Dermatas, N. Verdone, A. Polydera and L. Di Palma, *Nanomaterials application for heavy metals recovery from polluted water: The combination of nano zero-valent iron and carbon nanotubes. Competitive adsorption non-linear modeling*, Chemosphere 201 (2018), pp. 716–729.

[177] I.A.A. Terra, L.A. Mercante, R.S. Andre and D.S. Correa, *Fluorescent and colorimetric electrospun nanofibers for heavy-metal sensing*, Biosensors 7 (2017), p. 61. DOI: 10.3390/bios7040061

[178] A. Blanco, M.C. Monte, C. Campano, A. Balea, N. Merayo and C. Negro, *Chapter 5 - Nanocellulose for industrial use: Cellulose Nanofibers (CNF), Cellulose Nanocrystals (CNC), and Bacterial Cellulose (BC)*, in Micro and Nano Technologies, C.M. Hussain, ed., Elsevier, 2018, pp. 74–126.

[179] H.M. Saleh, S.M. El-Sheikh, E.E. Elshereafy and A.K. Essa, *Mechanical and physical characterization of cement reinforced by iron slag and titanate nanofibers to produce advanced containment for radioactive waste*, Constr. Build. Mater. 200 (2019), pp. 135–145.

[180] S. Thakkar and M. Misra, *Electrospun polymeric nanofibers: New horizons in drug delivery*, Eur. J. Pharm. Sci. 107 (2017), pp. 148–167.

[181] Z.R. Ege, A. Akan, F.N. Oktar, C.-C. Lin, B. Karademir and O. Gunduz, *Encapsulation of indocyanine green in poly(lactic acid) nanofibers for using as a nanoprobe in biomedical diagnostics*, Mater. Lett. 228 (2018), pp. 148–151.

[182] C.C. Piras, S. Fernández-Prieto and W.M. De Borggraeve, *Ball milling: A green technology for the preparation and functionalisation of nanocellulose derivatives*, Nanoscale Adv. 1 (2019), pp. 937–947.

[183] M.S. Nurul Atiqah, D.A. Gopakumar, F.A.T. Owolabi, Y.B. Pottathara, S. Rizal, N.A. Sri Aprilia et al., *Extraction of cellulose nanofibers via eco-friendly supercritical carbon dioxide treatment followed by mild acid hydrolysis and the fabrication of cellulose nanopapers*, Polymers 11 (2019), p. 1813. doi: 10.3390/polym11111813

[184] E.S. Madivoli, P.G. Kareru, A.N. Gachanja, S.M. Mugo, D.M. Sujee and K.M. Fromm, *Isolation of cellulose nanofibers from Oryza sativa residues via TEMPO mediated oxidation*, J. Nat. Fibers 19 (2022), pp. 1310–1322.

[185] T.J. Bondancia, L.J. Corrêa, A.J.G. Cruz, A.C. Badino, L.H.C. Mattoso, J.M. Marconcini et al., *Enzymatic production of cellulose nanofibers and sugars in a stirred-tank reactor: Determination of impeller speed, power consumption, and rheological behavior,* Cellulose 25 (2018), pp. 4499–4511.

[186] Q. Tarrés, H. Oliver-Ortega, S. Boufi, M. Àngels Pèlach, M. Delgado-Aguilar and P. Mutjé, *Evaluation of the fibrillation method on lignocellulosic nanofibers production from eucalyptus sawdust: A comparative study between high-pressure homogenization and grinding,* Int. J. Biol. Macromol. 145 (2020), pp. 1199–1207.

[187] P. Aditiawati, R. Dungani and C. Amelia, *Enzymatic production of cellulose nanofibers from oil palm empty fruit bunch (EFB) with crude cellulase of Trichoderma sp.,* Mater. Res. Express 5 (2018), p. 34005.

[188] Q. Liu, Y. Lu, M. Aguedo, N. Jacquet, C. Ouyang, W. He et al., *Isolation of high-purity cellulose nanofibers from wheat straw through the combined environmentally friendly methods of steam explosion, microwave-assisted hydrolysis, and microfluidization,* ACS Sustain. Chem. Eng. 5 (2017), pp. 6183–6191.

[189] A.B. Aghababai, A. Esmaeili and Y. Behjat, *Invent of a simultaneous adsorption and separation process based on dynamic membrane for treatment Zn(II), Ni(II) and, Co(II) industrial wastewater,* Arab. J. Chem. 14 (2021), p. 103231.

[190] J. Xue, T. Wu, Y. Dai and Y. Xia, *Electrospinning and electrospun nanofibers: Methods, materials, and applications,* Chem. Rev. 119 (2019), pp. 5298–5415.

[191] K.B.R. Teodoro, F.M. Shimizu, V.P. Scagion and D.S. Correa, *Ternary nanocomposites based on cellulose nanowhiskers, silver nanoparticles and electrospun nanofibers: Use in an electronic tongue for heavy metal detection,* Sensors Actuators, B Chem. 290 (2019), pp. 387–395.

[192] R. Chen, Y. Cheng, P. Wang, Q. Wang, S. Wan, S. Huang et al., *Enhanced removal of Co(II) and Ni(II) from high-salinity aqueous solution using reductive self-assembly of three-dimensional magnetic fungal hyphal/graphene oxide nanofibers,* Sci. Total Environ. 756 (2021), pp. 143871.

[193] A. Gedanken, *Ultrasonic processing to produce nanoparticles,* in Encyclopedia of Materials: Science and Technology, K.H.J. Buschow, R.W. Cahn, M.C. Flemings, B. Ilschner, E.J. Kramer, S. Mahajan, et al., eds., Elsevier, 2001, pp. 9450–9456.

[194] A. Huczko, *Template-based synthesis of nanomaterials,* Appl. Phys. A 376 (2000), pp. 365–376.

[195] F.F. Fang, Y.Z. Dong and H.J. Choi, *Effect of oxidants on morphology of interfacial polymerized polyaniline nanofibers and their electrorheological response,* Polymer 158 (2018), pp. 176–182.

[196] M.R. Yarandpour, A. Rashidi, R. khajavi, N. Eslahi and M.E. Yazdanshenas, *Mesoporous PAA/dextran-polyaniline core-shell nanofibers: Optimization of producing conditions, characterization and heavy metal adsorptions,* J. Taiwan Inst. Chem. Eng. 93 (2018), pp. 566–581.

[197] S.S. Paclijan, S.M.B. Franco, R.B. Abella, J.C.H. Lague and N.P.B. Tan, *Fresh and uncalcined solution blow spinning - Spun PAN and PVDF nanofiber membranes for methylene blue dye removal in water,* J. Membr. Sci. 7 (2021), pp. 173–184.

[198] L. Xia, H. Feng, Q. Zhang, X. Luo, P. Fei and F. Li, *Centrifugal spinning of lignin amine/cellulose acetate nanofiber for heavy metal ion adsorption,* Fibers Polym. 23 (2022), pp. 77–85

[199] C. Wan, Y. Jiao, S. Wei, L. Zhang, Y. Wu and J. Li, *Functional nanocomposites from sustainable regenerated cellulose aerogels: A review,* Chem. Eng. J. 359 (2019), pp. 459–475.

[200] A. Abraham, A.K. Mathew, H. Park, O. Choi, R. Sindhu, B. Parameswaran et al., *Pretreatment strategies for enhanced biogas production from lignocellulosic biomass,* Bioresour. Technol. 301 (2020), p. 122725.

[201] G. Banvillet, G. Depres, N. Belgacem and J. Bras, *Alkaline treatment combined with enzymatic hydrolysis for efficient cellulose nanofibrils production*, Carbohydr. Polym. 255 (2021), p. 117383.

[202] R. Casarano, P.A.R. Pires and O.A. El Seoud, *Acylation of cellulose in a novel solvent system: Solution of dibenzyldimethylammonium fluoride in DMSO*, Carbohydr. Polym. 101 (2014), pp. 444–450.

[203] J.C. do. Santos, L.Á.d. Oliveira, L.M. Gomes Vieira, V. Mano, R.T.S. Freire and T.H. Panzera, *Eco-friendly sodium bicarbonate treatment and its effect on epoxy and polyester coir fibre composites*, Constr. Build. Mater. 211 (2019), pp. 427–436.

[204] Y. Yue, J. Han, G. Han, G.M. Aita and Q. Wu, *Cellulose fibers isolated from energycane bagasse using alkaline and sodium chlorite treatments: Structural, chemical and thermal properties*, Ind. Crops Prod. 76 (2015), pp. 355–363.

[205] Y. Liu, X. Lv, J. Bao, J. Xie, X. Tang, J. Che et al., *Characterization of silane treated and untreated natural cellulosic fibre from corn stalk waste as potential reinforcement in polymer composites*, Carbohydr. Polym. 218 (2019), pp. 179–187.

[206] J. Xu, Z. Wu, Q. Wu and Y. Kuang, *Acetylated cellulose nanocrystals with highcrystallinity obtained by one-step reaction from the traditional acetylation of cellulose*, Carbohydr. Polym. 229 (2020), p. 115553.

[207] Y. Zhuang, J. Liu, J. Chen and P. Fei, *Modified pineapple bran cellulose by potassium permanganate as a copper ion adsorbent and its adsorption kinetic and adsorption thermodynamic*, Food Bioprod. Process. 122 (2020), pp. 82–88.

[208] A. Atiqah, M. Jawaid, M.R. Ishak and S.M. Sapuan, *Effect of alkali and silane treatments on mechanical and interfacial bonding strength of sugar palm fibers with thermoplastic polyurethane*, J. Nat. Fibers 15 (2018), pp. 251–261.

[209] N. F. Souza, J. A. Pinheiro, P. Silva, J.P.S. Morais, M.d.S.M. de Souza Filho, A.I.S. Brígida, C.R. Muniz and M. F. Rosa, *Development of chlorine-free pulping method to extract cellulose nanocrystals from pressed oil palm mesocarp fibers*, J. Biobased Mater. Bioenergy 9 (2015), pp. 372–379.

[210] K.F. Adekunle, *Surface treatments of natural fibres—A review: Part 1*, Open J. Polym. Chem. 05 (2015), pp. 41–46.

[211] P.L. Sruthi and P.H.P. Reddy, *Effect of alkali concentration on swelling characteristics of transformed kaolinitic clays*, Clays Clay Miner. 68 (2020), pp. 373–393. https://doi.org/10.1007/s42860-020-00081-x

[212] U.S. Gupta, M. Dhamarikar, A. Dharkar, S. Chaturvedi, A. Kumrawat, N. Giri et al., *Plasma modification of natural fiber: A review*, Mater. Today Proc. 43 (2020), pp. 451–457.

[213] H. Paudyal, B. Pangeni, K. Nath Ghimire, K. Inoue, K. Ohto, H. Kawakita et al., *Adsorption behavior of orange waste gel for some rare earth ions and its application to the removal of fluoride from water*, Chem. Eng. J. 195–196 (2012), pp. 289–296.

[214] H. Yu, L. Zheng, T. Zhang, J. Ren, W. Cheng, L. Zhang et al., *Adsorption behavior of Cd(II) on TEMPO-oxidized cellulose in inorganic/ organic complex systems*, Environ. Res. 195 (2021), p. 110848.

[215] A. Ali, A. Mannan, I. Hussain, I. Hussain and M. Zia, *Effective removal of metal ions from aqueous solution by silver and zinc nanoparticles functionalized cellulose: Isotherm, kinetics and statistical supposition of process*, Environ. Nanotechnol. Monit. Manag. 9 (2018), pp. 1–11.

[216] A. Daochalermwong, N. Chanka, K. Songsrirote, P. Dittanet, C. Niamnuy and A. Seubsai, *Removal of heavy metal ions using modified celluloses prepared from pineapple leaf fiber*, ACS Omega 5 (2020), pp. 5285–5296.

[217] M.J.K. Ahmed, M. Ahmaruzzaman and R.A. Reza, *Lignocellulosic-derived modified agricultural waste: Development, characterisation and implementation in sequestering pyridine from aqueous solutions*, J. Colloid Interface Sci. 428 (2014), pp. 222–234.

[218] X. Li, Y. Tang, Z. Xuan, Y. Liu and F. Luo, *Study on the preparation of orange peel cellulose adsorbents and biosorption of Cd²⁺ from aqueous solution*, Sep. Purif. Technol. 55 (2007), pp. 69–75.

[219] L.V.A. Gurgel, R.P. de Freitas and L.F. Gil, *Adsorption of Cu(II), Cd(II), and Pb(II) from aqueous single metal solutions by sugarcane bagasse and mercerized sugarcane bagasse chemically modified with succinic anhydride*, Carbohydr. Polym. 74 (2008), pp. 922–929.

[220] M. Hosseini, H. Zaki Dizaji, M. Taghavi and A.A. Babaei, *Preparation of ultra-lightweight and surface-tailored cellulose nanofibril composite cryogels derived from date palm waste as powerful and low-cost heavy metals adsorbent to treat aqueous medium*, Ind. Crops Prod. 154 (2020), p. 112696.

[221] M.H. Rahaman, M.A. Islam, M.M. Islam, M.A. Rahman and S.M.N. Alam, *Biodegradable composite adsorbent of modified cellulose and chitosan to remove heavy metal ions from aqueous solution*, Curr. Res. Green Sustain. Chem. 4 (2021), p. 100119.

[222] Q. Wu, M. Ren, X. Zhang, C. Li, T. Li, Z. Yang et al., *Comparison of Cd(II) adsorption properties onto cellulose, hemicellulose and lignin extracted from rice bran*, LWT 144 (2021), p. 111230.

[223] K. Taksitta, P. Sujarit, N. Ratanawimarnwong, S. Donpudsa and K. Songsrirote, *Development of tannin-immobilized cellulose fiber extracted from coconut husk and the application as a biosorbent to remove heavy metal ions*, Environ. Nanotechnol. Monit. Manag. 14 (2020), p. 100389.

[224] W.E. Oliveira, A.S. Franca, L.S. Oliveira and S.D. Rocha, *Untreated coffee husks as biosorbents for the removal of heavy metals from aqueous solutions*, J. Hazard. Mater. 152 (2008), pp. 1073–1081.

[225] P.F. Santos, J.B. Neris, F.H. Martínez Luzardo, F.G. Velasco, M.S. Tokumoto and R.S. da Cruz, *Chemical modification of four lignocellulosic materials to improve the Pb²⁺ and Ni²⁺ ions adsorption in aqueous solutions*, J. Environ. Chem. Eng. 7 (2019), p. 103363.

[226] M.B. Shakoor, N.K. Niazi, I. Bibi, M. Shahid, F. Sharif, S. Bashir et al., *Arsenic removal by natural and chemically modified water melon rind in aqueous solutions and groundwater*, Sci. Total Environ. 645 (2018), pp. 1444–1455.

[227] S.B. Khan, K.A. Alamry, H.M. Marwani, A.M. Asiri and M.M. Rahman, *Synthesis and environmental applications of cellulose/ZrO₂ nanohybrid as a selective adsorbent for nickel ion*, Compos. Part B Eng. 50 (2013), pp. 253–258.

[228] Z.M. Magriotis, S.S. Vieira, A.A. Saczk, N.A.V. Santos and N.R. Stradiotto, *Removal of dyes by lignocellulose adsorbents originating from biodiesel production*, J. Environ. Chem. Eng. 2 (2014), pp. 2199–2210.

[229] E. Misran, O. Bani, E.M. Situmeang and A.S. Purba, *Banana stem based activated carbon as a low-cost adsorbent for methylene blue removal: Isotherm, kinetics, and reusability*, Alexandria Eng. J. 61 (2021), pp. 1946–1955. https://doi.org/10.1016/j.aej.2021.07.022

[230] S. Manna, D. Roy, P. Saha, D. Gopakumar and S. Thomas, *Rapid methylene blue adsorption using modified lignocellulosic materials*, Process Saf. Environ. Prot. 107 (2017), pp. 346–356.

[231] J. Ponce, J.G.d.S. Andrade, L.N. dos Santos, M.K. Bulla, B.C.B. Barros, S.L. Favaro et al., *Alkali pretreated sugarcane bagasse, rice husk and corn husk wastes as lignocellulosic biosorbents for dyes*, Carbohydr. Polym. Technol. Appl. 2 (2021), p. 100061.

[232] T.C. Andrade Siqueira, I. Zanette da Silva, A.J. Rubio, R. Bergamasco, F. Gasparotto, E. Aparecida de Souza Paccola et al., *Sugarcane bagasse as an efficient biosorbent for methylene blue removal: Kinetics, isotherms and thermodynamics*, Int. J. Environ. Res. Public Health 17 (2020), p. 526.

[233] S.-L. Chan, Y.P. Tan, A.H. Abdullah and S.-T. Ong, *Equilibrium, kinetic and thermodynamic studies of a new potential biosorbent for the removal of basic blue 3 and Congo red dyes: Pineapple (Ananas comosus) plant stem*, J. Taiwan Inst. Chem. Eng. 61 (2016), pp. 306–315.

[234] D. Lin, Z. Liu, R. Shen, S. Chen and X. Yang, *Bacterial cellulose in food industry: Current research and future prospects*, Int. J. Biol. Macromol. 158 (2020), pp. 1007–1019.

[235] V.S. Munagapati, V. Yarramuthi, Y. Kim, K.M. Lee and D.-S. Kim, *Removal of anionic dyes (reactive black 5 and Congo red) from aqueous solutions using banana peel powder as an adsorbent*, Ecotoxicol. Environ. Saf. 148 (2018), pp. 601–607.

[236] A. Asfaram, M.R. Fathi, S. Khodadoust and M. Naraki, *Removal of direct red 12B by garlic peel as a cheap adsorbent: Kinetics, thermodynamic and equilibrium isotherms study of removal*, Spectrochim. Acta Part A Mol. Biomol. Spectrosc. 127 (2014), pp. 415–421.

[237] A. Alhujaily, H. Yu, X. Zhang and F. Ma, *Adsorptive removal of anionic dyes from aqueous solutions using spent mushroom waste*, Appl. Water Sci. 10 (2020), p. 183.

[238] J. Rashid, F. Tehreem, A. Rehman and R. Kumar, *Synthesis using natural functionalization of activated carbon from pumpkin peels for decolourization of aqueous methylene blue*, Sci. Total Environ. 671 (2019), pp. 369–376.

[239] L. Ma, C. Jiang, Z. Lin and Z. Zou, *Microwave-hydrothermal treated grape peel as an efficient biosorbent for methylene blue removal*, Int. J. Environ. Res. Public Health 15 (2018), pp. 239.

[240] H. Ao, W. Cao, Y. Hong, J. Wu and L. Wei, *Adsorption of sulfate ion from water by zirconium oxide-modified biochar derived from pomelo peel*, Sci. Total Environ. 708 (2020), p. 135092.

[241] D.R. Mulinari and M.L.C.P. da Silva, *Adsorption of sulphate ions by modification of sugarcane bagasse cellulose*, Carbohydr. Polym. 74 (2008), pp. 617–620.

[242] V. thi Quyen, T.-H. Pham, J. Kim, D.M. Thanh, P.Q. Thang, Q. Van Le et al., *Biosorbent derived from coffee husk for efficient removal of toxic heavy metals from wastewater*, Chemosphere 284 (2021), p. 131312.

[243] L.P. Lingamdinne, K.R. Vemula, Y.-Y. Chang, J.-K. Yang, R.R. Karri and J.R. Koduru, *Process optimization and modeling of lead removal using iron oxide nanocomposites generated from bio-waste mass*, Chemosphere 243 (2020), p. 125257.

[244] E. Zong, C. Wang, J. Yang, H. Zhu, S. Jiang, X. Liu et al., *Preparation of TiO₂/cellulose nanocomposites as antibacterial bio-adsorbents for effective phosphate removal from aqueous medium*, Int. J. Biol. Macromol. 182 (2021), pp. 434–444.

[245] I. Abdelfattah, A.A. Ismail, F. Al Sayed, A. Almedolab and K.M. Aboelghait, *Biosorption of heavy metals ions in real industrial wastewater using peanut husk as efficient and cost effective adsorbent*, Environ. Nanotechnol. Monit. Manag. 6 (2016), pp. 176–183.

[246] Y. Chen, H. Wang, W. Zhao and S. Huang, *Four different kinds of peels as adsorbents for the removal of Cd(II) from aqueous solution: Kinetics, isotherm and mechanism*, J. Taiwan Inst. Chem. Eng. 88 (2018), pp. 146–151.

[247] R. Jiang, J. Tian, H. Zheng, J. Qi, S. Sun and X. Li, *A novel magnetic adsorbent based on waste litchi peels for removing Pb(II) from aqueous solution*, J. Environ. Manage. 155 (2015), pp. 24–30.

[248] M. Akram, X. Xu, B. Gao, Q. Yue, S. Yanan, R. Khan et al., *Adsorptive removal of phosphate by the bimetallic hydroxide nanocomposites embedded in pomegranate peel*, J. Environ. Sci. 91 (2020), pp. 189–198.

[249] C. Fan and Y. Zhang, *Adsorption isotherms, kinetics and thermodynamics of nitrate and phosphate in binary systems on a novel adsorbent derived from corn stalks*, J. Geochemical Explor. 188 (2018), pp. 95–100.

[250] A. Nurain, P. Sarker, M.S. Rahaman, M.M. Rahman and M.K. Uddin, *Utilization of banana (Musa sapientum) peel for removal of Pb²⁺ from aqueous solution*, J. Multidiscip. Appl. Nat. Sci. 1 (2021), pp. 117–128.

[251] S. Mireles, J. Parsons, T. Trad, C.-L. Cheng and J. Kang, *Lead removal from aqueous solutions using biochars derived from corn stover, orange peel, and pistachio shell*, Int. J. Environ. Sci. Technol. 16 (2019), pp. 5817–5826.

[252] J. Jjagwe, P.W. Olupot, E. Menya and H.M. Kalibbala, *Synthesis and application of granular activated carbon from biomass waste materials for water treatment: A review*, J. Bioresour. Bioprod. 6 (2021), pp. 292–322. https://doi.org/10.1016/j.jobab.2021.03.003

[253] R. Davarnejad, S. Afshar and P. Etehadfar, *Activated carbon blended with grape stalks powder: Properties modification and its application in a dye adsorption*, Arab. J. Chem. 13 (2020), pp. 5463–5473.

[254] K.G. Akpomie and J. Conradie, *Banana peel as a biosorbent for the decontamination of water pollutants. A review*, Environ. Chem. Lett. 18 (2020), pp. 1085–1112.

[255] J.U. Ani, K.G. Akpomie, U.C. Okoro, L.E. Aneke, O.D. Onukwuli and O.T. Ujam, *Potentials of activated carbon produced from biomass materials for sequestration of dyes, heavy metals, and crude oil components from aqueous environment*, Appl. Water Sci. 10 (2020), pp. 1–11.

[256] G.L. Dotto and G. McKay, *Current scenario and challenges in adsorption for water treatment*, J. Environ. Chem. Eng. 8 (2020), p. 103988.

[257] D.R. Lobato-Peralta, E. Duque-Brito, A. Ayala-Cortés, D.M. Arias, A. Longoria, A.K. Cuentas-Gallegos et al., *Advances in activated carbon modification, surface heteroatom configuration, reactor strategies, and regeneration methods for enhanced wastewater treatment*, J. Environ. Chem. Eng. 9 (2021), p. 105626. https://doi.org/10.1016/j.jece.2021.105626

[258] M. El Gamal, H.A. Mousa, M.H. El-naas, R. Zacharia and S. Judd, *Separation and purification technology bio-regeneration of activated carbon: A comprehensive review*, 197 (2018), pp. 345–359.

[259] E. Gagliano, P.P. Falciglia, Y. Zaker, T. Karanfil and P. Roccaro, *Microwave regeneration of granular activated carbon saturated with PFAS*, Water Res. 198 (2021), p. 117121.

[260] B. Ferrández-Gómez, D. Cazorla-Amorós and E. Morallón, *Feasibility of electrochemical regeneration of activated carbon used in drinking water treatment plant. Reactor configuration design at a pilot scale*, Process Saf. Environ. Prot. 148 (2021), pp. 846–857.

[261] N.A. Bury, K.A. Mumford and G.W. Stevens, *The electro-Fenton regeneration of granular activated carbons: Degradation of organic contaminants and the relationship to the carbon surface*, J. Hazard. Mater. 416 (2021), p. 125792.

[262] C. Trellu, M. Gibert-Vilas, Y. Pechaud, N. Oturan and M.A. Oturan, *Clofibric acid removal at activated carbon fibers by adsorption and electro-Fenton regeneration – Modeling and limiting phenomena*, Electrochim. Acta 382 (2021), p. 138283.

[263] A. Larasati, G.D. Fowler and N.J.D. Graham, *Chemical regeneration of granular activated carbon: Preliminary evaluation of alternative regenerant solutions*, Environ. Sci. Water Res. Technol. 6 (2020), pp. 2043–2056.

[264] L. Chen, J. Yuan, T. Li, X. Jiang, S. Ma, W. Cen et al., *A regenerable N-rich hierarchical porous carbon synthesized from waste biomass for H₂S removal at room temperature*, Sci. Total Environ. 768 (2021), p. 144452.

[265] Y.H. Wang, S. Bayatpour, X. Qian, B. Frigo-Vaz and P. Wang, *Activated carbon fibers via reductive carbonization of cellulosic biomass for adsorption of nonpolar volatile organic compounds*, Colloids Surfaces A Physicochem. Eng. Asp. 612 (2021), p. 125908.

[266] Y. Hou, Y. Liang, H. Hu, Y. Tao, J. Zhou and J. Cai, *Facile preparation of multi-porous biochar from lotus biomass for methyl orange removal: Kinetics, isotherms, and regeneration studies*, Bioresour. Technol. 329 (2021), p. 124877.

[267] R. Araga, S. Soni and C.S. Sharma, *Fluoride adsorption from aqueous solution using activated carbon obtained from KOH-treated jamun (Syzygium cumini) seed*, J. Environ. Chem. Eng. 5 (2017), pp. 5608–5616.

[268] A. Batool and S. Valiyaveettil, *Chemical transformation of soya waste into stable adsorbent for enhanced removal of methylene blue and neutral red from water*, J. Environ. Chem. Eng. 9 (2021), p. 104902.

[269] S. Deng, Y. Nie, Z. Du, Q. Huang, P. Meng, B. Wang et al., *Enhanced adsorption of perfluorooctane sulfonate and perfluorooctanoate by bamboo-derived granular activated carbon*, J. Hazard. Mater. 282 (2015), pp. 150–157.

[270] P.C. Bhomick, A. Supong, M. Baruah, C. Pongener and D. Sinha, *Pine cone biomass as an efficient precursor for the synthesis of activated biocarbon for adsorption of anionic dye from aqueous solution: Isotherm, kinetic, thermodynamic and regeneration studies*, Sustain. Chem. Pharm. 10 (2018), pp. 41–49.

[271] B. Belhamdi, Z. Merzougui, H. Laksaci and M. Trari, *The removal and adsorption mechanisms of free amino acid l-tryptophan from aqueous solution by biomass-based activated carbon by H_3PO_4 activation: Regeneration study*, Phys. Chem. Earth, Parts A/B/C 114 (2019), p. 102791.

[272] J. Yuan, Y. Zhu, J. Wang, Z. Liu, J. Wu, T. Zhang et al., *The application of the modified durian biomass fiber as adsorbent for effective removal of ammonia nitrogen*, J. Iran. Chem. Soc. 19 (2022), pp. 435–445. https://doi.org/10.1007/s13738-021-02313-w

[273] W. Lyu, J. Li, L. Zheng, H. Liu, J. Chen, W. Zhang et al., *Fabrication of 3D compressible polyaniline/cellulose nanofiber aerogel for highly efficient removal of organic pollutants and its environmental-friendly regeneration by peroxydisulfate process*, Chem. Eng. J. 414 (2021), p. 128931. https://doi.org/10.1016/j.cej.2021.128931

[274] C. Pongener, P.C. Bhomick, A. Supong, M. Baruah, U.B. Sinha and D. Sinha, *Adsorption of fluoride onto activated carbon synthesized from Manihot esculenta biomass – Equilibrium, kinetic and thermodynamic studies*, J. Environ. Chem. Eng. 6 (2018), pp. 2382–2389.

[275] F. Xiao, Y. Cheng, P. Zhou, S. Chen, X. Wang, P. He et al., *Fabrication of novel carboxyl and amidoxime groups modified luffa fiber for highly efficient removal of uranium(VI) from uranium mine water*, J. Environ. Chem. Eng. 9 (2021), p. 105681.

[276] H. Nam, S. Wang and H.R. Jeong, *TMA and H_2S gas removals using metal loaded on rice husk activated carbon for indoor air purification*, Fuel 213 (2018), pp. 186–194.

[277] A. Sharma, Z.M. Siddiqui, S. Dhar, P. Mehta and D. Pathania, *Adsorptive removal of Congo red dye (CR) from aqueous solution by Cornulaca monacantha stem and biomass-based activated carbon: isotherm, kinetics and thermodynamics*, Sep. Sci. Technol. 54 (2019), pp. 916–929.

[278] L. Zhou, Q. Yu, Y. Cui, F. Xie, W. Li, Y. Li et al., *Adsorption properties of activated carbon from reed with a high adsorption capacity*, Ecol. Eng. 102 (2017), pp. 443–450.

[279] B.O. Fagbayigbo, B.O. Opeolu, O.S. Fatoki, T.A. Akenga and O.S. Olatunji, *Removal of PFOA and PFOS from aqueous solutions using activated carbon produced from Vitis vinifera leaf litter*, Environ. Sci. Pollut. Res. 24 (2017), pp. 13107–13120.

[280] W. Xiao, Z.N. Garba, S. Sun, I. Lawan, L. Wang, M. Lin et al., *Preparation and evaluation of an effective activated carbon from white sugar for the adsorption of rhodamine B dye*, J. Clean. Prod. 253 (2020), p. 119989.

[281] F. Guo, X. Jiang, X. Li, X. Jia, S. Liang and L. Qian, *Synthesis of MgO/Fe_3O_4 nanoparticles embedded activated carbon from biomass for high-efficient adsorption of malachite green*, Mater. Chem. Phys. 240 (2020), p. 122240.

[282] M. Erfani and V. Javanbakht, *Methylene blue removal from aqueous solution by a bio-composite synthesized from sodium alginate and wastes of oil extraction from almond peanut*, Int. J. Biol. Macromol. 114 (2018), pp. 244–255.

[283] C.Q. Teong, H.D. Setiabudi, N.A.S. El-Arish, M.B. Bahari and L.P. Teh, *Vatica rassak wood waste-derived activated carbon for effective Pb(II) adsorption: Kinetic, isotherm and reusability studies*, Mater. Today Proc. 42 (2019), pp. 165–171.

[284] V.H. Le, L.T.N. Huynh, T.N. Tran, T.T.N. Ho, M.N. Hoang and T.H. Nguyen, *Comparative desalination performance of activated carbon from coconut shell waste/carbon nanotubes composite in batch mode and single-pass mode*, J. Appl. Electrochem. 51 (2021), pp. 1313–1322.

[285] S. Chen, M. Zhou, H.F. Wang, T. Wang, X.S. Wang, H.B. Hou et al., *Adsorption of reactive brilliant Red X-3B in aqueous solutions on clay-biochar composites from bagasse and natural attapulgite*, Water 10 (2018), p. 703. https://doi.org/10.3390/w10060703

[286] L. Li, M. Wu, C. Song, L. Liu, W. Gong, Y. Ding et al., *Efficient removal of cationic dyes via activated carbon with ultrahigh specific surface derived from vinasse wastes*, Bioresour. Technol. 322 (2021), p. 124540.

[287] D.S.P. Franco, J. Georgin, M.S. Netto, D. Allasia, M.L.S. Oliveira, E.L. Foletto et al., *Highly effective adsorption of synthetic phenol effluent by a novel activated carbon prepared from fruit wastes of the Ceiba speciosa forest species*, J. Environ. Chem. Eng. 9 (2021), p. 105927.

[288] M.A. Adebayo, J.I. Adebomi, T.O. Abe and F.I. Areo, *Removal of aqueous Congo red and malachite green using ackee apple seed–bentonite composite*, Colloids Interface Sci. Commun. 38 (2020), p. 100311.

[289] A.J. Veeraragavan, R. Shanmugavel, N. Abraham, D. Subramanian and S. Pandian, *Kinetic studies validated by artificial neural network simulation for the removal of dye from simulated waste water by the activated carbon produced from Acalypha indica leaves*, Environ. Technol. Innov. 21 (2021), p. 101244.

[290] K.G.P. Nunes, L.W. Sfreddo, M. Rosset and L.A. Féris, *Efficiency evaluation of thermal, ultrasound and solvent techniques in activated carbon regeneration*, Environ. Technol. 42(26): (2021), pp. 4189–4200. doi: 10.1080/09593330.2020.1746839

[291] A. Jain, S. Kumari, S. Agarwal and S. Khan, *Water purification via novel nano-adsorbents and their regeneration strategies*, Process Saf. Environ. Prot. 152 (2021), pp. 441–454.

[292] E. Da'Na and A. Awad, *Regeneration of spent activated carbon obtained from home filtration system and applying it for heavy metals adsorption*, J. Environ. Chem. Eng. 5 (2017), pp. 3091–3099.

[293] N. Genç, E. Durna and E. Erkişi, *Optimization of the adsorption of diclofenac by activated carbon and the acidic regeneration of spent activated carbon*, Water Sci. Technol. 83 (2021), pp. 396–408.

4 Biomaterial for Wastewater Treatment

Keshava Joshi, Lokeshwari Navalgund and Vinayaka B. Shet

4.1 INTRODUCTION

Water demand is increasing worldwide every year by 1%–2% and is expected to continue increasing further mainly due to an increase in domestic and industrial sectors. The stress of water continues with population increase and in turn affects climate change (Sivakumar, 2011). Safe drinking water and sanitation are basic human rights as they sustain healthy livelihoods. Today we are in the age of pollutant removal and that too a significant portion of it is from industrial wastewater as industries consume a significant volume of water. Owing to the quick growth and expansion of industries in many places across the globe, the level of water contamination is increasing seriously and steadily and hence wastewater treatment technology remains a global concern (Carley and Christie, 2017). Wastewater is any form of water that has been contaminated and discharged in huge quantity, which has to be treated to reuse it to meet the demand of the society and minimize water scarcity problems. Today a zero discharge system is the cutting-edge technology in wastewater management and the process of reducing the contaminants needs better solutions (Tong and Elimelech, 2016). Wastewater treatment and recycling are being made mandatory for better water conservation by implementing different laws by regulatory agencies to move towards zero discharge for industrial effluents. Though there are many methods of treating wastewater like physico-chemical methods, biological methods based on different pollutants, size, characteristics, and need of the water quality, and so on, biomaterials play an important role in wastewater treatment (Rajasulochana and Preethy, 2016; Kumar et al., 2020). Hence, there is a need for exclusive secure innovative cost-effective technology that can be easily adopted in the present industrial system.

Biosorption is a process which has the ability to accumulate pollutants or bind the contaminants through a physico-chemical pathway and it doesn't discharge any sludge. Biosorption is a key possible solution as it is more economical and more eco-friendly than conventional methods. Biomaterials are made up of natural or synthetic materials usually with multiple components that interact with biological systems. The development of biomaterials is a growing field of research in different areas like medical research, biosensors, and wastewater treatment. The quantity of the impurities to be removed is dependent on the kinetic equilibrium and the composition of the surface of the biomaterial. A thorough understanding of the capability of different

DOI: 10.1201/9781032636368-4

biomaterials is necessary for the removal of all contaminants in wastewater. The recent advances in using various biomaterials as adsorbents are suitable for removing contaminants from wastewater instead of synthetic and costly materials (Rapo et al., 2020). This biosorption process with a variety of biomaterials available can become a more beneficial and effective technology to eliminate the contaminants and bridge the gap between laboratory results and industrial applications. The recent advances in biomaterials cover a wide range and as such we review the capability of biosorbents in removing organic and inorganic contaminants in wastewater. This chapter thus highlights the biosorption process and different biomaterials in wastewater treatment along with their challenges and applications.

4.1.1 BIOSORPTION PROCESS FOR WASTEWATER TREATMENT WITH REGENERATION

The choice of the technology for removal of contaminates from wastewater along with protecting the environment is a key challenging issue in today's millennium era. There are several conventional and non-conventional technologies like coagulation, flocculation, electro-dialysis, photocatalysis, membrane filtration, ion exchange, chemical precipitation, and so on, to protect and reuse the wastewater to meet today's water crisis and water demand (Azimi et al., 2017). But these technologies have various drawbacks like more energy consumption and sludge generation, fouling of membrane, incomplete removal of contaminant. Hence, today global research is bringing the technology in using natural materials and aiming to protect the ecosystem and improve the quality, which is referred to as green technology using plant material (Crini and Lichtfouse, 2018). So biosorption is one such phenomenon of green technology which involves the removal of a contaminant by interaction between a contaminate and microbial cells. Natural materials or waste materials from agriculture and industrial activities are considered as cheaper and safer environmental adsorbents (Dawood and Sen, 2014; Saleh, 2021). The availability of such resources is a factor of concern in selecting the biomass or biosorbent for the cleaning purpose. These biosorbents are prepared from naturally available resources or from the waste resources or biomass of algae, fungi, bacteria, moss, plants, etc., which are inactivated and pretreated by washing with alkali and drying and preparing the granules (Kanamarlapudi et al., 2018)

Biosorption is a subcategory of adsorption process which is quick and involves the biomaterial as adsorbent in fixing the contaminants in wastewater on to the surface of the biomass. Biosorption involves a two-phase method in which the solid biosorbent dissolves in a liquid-phase solvent having contaminants or involves the binding of the sorbate on to the biosorbent (Godage and Gionfriddo, 2020). The attraction capacity of the biosorbent attracts the contaminant in the solvent by different bonding mechanisms like electrostatic force, van der Waals force, ion exchange, reduction, precipitation, etc., till equilibrium is reached. These biosorbents contains chemical and functional groups like amine, amide, sulphonate, carbonyl, carboxyl, etc., that can attract and sequester the components present in wastewater (Kanamarlapudi et al., 2018). Biosorption is accomplished through varied choice of pH, temperature, concentration, dosage, and other challenging factors. The degree of sorbent affinity to adsorbate depends on chemical characteristics, molecular weight, the oxidation and reduction state of the pollutant, and the structure and nature of the biosorbent (Gadd, 2009). Biosorption is a surface phenomenon and its separation is based on the selective sorption that is based on the isotherms and the thermodynamic and kinetic selectivity of the contaminants by the biosorbent owing to the specific interaction between the surface of the biosorbent and the contaminant. Biosorption has significant potential; it is a new and energy saving technology for cleaning the environment because of the use of low-cost material, minimum operational costs, sludge minimization, biosorbent regeneration, recovery of the metals and other useful material absorbed, high efficiency, no toxic effects, and no requirements for nutrients (Fomina and Gadd, 2014).

In biosorption, the biomaterial is adapted by either physical or chemical treatment and is applied directly or immobilized on different support systems. Various parameters like dosage, concentration, pH, temperature, speed of agitation, surface area, etc., influence the rate of biosorption. Immobilization is a physical attachment of the cell to a particular area without the loss of biological activity and is known as bioencapsulation or entrapment (Willaert, 2007). Immobilization has the advantage in that the technique protects the cell or enzyme from contaminants, separation from the product is easy, cell stability against pH, temperature, and impurities is enhanced, capacity is increased, resistance to chemical and mechanical strength improves, effortless separation of biomass from wastewater containing contaminants is facilitated, and a non-destructive cell recovery is possible. Immobilization of the biosorbent is carried out by either entrapment or encapsulation techniques. Suitable immobilization matrixes like calcium or sodium alginate, polyacrylamide, silica, polysulphone, and polyurethane which are cheap and feasible (Xu et al., 2011) are used to operate because of their chemical resistance and the good mechanical strength of the bioparticles. Many researchers are working towards dead biomass instead of living biomass especially in continuous system of biosorption as living biomass has drawbacks like low strength, little rigidity, loss of biosorbent, and difficulty in separation (Fomina and Gadd, 2014; Kanamarlapudi et al., 2018). Hence many researchers (Umali et al., 2006; Mohapatra et al., 2019; Tavana et al., 2020) are using dead biomass for achieving higher efficiency in the biosorption process as it overcomes the difficulties of living cells like no nutrients, less impact of pH and temperature, easy recovery of ions, no toxicity, easy regeneration and reuse.

Currently many have reported (Acheampong et al., 2013; Sreenivas et al., 2014; Cardoso et al., 2018; Hammo et al., 2021) the application of study of biomaterial in a fixed bed column, which offers advantage of large-scale process of biosorption as it involves the continuous process. Regenerating and reusing the biosorbent in the column studies after recovering the useful material is more economical and viable. The biosorbent particles are packed in a column, which act as a bed device for continuous contaminant removal and for this various bioreactors are available to operate on a large scale. In this process, once the packed bed is saturated the biomass packed inside can be regenerated with acids or hydroxide solutions, which is known as desorption. During regeneration useful materials can be recovered and regenerated biomaterial can be reused for further biosorption experiments, which in turn help in reducing the residues. Many researchers have worked with recovery of heavy metals (Khan et al., 2012; Acheampong et al., 2013; Mishra et al., 2016), organic compounds

(Akar et al., 2013), etc., in column studies and also regenerated the biomaterial for further biosorption process and evaluated its efficiency and effectiveness.

4.2 BIOMATERIAL: SELECTION, SYNTHESIS, AND CHARACTERIZATION

Biosorption is receiving a lot of attention nowadays because of a wide range of materials available as biosorbents. In the adsorption process, though there are large materials (adsorbents) available with their high decontamination capability, still the challenging issue is that they are not biodegradable. Hence, natural adsorbents called biosorbents have been given priority over the synthetic ones available which are of moderate cost and are a competent alternative for wastewater treatment. These biomaterials are available easily in nature and hence they are used to remove pollutants from wastewater. These biological materials can be dead biomass or living biomass or microorganisms and biomass coming from agricultural and food wastes. The choice and selection of the biosorbent or biomaterial depends on water solubility, surface function, size and distribution, large availability, biosorption capacity, rejuvenation and reclaim of biosorbent. Also, the choice of any biomaterial needs to be looked at with minimum or no activation needed; it should be easily available at the point source and be available in sufficient quantity and also be non-reactive with water.

In the biosorption process the material can be used directly to remove the pollutants or it has to be synthesized from the natural material as a nanomaterial or biochar and hence used to remove the pollutants. Synthesis of biomaterial is by biological method or green synthesis from bacteria, fungi, algae, plant extracts like leaves, fruits, flower, root, etc. Characterization of biomaterial is to understand the material in detail with its physical, chemical, mechanical, surface characterization. Selection of the characterization tool is also a challenging issue in studying the biomaterials as they are utilized for a variety of different applications. Different equipments used for characterization of biomaterial are briefly discussed below. Figure 4.1 presents the general procedure of biosorption and the different characterization techniques. These techniques are briefly explained below.

UV spectrophotometer: It is one of the simplest, fastest, and inexpensive methods to find out the concentration of an element in a given solution. It measures the intensity of light passing through a sample solution as absorbance in a cuvette by setting the known wavelength of the element and compares it with intensity of light in blank reading. The main components of the spectrophotometer are the source, the sample holder, and a suitable detector.

Fourier transform infrared: It is a technique used to identify pure substance and impurities in the composition of materials and is an analytical tool to evaluate a wide range of materials when IR radiation is passed through the sample. It provides information on the functional groups (hydroxyl, carbonyl, ether, ester, amides, etc.) present on the surface of biosorbents and the nature of the bond involved in biosorption as it generates spectra and patterns for different chemical structures or molecules.

X-ray diffraction (XRD): It is a methodology used to find out the crystalline structure of the sample and doesn't provide any information on the nature of the chemical. It utilizes X-ray scattering phenomena (array of atoms gives rise to a definite

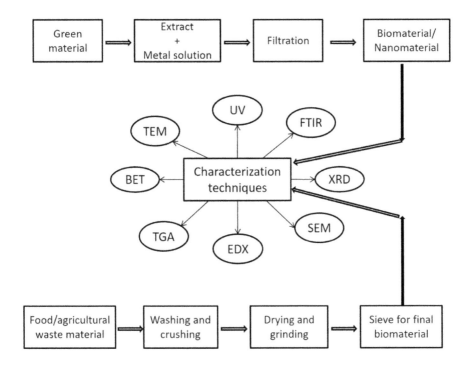

FIGURE 4.1   General outline of biosorption and different characterization techniques.

diffraction pattern to give a qualitative image) to determine the crystalline structure or atomic arrangement within the crystal lattice. XRD is generally a bulk characterization and non-destructive technique which can be operated at room temperature and pressure.

Scanning electron microscopy (SEM): In this method the surface image with the microstructure and morphology of the given material is obtained. The electron beam is passed through a sample to read the surface and interpret it to obtain the surface image of the material. The change in the morphological and structure arrangements before and after biosorption is qualitatively analysed for a biosorbent. It is one of the powerful and versatile tools to determine the biocompatibility of biomaterials and for measuring the surface properties of materials by understanding the interface of the biological layers.

Energy-dispersive X-ray (EDX): It is one of the tools to identify elements through chemical identification and their concentration and it also provides the data of different elements present in the biosorbent. XRD provides phase identification whereas EDX provides chemical identification.

Thermo-gravimetric analysis (TGA): It is primarily used to measure the material thermal stability, where the weight of sample is measured over time as the temperature changes. Thermal analysis is required in the biosorption process for determining the maximum temperature that is required for the biosorbent to maintain its properties.

Brunauer, Emmett, and Teller (BET): It is a significant investigation method for measuring the specific surface area of the substances and it utilizes the probing gas

(N_2), which is an adsorbate that does not chemically react with the biosorbent. This analysis predicts the rate of dissolution as it is proportional to the specific surface area and in turn it helps in predicting the bioavailability. The analysis also depicts the monolayer or multilayer biosorption of molecules on solid surfaces.

Transmission electron microscopy (TEM): In this method a microscope which uses a high-voltage electron beam to envisage specimen and generate an extremely exaggerated image is used. Light microscope can magnify the image 2,000 times, whereas TEM can magnify the image up to two million times. TEM is used to identify and characterize viruses, bacteria, protozoa, and fungi and also learn the morphology of cells and their organelles.

4.3 TYPES OF BIOMATERIAL AND THEIR APPLICATIONS

A wide range of biomaterials is available in nature for the biosorption process. There are different types of materials that can be used as biosorbents, as shown in the Figure 4.2. Today the choice and identification of the biosorbent for different applications from the large spectrum available is one of the major challenges in the field of wastewater treatment. These biosorbents can be living or dead biomass from natural sources like plant, animals, and organisms or from industrial and agricultural wastes. Table 4.1 provides the literature on different applications of biomaterials in wastewater treatment. These biomaterials are derived from nature or synthesized in the laboratory by different approaches using polymers, ceramics, composites, nanomaterials. The use of these different biomaterials has been an area of interest for many researchers in the past few decades. They are great alternatives in water purification technology because of the vast availability in quantity and quality, effective cost, better performance; also, it's a greener and cleaner technology.

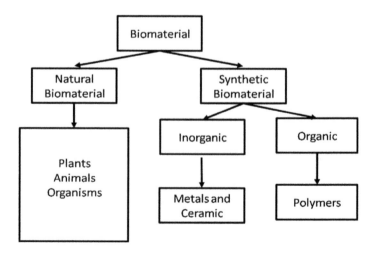

FIGURE 4.2 The different types of biomaterials.

TABLE 4.1

List of few applications of different biomaterial used in wastewater treatment

Sl. no.	Biomaterial/organism	Application	References
1	Mukiamaderaspatna plant extract and green synthesis of silver nanocomposite hydrogel (SNC)	Methylene blue (MB) dye removal from aqueous solution. The maximum Langmuir monolayer adsorption capacity was calculated as 213.7 mg of MB dye per g of SNC. The pseudo-first-order, pseudo-second-order, and Elovich kinetic models were used to fit the adsorption kinetics data.	Karthiga Devi et al. (2016)
2	Scenedesmus strain	85.5% reduction in total nitrogen, 83.2% reduction in phosphorus, and 89.3% reduction in chemical oxygen demand in municipal wastewater. Kinetic study showed best fitness of pseudo-second-order kinetic model and experimental data was best fitted with Freundlich adsorption isotherm model.	Maqbool et al. (2019)
3	Iron nanoparticle synthesized from tangerine peel extract	The maximum removal of cadmium ions (90%) occurred at pH of 4 and adsorbent dose of 0.4 g/100 mL. Adsorption of cadmium ions by synthesized iron oxide nanoparticles followed Freundlich adsorption model and pseudo-second-order equation.	Ehrampoush et al. (2015)
4	Magnetic C. selloana flower spikes	Malachite green (MG) and MB dyes removal from aqueous solution. Fitted Langmuir isotherm model with $R = 0.99$.	Parlayıcı and Pehlivan (2021)
5	Biochar	The performance of biochar amendment in constructed wetlands/biofilters with respect to the removal of nitrogen, phosphorus, organic contaminants, heavy metals.	Deng et al. (2021)
6	Granular biomass	Applied for reduction of uranium. Biosorption was observed to be rapid (<1 h) in acidic pH range (1–6) compared to that at pH 7.0 or above. Almost complete removal of uranium was observed in the range 6–100 mg L^{-1} in less than 1 h. Redlich-Peterson model gave the best fit when the experimental data were analysed using different adsorption isotherm equations.	Nancharaiah et al. (2006)
7	Synthesized zero-valent iron nanoparticles (ZVINs) using Calotropis gigantea (CG) flower extract	Applied for MB dye and analine reduction. Langmuir adsorption model was a good fit to the adsorption of MB and aniline on the surface of sorbent than the Freundlich model, and the adsorption process follows pseudo-second-order kinetics.	Sravanthi et al. (2018)

(continued)

TABLE 4.1
(Continued)

Sl. no.	Biomaterial/organism	Application	References
8	Rice straw and palm oil midrib	The calculated maximum dye adsorption capacities onto rice straw and palm oil midrib activated carbon were 55.86 and 69.44 mg g^{-1}, respectively. Adsorption using biomass-based activated carbon offers a good technique for textile wastewater treatment as it can remove up to 95% of the colour intensity besides reducing other pollutants such as COD, nitrate, and phosphate.	Firdaus et al. (2017)
9	Fungal biomasses of *Mucor rouxii* and *Absidia coerulea* along with chitosan and walnut shell media	The study demonstrated that the removal efficiencies by *M. rouxii* for these oils were in the 77%–93% range at a pH of 5.0. The adsorption capacities for standard mineral oil, vegetable oil, and cutting oil were 77.2, 92.5, and 84 mg g^{-1} of biomass, respectively.	Srinivasan and Viraraghavan (2010)
10	Novel actinobacterium Zhihengliuella sp.	Removal of toxic heavy metals including Cr, Cd, Mn, Ni, Pb, Zn, and Fe.	
11	Macrofungus (Lactariusscrobiculatus) biomass	The maximum biosorption capacity of *L. scrobiculatus* was found to be 56.2 mg g^{-1} for Pb(II) and to be 53.1 mg g^{-1} for Cd(II). The recovery of the metal ions from *L. scrobiculatus* biomass was found as higher than 95% using 1 M HCl and 1 M HNO$_3$. Furthermore, the reusability of the biosorbent was determined after six consecutive sorption-desorption cycles.	Anayurt et al. (2009)
12	Mushroom biomass Lepiotahystrix	Adsorption capacities were found to be 3.9 and 8.9 mg g^{-1} at a contact time of 25–40 min and initial metal ion concentration of 300–500 mg g^{-1} for Pb and Cu, respectively. The biosorption process follows second-order kinetics and fits the Langmuir isotherm model.	Kariuki et al. (2017)
13	Aquatic weeds (mangrove leaves and water lily)	In case of Cr(VI), mangrove leaves showed maximum removal/reduction capacity (8.87 mg g^{-1}) followed by water lily (8.44 mg g^{-1}).	Elangovan et al. (2008)
14	*Moringa oleifera* seeds	Biosorption of Cd(II), Cr(III), and Ni(II) on unmodified shelled *Moringa oleifera* seeds.	Sharma et al. (2007)
15	Powder of steam of *Acacia nilotica*	The biomass of *A. nilotica* was found to be effective for the removal of As with 95% sorption efficiency at a concentration of <200 μg L^{-1} of As solution, and thus uptake capacity is 50.8 mg As per g of biomass.	Baig et al. (2010)

	Biomaterial	Description	Reference
16	Green algae waste biomass	Biosorption of Pb(II), Cd(II), and Co(II). The results indicate that the Langmuir model provides the best correlation of experimental data, and the pseudo-second-order kinetic fit the best.	Bulgariu and Bulgariu (2012)
17	Apricot (*Prunus armeniaca* L.) shells (SHM)	The SHM showed high removal efficiency towards multiple metal ions. The amounts of Fe, Pb, Cu, and Cr ions were reduced by 97%, 87%, 81%, and 80%, respectively, while Ni and Zn amounts were reduced by 33% and 14%.	Sostaric et al. (2018)
18	Chitin, shrimp waste	The biocoagulant composed of demineralized shrimp waste (DMSW), with a dosage of 200 mg DW per L, presented the best results for both removals; the turbidity reached more than 95% and the organic matter was 80%.	Frantz et al. (2020)
19	*Thamnidium elegans* (*T. elegans*) cells	A quadratic model was built for the batch process and optimum values of variables were recorded as pH 2.0, biosorbent amount 0.06 g, and contact time 39.3 min with the removal yield of 99.42%. *T. elegans* cells showed great potential for the treatment of coloured real wastewaters.	Akar et al. (2017)
20	Fungus Phomopsis sp.	Results indicate that after 24 h contact time, up to 870 μmol g^{-1} of lead, 390 μmol g^{-1} of copper, 230 μmol g^{-1} of cadmium, 150 μmol g^{-1} of zinc, and 110 μmol g^{-1} of nickel ions are adsorbed into the material. After approximately 10 min, about 70% of the overall adsorption process has already been completed and also extended to a real wastewater effluent, confirming the potential of this biomaterial as a depolluting agent.	Saiano et al. (2005)
21	MnO$_2$@biomass carbon (CMn)	The highest adsorption capacity of CMn for amoxicillin was 16.094 mg g^{-1} in a neutral medium. The adsorption isotherms were fitted using the Langmuir model. The kinetics of AMX adsorption was determined to follow the pseudo-first-order model. Recycle experiments indicated that CMn exhibited high adsorption efficiency (>80%) after four cycles.	Ren et al. (2020)
22	Synthesized sawdust and sugarcane molasses with styrene-maleic acid copolymer	Adsorption efficiency of 90 was observed in 2 h using SM-SMA composites for congo red dye. The equilibrium data was analysed using Langmuir, Freundlich, Temkin, and Dubinin-Radushkevich isotherm models.	Gonte et al. (2014)
23	Brown-rot fungus *Lentinus edodes*	Langmuir simulation of the biosorption showed that the maximum uptake of cadmium was 5.58 mmol g^{-1} in weak acid condition, which was much higher than for many other biosorbents.	Chen et al. (2008)

(continued)

TABLE 4.1
Continued

Sl. no.	Biomaterial/organism	Application	References
24	Palm-date stones	The ability of palm-date stones to adsorb basic violet 3 (BV3) and basic red 2 (BR2) from wastewaters is assessed through kinetic, thermodynamic, and equilibrium investigations. Kinetic and dynamic studies showed that 77% and 93% of BR2 and BV3 were removed within a contact time of 15 min, which proves the applicability of the process at an industrial scale.	Wakkel et al. (2019)
25	Green synthesized polyaniline-Scenedesmus biocomposite	The physico-chemical analysis of municipal wastewater after microalgae growth depicted 85.5% reduction in total nitrogen, 83.2% reduction in phosphorus, and 89.3% reduction in chemical oxygen demand, which clearly reveals that this strain can be efficiently used to avoid eutrophication caused by municipal wastewaters.	Maqbool et al. (2019)
26	Hybrid biomaterials based on methylotrophic yeast *Ogataea polymorpha*	The biofilter based on the organic silica matrix encapsulated in the methylotrophic yeast *Ogataea polymorpha* BKM Y-2559 has an oxidizing power of three times more than the capacity of the aeration tanks used at the chemical plants during methyl alcohol production.	Kamanina et al. (2016)
27	Agrocybecylindracea substrate-Fe_3O_4	Optimized by the Taguchi method, Cr(VI) removal percentage was up to 73.88 at 240 min, 40°C, pH 3, Cr(VI) concentration 200 mg L^{-1}, dosage 12 g L^{-1}, rpm 200.	Wang et al. (2018)
28	Raw shrimp shell (RSS), a biomaterial of animal origin	To evaluate the effectiveness of crushed shrimp shell particles in order to remove nitrate ions from municipal wastewater effluents with a significant purification yield of 79%.	Abali et al. (2021)
29	Hen feathers	The optimum conditions for removal of MB were found to be pH 7.0, biosorbent dose = 1.0 g, and initial dye concentration = 50 mg L^{-1}. The maximum monolayer sorption capacity was determined as 134.76 mg g^{-1} at 303 K.	Chowdhury and Saha (2012)
30	Sulphate-reducing bacteria	The parameters affecting Cd^{2+} removal of this system were obtained: the initial Cd^{2+} concentrations 150 mg L^{-1}, optimum pH 7.0, optimum temperature 37°C, optimum time 48 h, respectively.	Wu et al. (2017)
31	Cactus *Opuntia ficus Indica*	The dosage of cactus added was correlated with turbidity of wastewater. Depending upon turbidity of wastewater, turbidity removal varies from 70% to 90% for dosage 30 to 100 mg L^{-1} and cactus has potential for wastewater treatment.	Deshmukh and Hedaoo (2019)
32	Aerobic granules	Removal of uranium was observed in the range 6–100 mg l^{-1} in less than 1 h. The Redlich-Peterson model gave the best fit when the experimental data were analysed using different adsorption isotherm equations.	Nancharaiah et al. (2006)

	Biomaterial	Description	Reference
33	*Prunus amygdalus* L. (almond) shell	The adsorption kinetics conformed to the pseudo-second-order kinetic model. The equilibrium data pointed out an excellent fit to the Langmuir isotherm model with maximum monolayer adsorption capacity of 41.34 mg g^{-1} at 293 K for methyl orange removal.	Deniz (2013)
34	Olive stone	Simultaneous efficient removal of chromium, copper, and nickel in a fixed column reactor.	Martín-Lara et al. (2014)
35	Coir pith	The coir pith exhibited the highest uptake capacity for uranium at 317 K, at the final solution pH value of 4.3 and at the initial uranium concentration of 800 mg L^{-1}.	Parab et al. (2005)
36	Activated coconut shell carbon	Biosorption of zinc(II) and desorption studies to recover the metal.	Amuda et al. (2007)
37	Aloe vera leaf powder	A small amount of the adsorbent (0.3 g L^{-1} water) could remove as much as 74.6% of Pb in 30 min from a solution. Uptake capacity also increases by modifying the aloe vera by treating it with H_3PO_4. With the help of this modification uptake can be increased upto 96.2%.	Malik et al. (2015)
38	*Thamnidium elegans* cells	This study describes the potential use of *Thamnidium elegans* (*T. elegans*) cells for the biosorptive treatment of real industrial wastewater for dye removal with an efficiency of 97%.	Akar et al. (2017)
39	*Azadirachta indica* leaf powder	The nano biomaterial exhibits sorption efficiency of Cd(II) as 96.23% and Ni(II) as 84.12% and stability in terms of regeneration cycles.	Goyal and Masram (2016)
40	Fruit shell of gulmohar (*Delonix regia*)	The maximum biosorption capacity of fruit shell of gulmohar to remove Cr(VI) was 12.28 mg g^{-1}. The study showed that the abundant and inexpensive fruit shell of gulmohar biosorbent has a potential application in the removal of Cr(VI) from electroplating wastewater and its conversion into less or non-toxic Cr(III).	Prasad and Abdullah (2010)
41	Animal bone carbon	Treatment of industrial effluent. The equilibrium isotherm equation check of Freundlich gave $n = 0.985$; $K = 1.798$.	Mohammed et al. (2021)
42	Rice husk	Adsorption experiments analyse the increased removal efficiency of rice husk composites for dye, showing removal of 81% and 77% on oleic and stearic acid modification, respectively.	Rafique and Zulfiqar (2014)
43	Magnetite-coated pine cone biomass	Removal of arsenic from wastewater.	Pholosi et al. (2018)
44	*Zizyphus mauritiana* seeds	Maximum adsorption capacities of aspirin were found to be 8.95 mg g^{-1} with *Zizyphus mauritiana* seeds.	N'diaye and Kankou (2020)

(continued)

TABLE 4.1
(Continued)

Sl. no.	Biomaterial/organism	Application	References
45	Chitin	Kinetic data and equilibrium removal isotherms were measured for removal of cadmium.	Benguella and Benaissa (2002)
46	Lignocellulosic waste	Biosorption mechanism of dye removal, desorption, and usability in real conditions were explored. The removal process followed Langmuir isotherm and pseudo-second-order kinetics.	Akar et al. (2013)
47	Phosphated cellulose	The efficient drug ranitidine sorption has a low dependence on temperature, with maximum values of 85.0, 82.0 mg, and 85.7 $mg \cdot g^{-1}$ for Cel-P10 at 298, 308, and 318 K, respectively. The best sorption occurred at pH = 6 with a saturation time of 210 min.	Bezerra et al. (2014)
48	Nano zero-valent iron	Nano zero-valent iron (nZVI) is widely used to remediate groundwater and wastewater from heavy metals and stable organic pollutants.	Chedia et al. (2015)
49	Laccase, a multicopper oxidoreductase enzyme	Laccase, a multicopper oxidoreductase enzyme, has shown great potential in oxidizing a large number of phenolic and non-phenolic emerging contaminants.	Ba et al. (2013)
50	Inorganic hybrid hydrogel	The study revealed the efficient adsorption of the anionic and neutral dyes, while the adsorption of the cationic dye was much lower.	Upadhyay et al. (2020)

4.3.1 NATURAL BIOMATERIALS

The biomaterials used from nature like plants, animals, organisms like bacteria, fungi in wastewater purification are called natural biomaterials. These are derived from different parts of plants and animals. These natural materials can also be from industrial and agricultural wastes like sewage sludge, by-products like bagasse from the sugar industry, rice husk, orange peels, egg shells, etc. Natural biomaterials have shown quite a few advantages over synthetic biomaterials like vast availability, effective cost, and safe disposal.

The chitosan-based adsorbents are natural amino polymers, which have a good potential ability to bind the pollutants present in wastewater and they are ecofriendly, cost-effective, biocompatible, and biodegradable. Today many researchers are using cactus as a natural material and substitute for synthetic material in wastewater treatment for reducing turbidity, Chemical oxygen demand (COD), and heavy metals. Though plants are available as a plentiful source of cellulose, that is, vegetable cellulose (VC), still researchers are using quiet a lot of different bacteria as a potential source of bacterial cellulose (BC). These are highly porous with small pore size and are perfect for fine filtration; also, cellulose membranes have proven to be efficient as filters for reducing various contaminates in wastewater. The biosorbent from different plant sources as vegetable extracts, seeds, stem, leaves, aloe vera, tamarind, date palm, lime tree, etc., are also being studied as efficient biosorbents by many researchers. Extracellular polysaccharides (EPS) mainly composed of proteins, carbohydrates, lipids, and humic substances are biopolymers with a high molecular weight complex. Because of its outstanding unique structural and functional groups like uronic acids, sulphated units and phosphate groups are employed on a larger scale for remediation of heavy metals oil and hydrocarbons. Beacuse of EPS ecological and commercial values bacterial EPS have been considered as sustainable and cost-effective alternatives to economical bioremediation of the environment. Many researchers have also used bacteria, yeast, fungi, and algae as biosorbents which are capable of tolerating adverse conditions. Experiments were performed using both living and dead organisms and dead organisms rather than living biomass have shown efficient binding of ions and also proven to be economical. The usage of algae as a biosorbent has gained more importance owing to advantages such as limited necessity of nutrients, good sorption capacity, low sludge, high surface area, and more potential for regeneration. The cell wall of algae has a high content of cellulose and acids, which enhances the sorption rate, and is thus economical and ecofriendly. The cell wall of fungi also has a good binding property as it has proteins, lipids, and polyphosphates and hence is more ecofriendly and biosorbent. All these biomasses are obtained very easily and at a lower cost as by-products from industries and also from low-cost growth media.

4.3.2 SYNTHETIC BIOMATERIALS

Synthetic biomaterials are classified as organic, which mainly includes polymers and inorganic materials including metals, ceramics, and metal oxides.

Polymers are considered to be very good organic biosorbents because of their chemical and mechanical stability, high flexibility, and good surface area. These biosorbents are used to remove both organic and inorganic elements. The composites of polymers can be prepared very easily and have many applications and good regeneration capacity because of the enhanced properties due to composites. Today biopolymers are considered to be more capable and suitablebiosorbents for reducing the pollutants in wastewater as they are more easily manipulated, modified, and sustainable. Hydrogels are also another category of polymers having hydrophilic groups like hydroxyl, carboxyl, and amide that swell in water. The polymer hydrogel can be synthesized by different methods like microemulsion, freeze drying with different geometries in the form of film, ring, bead, fibre depending on the applications. Hydrogels are called smart materials as it can change the shape when any change occurs in the environment. Hence, hydrogels are exhibiting greater performance as biosorbents in a wide variety of applications in wastewater treatment.

Carbonaceous materials in the form of organic and inorganic composites, nanomaterial, etc., have extensively been used in diverse applications for wastewater treatment. Different carbon materials like activated carbon, carbon nanotubes, graphite, grapheme, carbon mesosphere, metal organic frame works (MOFs) have been used for different applications. The excellent properties like enhanced thermal, chemical and mechanical stability, large surface area, high solubility, low cost, high hydrophobic have received much attention of researchers in the area of biosorption. Polymer and grapheme composites have been considered for application as a biosorbent as they have enhanced properties like selectivity, chlorine resistance, mechanical stress, fouling resistance, which enhances the rate of biosorption.

The application of activated carbon produced from different agro-wastes in combination with polymers has proved to be a promising technology in the biosorption process as it has high surface area and good porosity. It is a low-cost and sustainable technology as the agro-waste is being utilized for environmental applications.

Many researchers are working with green synthesized metal nanoparticles from various biological materials like plant and leaf extracts, bacteria, fungi, etc. The commonly used green synthesized metal nanoparticles for biosorption are TiO_2, ZnO, CuO, Fe_3O_4, and silver nanoparticles. This technology is considered promising as compared to conventional treatment methods as this is more convenient, the process for formation of nanoparticles is simple, and it is environmentally friendly to minimize the toxicity and formation of harmful by-products.

4.4 VARIOUS FACTORS CONTROLLING POLLUTANT REMOVAL AND REGENERATION AND REUSE OF BIOSORBENTS

The discharging of all pollutants in the form of organic or inorganic toxins has led to a serious environmental issue. Biosorption is one of the widely used technologies and the biosorbents used here are considered to be emerging, greener, cost-effective, and efficient alternatives. The process of biosorption is influenced by various parameters that control the removal of different pollutants, as indicated in Figure 4.3. The figure also shows the process of biosorption and the different parameters influencing the biosorption process and regeneration and desorption process, which are discussed below in brief.

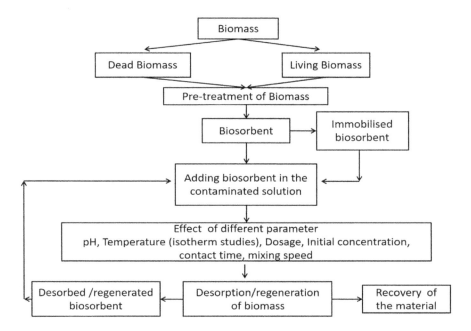

FIGURE 4.3 Experimental procedure of biosorption and effect of different parameters on biosorption.

4.4.1 pH

It is one of the significant parameters in the biosorption method as it affects the solubility of the pollutants and the degree of the biosorbent during the process of reaction. Biomass is regarded as a complex ion exchange biosorbent as it has active binding sites. The extent of displacement of the proton and bond formation is determined by the pH value. The variation in the pH values can modify the chemical state of the functional groups and has an impact on the characteristics and availability of the ions that are responsible for biosorption. Generally, the pH varies from 2.5 to 7.00 pH for any pollutant in the biosorption process.

4.4.2 Temperature (Thermodynamic Studies)

The rate of biosorption has an impact on the temperature of the wastewater as bacteria grow across a wide range of temperatures. The optimum temperature for any biosorption process is between 20°C and 40°C. Higher temperatures above 45°C damage the microorganism or the cells and in turn this affects the rate of biosorption. Based on the preferred range of temperatures microorganisms are classified as psychrophiles (15°C–20°C), mesophiles (20°C–40°C), and thermophiles (50°C–60°C).

Most of the biosorption processes are exothermic in process and the biosorption capacity or ion removal rate increases with a decrease in temperature. The optimal temperature is usually 25°C in most of the cases. The uptake of the pollutant by biomass is calculated using the thermodynamic equations. The change in Gibbs free energy of biosorption is given as $\Delta G^\circ = -RT \ln K_e$, where ΔG is the Gibbs free energy change for biosorption in J mol^{-1}, R is the universal gas constant in J mol^{-1} K^{-1}, T is the temperature in K, K_e is the equilibrium constant. From thermodynamics, we have the Gibbs Helmholtz equation $\Delta G^\circ = \Delta H^\circ - T\Delta S^\circ$; after plotting T against ΔG°, we get a straight line with slope $-\Delta S^\circ$ entropy and intercept ΔH° enthalpy. If $\Delta G > 0$ the reaction is non-spontaneous and endothermic, if $\Delta G < 0$ the reaction is spontaneous and exothermic, and if $\Delta G = 0$ the reaction is at equilibrium.

4.4.3 TIME INTERVAL/CONTACT TIME

The important factor that is considered in the biosorption process is the time allotted for the process to take place and biosorption is a rapid process. The study of the contact time or time interval durations is important in the development of feasible commercial processes. Shake flask experiments are carried out for a known time to obtain the equilibrium time, which is a significant parameter for the cost valuation of wastewater treatment. Though contact time is a limiting factor for biosorption processes, experimental conditions show that increasing the contact time allows biosorbent material to get exposed to the maximum biosorption capability. When a biosorbent reaches its maximum biosorption capability its binding sites become completely saturated and have no further effect and hence reach equilibrium.

4.4.4 INITIAL CONCENTRATION STUDIES

The rate of biosorption is greatly affected by the initial concentration of the pollutant and many researchers (Zhou et al., 2007; Bankar & Nagaraja, 2018) have reported that the increase in the initial concentration increases the rate of biosorption. With increase in the initial concentration the mass transfer driving force, that is, number of collisions of molecules, becomes larger, resulting in higher biosorption. It is observed that 100% biosorption takes place at lower concentration, as most of the ions existing in the solution interact and bind at the site. As the concentration is increased, few ions are left unabsorbed in the solution due to the saturation of the binding sites.

4.4.5 SURFACE AREA

The surface area plays a very significant part in the biosorption process as it depends on the external area available per unit volume of an object. The SEM analysis of a biosorbent shows the surface area available and the extent of biosorption process and its capacity. Higher the surface area, higher would be the rate of biosorption.

4.4.6 ISOTHERM STUDIES

A biosorption isotherm describes how the pollutant in the solution phase gets transmitted to the biosorbent and reaches equilibrium condition and is usually illustrated by two different models: Freundlich and Langmuir. These isotherms are plotted by considering the rate of biosorption and the saturation limit reached. It indicates the relationship between the equilibrium pollutant in the liquid phase to the biosorbent amount in the solid phase. The Langmuir isotherm provides the equilibrium relation with respect to monolayer adsorbate formation on the outer surface of the adsorbent. The Freundlich isotherm provides the equilibrium relation with multilayer adsorption properties. The amount of ion sorbed is determined by the initial concentrations using the equation $q_e = \frac{(C_i - C_e)V}{M}$ and then by applying it to the isotherm models in the linear form expressed as Langmuir model $\frac{C_e}{q_t} = \frac{1}{Q_m k_L} + \frac{C_e}{Q_m}$ and Freundlich model $\log q_e = \log k_F + \frac{1}{n} \log C_e$, where q_e and Q_m are ion sorbed at equilibrium and maximum capacity in mg g^{-1}, respectively, k_L and k_F are Langmuir and Freundlich equilibrium constants relating to the biosorption capacity, and n is the empirical parameter showing the intensity of biosorption.

4.4.7 KINETIC STUDIES

In the biosorption process, kinetic studies are done to comprehend the quality of the biosorbent by knowing the number of ions absorbed by the biomass at different contact times. Chemical kinetics indicates the kind of reaction taking place along with factors affecting the rate of reaction. The nature of sorption depends on the chemical and physical characteristics of the biosorbent. The quality of the biosorbent is assessed by knowing the rate at which the ion is being attracted by the biosorbent and also to what extent it is retained with or without immobilized form by fitting into the model. The rate of ion uptake is change in the initial amount of the ion in the solution to that retained in the solution after biosorption.

Here both biosorption and desorption procedures are dependent on time, and it is necessary to assess the rate of biosorption for the design and regeneration of the biosorbent. Kinetic biosorption studies provide the information about optimal settings (contact time, temperature, adsorbent dosage, and pH), the mechanism of sorption, and possible rate-limiting step for physical and chemical sorption.

The kinetics of biosorption is derived from the general kinetic law and is classified in two ways. The two most accepted kinetic models Langergren pseudo-first- and second-order models were used to compare the performance of the two proposed models: $q_t = q_e(1 - e^{-k_1 t})$ and $q_t = \frac{k_2 q_e^2 t}{1 + k_2 q_e t}$. Here, q_e and q_t (mg g^{-1}) are the adsorptive removal capacity at equilibrium and variable time (t), respectively, and k_1 and k_2 are the rate constants for the first- and second-order kinetic models. The linear form of the model is expressed as $\ln(q_e - q_t) = \ln q_e - k_1 t$ and $\frac{t}{q_t} = \frac{1}{k_2 q_e^2} + \frac{1}{q_e} t$. Here, a plot of $\ln(q_e - q_t)$ versus time (t) offers the estimation of q_e and k_1 from the intercept and the slope of the plot, individually. Similarly, a plot of (t/q_t) versus t offers values of k_2 and q_e from the intercept and slope, individually. The finest model fit is chosen based on the best regression value obtained and the correlation coefficient in the plot.

4.4.8 DESORPTION STUDIES

The biosorbent that has been exhausted after the biosorption process can be reused after recovering the ion or metal adsorbed. Such process is called desorption and it can be carried out using desorbing agents like HCl, NaOH, EDTA, etc. The choice of the recovery agent is more important in desorption studies, as the recovery agent used in the process should not harm the physical properties of the biosorbent and should retain its original shape and structure to provide maximum efficiency in binding. The desorption efficiency is given as $\text{Desorption efficiency} = \dfrac{C_{\text{desorption}}}{C_{\text{adsorption}}} \times 100$

where $C_{\text{desorption}}$ (mol L^{-1}) and $C_{\text{adsorption}}$ (mol L^{-1}) are, respectively, the concentrations of desorbed and adsorbed ions in equilibrium by the biosorbent.

The important concept in the biosorption process is not only metal removal, but also the recovery of that metal using the desorption process due to cost considerations and its impact on the environment. The stability and potential recovery of biomass is assessed through the desorption cycles.

4.5 CHALLENGES AND APPLICATION OF BIOMATERIAL IN WASTEWATER TREATMENT

Selection of biowaste material is one of the challenging issues based on its sufficient availability, minimum preparations needed for biosorbent, minimum or no activation, non-reactive regeneration of the biosorbent. Biomaterials are with less sorption efficiency and stability, and with commercial scale applications, need to be investigated.

A lot of investigation in the area of biomaterials and composites is desired in order to enhance environmental sustainability and advancement in green technology with increase in biocompatible materials and to present the technology for commercial applications of wastewater treatment.

Today many industries are moving towards zero discharge concepts by utilizing the waste and by-products in one or the other form. Biosorbents are also one such application of such waste materials discharged from industries. It solves the dual problem of the utilization of the waste and the treatment of the effluent using this waste. The utilization of the biosorbent from industrial waste is increasing at a speed of 5%. Still many challenging issues are to be looked into by many researchers to increase this rate from 5% to a higher rate of consumption. Another challenging issue is the continuous studies of the biosorption process in industrial practice by using living biomass in column studies. High pressures can disrupt the biomass and hence the rate of biosorption decreases. Many researchers are working with dead biomass and converting agricultural waste into activated form to overcome these problems.

4.6 CONCLUSION

Biosorption is an alternative technique of advance wastewater treatment for reduction of pollutants due to its ecofriendliness and sustainability. The current chapter highlights the importance of biomaterials and their abundant availability. The selection of the biomaterial is important due to their structural, physiochemical, and biological features. There are wide varieties of biomaterials available naturally which can be used as free cells or dead biomass for making industrial wastewater treatment more effective, cheap, ecofriendly to have greener and cleaner technology. The possibility of reuse or of regenerating the biomass essentially symbolizes the key element in the area of wastewater treatment. Though many have investigated and worked in the area of biosorption with different biomaterials and using different combinations and mechanisms, much work remains to be done to enhance the efficiency for larger-scale applications. The process becomes more cost-effective by using biodegradable material and also by regenerating the material for the reuse. Therefore, the constructive efforts in the direction of improvement in the area of understanding the binding capacity, increasing the stability of biosorbents to improve mechanical strength, the regeneration capacity of the biomaterial, scaling up of the bioreactor with biosorption process are the keen targets of the research community.

REFERENCES

Abali, M. H., Zaghloul, A., Sinan, F., & Zerbet, M. (2021). Removal of nitrate ions by adsorption onto micro-particles of shrimp-shells waste: Application to wastewater of infiltration-percolation process of the city of Agadir (Morocco). *Materials Today: Proceedings*, 37, 3898–3904.

Acheampong, M. A., Pakshirajan, K., Annachhatre, A. P., & Lens, P. N. (2013). Removal of Cu(II) by biosorption onto coconut shell in fixed-bed column systems. *Journal of Industrial and Engineering Chemistry*, 19(3), 841–848.

Akar, S. T., Yilmazer, D., Celik, S., Balk, Y. Y., & Akar, T. (2013). On the utilization of a lignocellulosic waste as an excellent dye remover: Modification, characterization and mechanism analysis. *Chemical Engineering Journal*, 229, 257–266.

Akar, T., Sayin, F., Turkyilmaz, S., & Akar, S. T. (2017). The feasibility of Thamnidium elegans cells for color removal from real wastewater. *Process Safety and Environmental Protection*, 105, 316–325.

Amuda, O. S., Giwa, A., & Bello, I. A. (2007). Removal of heavy metal from industrial wastewater using modified activated coconut shell carbon. *Biochemical Engineering Journal*, 36(2), 174–181.

Anayurt, R. A., Sari, A., & Tuzen, M. (2009). Equilibrium, thermodynamic and kinetic studies on biosorption of Pb(II) and Cd(II) from aqueous solution by macrofungus (Lactariusscrobiculatus) biomass. *Chemical Engineering Journal*, 151(1–3), 255–261.

Azimi, A., Azari, A., Rezakazemi, M., & Ansarpour, M. (2017). Removal of heavy metals from industrial wastewaters: a review. *ChemBioEng Reviews*, 4(1), 37–59.

Ba, S., Arsenault, A., Hassani, T., Jones, J. P., & Cabana, H. (2013). Laccase immobilization and insolubilization: From fundamentals to applications for the elimination of emerging contaminants in wastewater treatment. *Critical Reviews in Biotechnology*, 33(4), 404–418.

Baig, J. A., Kazi, T. G., Shah, A. Q., Kandhro, G. A., Afridi, H. I., Khan, S., & Kolachi, N. F. (2010). Biosorption studies on powder of stem of Acacia nilotica: Removal of arsenic from surface water. *Journal of Hazardous Materials*, 178(1–3), 941–948.

Bankar, A., & Nagaraja, G. (2018). Recent trends in biosorption of heavy metals by Actinobacteria. In *New and Future Developments in Microbial Biotechnology and Bioengineering* (pp. 257–275). Elsevier.

Benguella, B., & Benaissa, H. (2002). Cadmium removal from aqueous solutions by chitin: Kinetic and equilibrium studies. *Water Research*, 36(10), 2463–2474.

Bezerra, R. D., Silva, M. M., Morais, A. I., Osajima, J. A., Santos, M. R., & Airoldi, C. (2014). Phosphated cellulose as an efficient biomaterial for aqueous drug ranitidine removal. *Materials*, 7(12), 7907–7924.

Bulgariu, D., & Bulgariu, L. (2012). Equilibrium and kinetics studies of heavy metal ions biosorption on green algae waste biomass. *Bioresource Technology*, 103(1), 489–493.

Cardoso, S. L., Costa, C. S. D., da Silva, M. G. C., & Vieira, M. G. A. (2018). Dealginated seaweed waste for Zn(II) continuous removal from aqueous solution on fixed-bed column. *Journal of Chemical Technology & Biotechnology*, 93(4), 1183–1189.

Carley, M., & Christie, I. (2017). *Managing sustainable development*. Routledge.

Chedia, R., Jalagonia, N., Kuchukhidze, T., Sanaia, E., Kvartskhava, G., Gabunia, V., & Marquis, F. (2015, December). Impregnation of nano zero-valent iron in biomaterials for remediation of wastewater. In *2015-Sustainable Industrial Processing Summit* (Vol. 9, pp. 109–120). Flogen Star Outreach.

Chen, G., Zeng, G., Tang, L., Du, C., Jiang, X., Huang, G.,... & Shen, G. (2008). Cadmium removal from simulated wastewater to biomass byproduct of Lentinus edodes. *Bioresource Technology*, 99(15), 7034–7040.

Chowdhury, S., & Saha, P. D. (2012). Biosorption of methylene blue from aqueous solutions by a waste biomaterial: Hen feathers. *Applied Water Science*, 2(3), 209–219.

Crini, G., & Lichtfouse, E. (2018). Wastewater treatment: An overview. *Green Adsorbents for Pollutant Removal*, vol. 18, 1–21.

Dawood, S., & Sen, T. (2014). Review on dye removal from its aqueous solution into alternative cost effective and non-conventional adsorbents. *Journal of Chemical and Process Engineering*, 1(104), 1–11.

Deng, S., Chen, J., & Chang, J. (2021). Application of biochar as an innovative substrate in constructed wetlands/biofilters for wastewater treatment: Performance and ecological benefits. *Journal of Cleaner Production*, vol. 293, 126156.

Deniz, F. (2013). Adsorption properties of low-cost biomaterial derived from Prunus amygdalus L. for dye removal from water. *The Scientific World Journal*, 2013. 961671, 1–8.

Deshmukh, S. O., & Hedaoo, M. N. (2019). Wastewater treatment using bio-coagulant as cactus Opuntia ficus indica. *Carbon*, 29(54.80), 53–30.

Ehrampoush, M. H., Miria, M., Salmani, M. H., & Mahvi, A. H. (2015). Cadmium removal from aqueous solution by green synthesis iron oxide nanoparticles with tangerine peel extract. *Journal of Environmental Health Science and Engineering*, 13(1), 1–7.

Elangovan, R., Philip, L., & Chandraraj, K. (2008). Biosorption of chromium species by aquatic weeds: Kinetics and mechanism studies. *Journal of Hazardous Materials*, 152(1), 100–112.

Firdaus, M. L., Krisnanto, N., Alwi, W., Muhammad, R., & Serunting, M. A. (2017). Adsorption of textile dye by activated carbon made from rice straw and palm oil midrib. *Aceh International Journal of Science and Technology*, 7(1), 1–7.

Fomina, M., & Gadd, G. M. (2014). Biosorption: Current perspectives on concept, definition and application. *Bioresource Technology*, 160, 3–14.

Frantz, T. S., de Farias, B. S., Leite, V. R. M., Kessler, F., Cadaval Jr, T. R. S. A., & de Almeida Pinto, L. A. (2020). Preparation of new biocoagulants by shrimp waste and its application in coagulation-flocculation processes. *Journal of Cleaner Production*, 269, 122397.

Gadd, G. M. (2009). Biosorption: Critical review of scientific rationale, environmental importance and significance for pollution treatment. *Journal of Chemical Technology & Biotechnology: International Research in Process, Environmental & Clean Technology*, 84(1), 13–28.

Godage, N. H., & Gionfriddo, E. (2020). Use of natural sorbents as alternative and green extractive materials: A critical review. *Analytica Chimica Acta*, 1125, 187–200.

Gonte, R. R., Shelar, G., & Balasubramanian, K. (2014). Polymer–agro-waste composites for removal of Congo red dye from wastewater: Adsorption isotherms and kinetics. *Desalination and Water Treatment*, 52(40–42), 7797–7811.

Goyal, P., & Masram, D. T. (2016). Azadirachta indica nano biomaterial: A green economical biomaterial for removal of Cd(II) and Ni(II) from waste water. *Materials Today: Proceedings*, 3(6), 1470–1476.

Hammo, M. M., Akar, T., Sayin, F., Celik, S., & Akar, S. T. (2021). Efficacy of green waste-derived biochar for lead removal from aqueous systems: Characterization, equilibrium, kinetic and application. *Journal of Environmental Management*, 289, 112490.

Kamanina, O. A., Lavrova, D. G., Arlyapov, V. A., Alferov, V. A., & Ponamoreva, O. N. (2016). Silica sol-gel encapsulated methylotrophic yeast as filling of biofilters for the removal of methanol from industrial wastewater. *Enzyme and Microbial Technology*, 92, 94–98.

Kanamarlapudi, S. L. R. K., Chintalpudi, V. K., & Muddada, S. (2018). Application of biosorption for removal of heavy metals from wastewater. *Biosorption*, 18, 69.

Kariuki, Z., Kiptoo, J., & Onyancha, D. (2017). Biosorption studies of lead and copper using rogers mushroom biomass 'Lepiotahystrix'. *South African Journal of Chemical Engineering*, 23(1), 62–70.

Karthiga Devi, G., Senthil Kumar, P., & Sathish Kumar, K. (2016). Green synthesis of novel silver nanocomposite hydrogel based on sodium alginate as an efficient biosorbent for the dye wastewater treatment: Prediction of isotherm and kinetic parameters. *Desalination and Water Treatment*, 57(57), 27686–27699.

Khan, M. A., Ngabura, M., Choong, T. S., Masood, H., & Chuah, L. A. (2012). Biosorption and desorption of nickel on oil cake: Batch and column studies. *Bioresource Technology*, 103(1), 35–42.

Kumar, A., Sharma, G., Naushad, M., Ala'a, H., Garcia-Penas, A., Mola, G. T.,… & Stadler, F. J. (2020). Bio-inspired and biomaterials-based hybrid photocatalysts for environmental detoxification: A review. *Chemical Engineering Journal*, 382, 122937.

Malik, R., Lata, S., & Singhal, S. (2015). Removal of heavy metal from wastewater by the use of modified aloe vera leaf powder. *International Journal of Basic and Applied Chemical Sciences*, 5(2), 6–17.

Maqbool, M., Bhatti, H. N., Sadaf, S., Zahid, M., & Shahid, M. (2019). A robust approach towards green synthesis of polyaniline-Scenedesmus biocomposite for wastewater treatment applications. *Materials Research Express*, 6(5), 055308.

Martín-Lara, M. A., Blázquez, G., Trujillo, M. C., Pérez, A., & Calero, M. (2014). New treatment of real electroplating wastewater containing heavy metal ions by adsorption onto olive stone. *Journal of Cleaner Production*, 81, 120–129.

Mishra, A., Tripathi, B. D., & Rai, A. K. (2016). Packed-bed column biosorption of chromium(VI) and nickel(II) onto Fenton modified Hydrilla verticillata dried biomass. *Ecotoxicology and Environmental Safety*, 132, 420–428.

Mohammed, N., Nadjia, H., Amina, G., Yamina, B., & Souhila, O. (2021). Study and characterization of a biomaterial: Animal bone. Application to the treatment of an industrial effluent. In *Advances in Green Energies and Materials Technology* (pp. 139–145). Springer, Singapore.

Mohapatra, R. K., Parhi, P. K., Pandey, S., Bindhani, B. K., Thatoi, H., & Panda, C. R. (2019). Active and passive biosorption of Pb(II) using live and dead biomass of marine bacterium Bacillus xiamenensis PbRPSD202: Kinetics and isotherm studies. *Journal of Environmental Management*, 247, 121–134.

N'diaye, A. D., & Kankou, M. S. A. (2020). Adsorption of aspirin onto biomaterials from aqueous solutions. *Cellulose*, 11, 41–43.

Nancharaiah, Y. V., Joshi, H. M., Mohan, T. V. K., Venugopalan, V. P., & Narasimhan, S. V. (2006). Aerobic granular biomass: A novel biomaterial for efficient uranium removal. *Current Science*, Vol. 91, 503–509.

Parab, H., Joshi, S., Shenoy, N., Verma, R., Lali, A., & Sudersanan, M. (2005). Uranium removal from aqueous solution by coir pith: Equilibrium and kinetic studies. *Bioresource Technology*, 96(11), 1241–1248.

Parlayıcı, Ş., & Pehlivan, E. (2021). Biosorption of methylene blue and malachite green on bio-degradable magnetic Cortaderiaselloana flower spikes: Modeling and equilibrium study. *International Journal of Phytoremediation*, 23(1), 26–40.

Pholosi, A., Naidoo, B. E., & Ofomaja, A. E. (2018). Clean application of magnetic biomaterial for the removal of As(III) from water. *Environmental Science and Pollution Research*, 25(30), 30348–30365.

Prasad, A. G. D., & Abdullah, M. A. (2010). Biosorption of Cr(VI) from synthetic wastewater using the fruit shell of gulmohar (Delonix regia): Application to electroplating wastewater. *BioResources*, 5(2), 838–853.

Rafique, U., & Zulfiqar, S. (2014). Synthesis and technological application of agro-waste composites for treatment of textile waste water. *International Journal of Agriculture Innovations and Research*, 3(3), 842–845.

Rajasulochana, P., & Preethy, V. (2016). Comparison on efficiency of various techniques in treatment of waste and sewage water–A comprehensive review. *Resource-Efficient Technologies*, 2(4), 175–184.

Rapo, E., Aradi, L. E., Szabo, A., Posta, K., Szep, R., & Tonk, S. (2020). Adsorption of remazol brilliant violet-5R textile dye from aqueous solutions by using eggshell waste biosorbent. *Scientific Reports*, 10(1), 1–12.

Ren, L., Zhou, D., Wang, J., Zhang, T., Peng, Y., & Chen, G. (2020). Biomaterial-based flower-like MnO_2@carbon microspheres for rapid adsorption of amoxicillin from wastewater. *Journal of Molecular Liquids*, 309, 113074.

Saiano, F., Ciofalo, M., Cacciola, S. O., & Ramirez, S. (2005). Metal ion adsorption by Phomopsis sp. biomaterial in laboratory experiments and real wastewater treatments. *Water Research*, 39(11), 2273–2280.

Saleh, T. A. (2021). Protocols for synthesis of nanomaterials, polymers, and green materials as adsorbents for water treatment technologies. *Environmental Technology & Innovation*, vol. 24, 101821.

Sharma, P., Kumari, P., Srivastava, M. M., & Srivastava, S. (2007). Ternary biosorption studies of Cd(II), Cr(III) and Ni(II) on shelled Moringa oleifera seeds. *Bioresource Technology*, 98(2), 474–477.

Sivakumar, B. (2011). Global climate change and its impacts on water resources planning and management: Assessment and challenges. *Stochastic Environmental Research and Risk Assessment*, 25(4), 583–600.

Sostaric, T. D., Petrovic, M. S., Pastor, F. T., Loncarevic, D. R., Petrovic, J. T., Milojkovic, J. V., & Stojanović, M. D. (2018). Study of heavy metals biosorption on native and alkali-treated apricot shells and its application in wastewater treatment. *Journal of Molecular Liquids*, 259, 340–349.

Sravanthi, K., Ayodhya, D., & Swamy, P. Y. (2018). Green synthesis, characterization of biomaterial-supported zero-valent iron nanoparticles for contaminated water treatment. *Journal of Analytical Science and Technology*, 9(1), 1–11.

Sreenivas, K. Á., Inarkar, M. Á., Gokhale, S. V., & Lele, S. S. (2014). Re-utilization of ash gourd (Benincasahispida) peel waste for chromium(VI) biosorption: Equilibrium and column studies. *Journal of Environmental Chemical Engineering*, 2(1), 455–462.

Srinivasan, A., & Viraraghavan, T. (2010). Oil removal from water using biomaterials. *Bioresource Technology*, 101(17), 6594–6600.

Tavana, M., Pahlavanzadeh, H., & Zarei, M. J. (2020). The novel usage of dead biomass of green algae of Schizomerisleibleinii for biosorption of copper(II) from aqueous solutions: Equilibrium, kinetics and thermodynamics. *Journal of Environmental Chemical Engineering*, 8(5), 104272.

Tong, T., & Elimelech, M. (2016). The global rise of zero liquid discharge for wastewater management: Drivers, technologies, and future directions. *Environmental Science & Technology*, 50(13), 6846–6855.

Umali, L. J., Duncan, J. R., & Burgess, J. E. (2006). Performance of dead Azolla filiculoides biomass in biosorption of Au from wastewater. *Biotechnology Letters*, 28(1), 45–50.

Upadhyay, A., Narula, A., & Rao, C. P. (2020). Copper-based metallogel of bovine serum albumin and its derived hybrid biomaterials as aerogel and sheet: Comparative study of the adsorption and reduction of dyes and nitroaromatics. *ACS Applied Bio Materials*, 3(12), 8619–8626.

Wakkel, M., Khiari, B., & Zagrouba, F. (2019). Textile wastewater treatment by agro-industrial waste: Equilibrium modelling, thermodynamics and mass transfer mechanisms of cationic dyes adsorption onto low-cost lignocellulosic adsorbent. *Journal of the Taiwan Institute of Chemical Engineers*, 96, 439–452.

Wang, C., Liu, H., Liu, Z., Gao, Y., Wu, B., & Xu, H. (2018). Fe_3O_4 nanoparticle-coated mushroom source biomaterial for Cr(VI) polluted liquid treatment and mechanism research. *Royal Society Open Science*, 5(5), 171776.

Willaert, R. (2007). Cell immobilization and its applications in biotechnology: Current trends and future prospects. In *Fermentation Microbiology and Biotechnology* (2nd ed., pp. 287–332). CRC Press, Boca Raton, FL.

Wu, M., Yan, X., Liu, K., & Deng, L. (2017). Application of activated biomaterial in the rapid start-up and stable operation of biological processes for removal cadmium from effluent. *Water, Air, & Soil Pollution*, 228(1), 1–12.

Xu, J., Song, X. C., Zhang, Q., Pan, H., Liang, Y., Fan, X. W., & Li, Y. Z. (2011). Characterization of metal removal of immobilized Bacillus strain CR-7 biomass from aqueous solutions. *Journal of Hazardous Materials*, 187(1–3), 450–458.

Zhou, M., Liu, Y., Zeng, G., Li, X., Xu, W., & Fan, T. (2007). Kinetic and equilibrium studies of Cr (VI) biosorption by dead Bacillus licheniformis biomass. *World Journal of Microbiology and Biotechnology*, 23, 43–48.

5 Biomaterials to Bioethanol

A Practical Approach

Vinayaka B. Shet, Keshava Joshi,
Lokeshwari Navalgund, Sandesh. K, Ujwal. P
and Nabisab Mujawar Mubarak

5.1 INTRODUCTION

5.1.1 PREAMBLE

Today there is an increased concern regarding the release of greenhouse gases, which is causing global warming at an alarming rate due to the enhanced consumption of fossil fuels, thereby leading to various environmental complications. Thus, the progressive reduction of fossil fuels has contributed towards the production of renewable energy from bioresources. Most of the countries are importing crude oil to meet fuel requirements. To reduce the usage of crude oil and financial burden, importance is given for bioethanol production as a lucrative alternative [1]. The continuous increase in demand for fossil fuel has resulted in a steep rise in price and environmental problems. The demand can be met by developing alternative fuel from renewable resources. Increasing crude oil prices and the requirement for sustainable energy sources has led to the production of biofuels at a competitive cost [2]. Advances in synthetic biology and metabolic engineering have paved the way to achieve sustainable production of biofuels [3]. Presently, the world is facing climate change and there is a need for meeting the expanding energy needs while limiting greenhouse gas emissions. During the December 1997 United Nations Framework Convention on Climate Change (UNFCCC), developed countries agreed on and endorsed using renewable energy as per the Kyoto Protocol by step-by-step replacement of fossil fuel to reduce the net CO_2 emission [4]. Renewable resources are utilized to replace fossil fuel in order to reduce the CO_2 emission; ethanol is blended with petrol or gasoline and used as fuel. Various biomaterials containing starch, sucrose, cellulose and lignocellulose have been explored for the production of bioethanol. Abundantly available underutilized biomaterials that do not compete with food are explored for the research and development of economic bioethanol production [5, 6].

Structurally biomaterials possess polysaccharide networks; it is challenging to break down and exhibit non-availability of the fermentable sugars to microorganisms. Biomaterials having a complex structure pose a challenge during the conversion

DOI: 10.1201/9781032636368-5

of biomaterials into simple sugars, thus making the production of bioethanol commercially non-competitive [7]. Different strategies and technologies are under development for the conversion of biomaterials for releasing sugar. Emphasis is to develop the process and scale up bioethanol production from biomaterials by approving refineries under oil manufacturing companies for commercialization.

5.1.2 Overview of Bioethanol

Ethyl alcohol has been used as a fuel for decades since it is highly inflammable. It is soluble in water and appears colourless. As per the USA 1992 energy policy act, ethyl alcohol is considered as an alternative fuel and currently ethyl alcohol is used along with conventional fuel to power motor vehicles by blending with petrol. Ethanol derived from biomaterials is considered as renewable fuel. Ethanol is a two-carbon hydrocarbon with hydroxyl groups, which are hydrophilic and capable of hydrogen bonding with water molecules. Ethanol is a monohydric primary alcohol; it is a colourless liquid with slight odour and is volatile in nature. The burning of ethanol is not always visible and it has a smokeless blue flame. The melting and boiling point of ethanol are −117.3°C and 78.5°C, respectively. Ethanol and water mixes well in all proportions and thus they are difficult to separate out. It is more viscous and less volatile in nature due to the hydroxyl group participating in hydrogen bonding. Ethanol forms a constant boiling mixture; the boiling point of the binary azeotrope 78.15°C (ethanol-water) is below that of pure ethanol; hence absolute ethanol recovery by simple distillation is not possible.

Ethanol is produced through chemical and biochemical (fermentation) processes. The fermentation process is economic and eco-friendly. The fermentation process takes place at room temperature and atmospheric pressure. Since underutilized biomaterials are available abundantly and it is economic to produce ethanol by saccharification to release simple sugars followed by microbial fermentation, research is being carried out to utilize different types of biomaterials for the production of bioethanol. Bioethanol refers to ethanol that is produced from biomass. Bioethanol is a sustainable energy resource due to its long-term environmental and economic advantages. It is also used as an "octane booster" in gasoline. Bioethanol production through the fermentation process is governed by various parameters such as agitation rate, pH, inoculum size, temperature, sugar concentration and time. Yeast metabolism plays a critical role in converting hexose and pentose sugars into bioethanol. The quantity and rate of bioethanol production is enhanced by yeast. Bioethanol can be classified into four broad categories on the basis of sources from which it is produced: First-generation bioethanol is produced from starchy sources like cassava, potato, wheat, rice, maize, sweet potato, barley, corn, wheat and waste food grains, using well-defined technology, and it is easy to produce but poses conflict, viz. food vs. fuel. However, it helps the agricultural sector due to the increased demand for food crops and leads to an increase in the price of food crops. Sugars and starch materials can also be used in the production of bioethanol. Sugary materials utilized for first-generation bioethanol production are sweet sorghum, sugar beet, fruit and confectionery industrial waste. Second-generation biofuels are sourced from cellulosic energy crops, agricultural residue, grass, straw and wood. They are harder to extract as sugars when compared to first-generation biofuels but are presently considered the most reliable source for biofuel production due to the availability of feedstock and since they don't compete with food material. Lignocellulosic feedstocks like cotton stalks, coconut residue, bagasse, rice straw, corn cobs, wheat straw, rice husk, jute sticks, waste paper, paper pulp, sawdust, sewage treatment sludge, wood chips and bark are used for the production of bioethanol. Lignin, cellulose and hemicelluloses constitute the major components of lignocellulosic residues [2]. Liquefaction and hydrolysis of feedstock are carried out to release sugar monomers. Pretreatment of the feedstock is undertaken to release simple sugars. In lignocellulosic materials, during the pretreatment, deformation of the rigid components and degradation of the crystallinity takes place.

Third-generation biofuels are sourced from algal biomass, mainly composed of microalgae. Microalgae can provide several types of biofuels ranging from biodiesel and methane to bio-hydrogen. There are many advantages of producing biofuel from algae: algal cultivation can cover a larger area than food crops, with a short harvesting cycle and hence higher growth rate. Thus, good agricultural land is not required for biomass production. Although this technology looks promising, it is still in its nascent stage. Moreover, biofuel produced from this generation of biomass is less stable and highly unsaturated. Since algae can prevent the competition of feedstock with agricultural plant, public acceptance is also positive. Algae contain a low level of hemicelluloses and lignin, which makes it suitable for bioethanol production. Biomass is hydrolysed to convert the polysaccharide into free monomer molecules, which can be fermented to bioethanol. Fourth-generation biofuels are developed by adopting advanced biochemistry or revolutionary processes combining genetically modified feedstock, which is engineered to capture more carbon and helps in the efficient production of fuel. While the key strategy is capturing followed by sequestration of CO_2, some prefer the 'synthetic genomics' plan to combine the fuel processing and feedstock growth processes by organisms that intake CO_2 and release sugars.

Cetane number has an influence on engine startability, including starting at low temperatures, emissions, peak cylinder pressure, engine durability and combustion noise. The flash point determines the lowest temperature at which a fuel ignites when exposed to an ignition source. Vapour pressure is defined as the pressure exerted by a vapour over a liquid in a container at a specified temperature. The calorific value of fuel has an influence on power output of the engine and fuel consumption. The average energy value of petrol is about 41.3 MJ kg^{-1}, while for bioethanol it is 26.4 MJ kg^{-1}.

5.1.3 NEED FOR BIOETHANOL

Biofuels are being used for renewable replacement of non-renewable fuels due to environmental and economic benefits. Renewable energy is progressively contributing more to the total world energy demand and it is being considered for sustainable development. Global production capacity for the year 2015 was estimated to be 225 million gallons per year for advanced biofuels. Countries like France, Sweden, the United Kingdom, Brazil, the Netherlands, Canada, China and the United States have contributed 390 million gallons per year. The United States and Brazil produced biofuels from corn and sugarcane, respectively, to meet 70% of the global supply in 2015 [8]. In Asia, biofuel production is from cassava, wheat, corn and sugarcane with investment also occurring in Jatropha, rapeseed, soybean and palm. The biomaterial and region-based diversification may be favourable to the formation of an international biofuel production. Bioethanol is identified as an alternative biofuel to crude oil worldwide due to its significant contribution in the reduction of environmental pollution and a reduction in the consumption of crude oil.

5.2 BIOMATERIALS AS POTENTIAL FEEDSTOCK FOR BIOETHANOL

Agriculture, horticulture, non-agricultural and forestry residues are incinerated causing air pollution. These residues can be channelized for bioethanol production. The residue is mainly composed of starch and lignocellulose. Kitchen waste generated on a day-to-day basis is also a potential biomaterial. Processing, pretreatment and fermentation are the steps involved in conversion of biomaterials to bioethanol.

5.2.1 STARCH

Starch is the storage polysaccharide made up of glucose units. Glycosidic bonds link glucose units to form polymeric structures. The glucose generated during photosynthesis is stored as starch either in grain (seed) or in tuber.

5.2.2 LIGNOCELLULOSE

Lignocellulosic materials are the building blocks of all plants. Lignocellulose is composed of carbohydrate polymer cellulose (30%–50%), hemicellulose (15%–30%) and branched polymer lignin (12%–35%) (Table 5.1) [9].

5.2.2.1 Cellulose

The approximate molecular weight of cellulose ranges from 10,000 to 150,000 Da, and it is a fibrous polysaccharide present in plant cell walls. Cellulose is made by joining hexose sugar, D-glucose by β-1-4-glucosidic bonds crystalline, insoluble. Cellulose strands are held together to form cellulose fibrils which are linked by

---

TABLE 5.1

Composition of lignocellulosic feedstock

Feedstock	Cellulose (%)	Lignin (%)	Hemicellulose (%)	Ash (%)	References
Corn stover	38–40	7–21	28	3.6–7	[10]
Sorghum stalks	27	11	25	–	[11]
Rice straw	28–36	12–14	23–28	14–20	[12]
Wheat straw	33–38	17–19	26–32	6–8	[11]
Coir	36–43	41–45	0.15–0.25	2.7–10.2	[10]
Leaflets	23.83	38.68	–		[10]

hydrogen bonds. Cellulose fibres are connected by inter- and intramolecular hydrogen bonds. A major bottleneck is to disarray the lignin because of its rigid structure. It is the second most available abundant polymer on earth. Cellulose is used as a nutritional source by a wide range of microorganisms and herbivores.

5.2.2.2 Hemicellulose

Hemicellulose is a heterogeneous polymer composed of branched and straight chain polysaccharides present in the secondary cell wall. The molecular weight of hemicelluloses is less than that of cellulose and it can be easily hydrolysed. During chemical treatment, initially, the side chain of the hemicellulose reacts followed by its backbone. Lignin and cellulose fibres are connected by hemicelluloses to provide a rigid network. Due to the lack of crystalline structure, it can be easily degraded to a simple sugar because of a low degree of polymerization. It may contain hexose, pentose and uronic acid. Hemicellulose is a hetero-polymer of D-glucose, D-xylose, D-galactose, L-arabinose and D-mannose. Xylans are the most abundantly available hemicellulose. Hemicellulose is sensitive to hydrolysis conditions; hence, the factors affecting hydrolysis have to be controlled to avoid formation of furfurals and hydroxymethyl furfurals to prevent formation of fermentation inhibitors.

5.2.2.3 Lignin

It is the aromatic polymer of phenylpropane units, and is hydrophobic in nature having a crosslinked structure present in the plant cell wall. The purpose of the lignin is to provide support. It is an amorphous polyphenol having heterogeneity through coupling of 4-hydrophenylpropanoids. Lignin is an amorphous form and holds together cellulose and hemicellulose fibres. It is a recalcitrant substance in the lignocellulosic biomass and prevents biomass from microbial and chemical degradation. It has a complex arrangement of phenolic compounds.

5.2.3 SOURCES OF BIOMATERIALS

To understand the process of converting biomaterials into bioethanol, few examples of starchy and lignocellulosic residues are discussed in the current section.

5.2.3.1 Pongamia De-oil Seedcake

The *Pongamia pinnata* generally known as Karanja or Honge is a non-edible, oil-rich, fast-growing leguminous tree. The trees of *P. pinnata* grow in the tropical region of Asia, Australia, islands of Indian Ocean and are extensively spread in the majority of Indian states and to grow well it prefers less rainfall. It can fix atmospheric nitrogen and survive in drought conditions. Seed yield is good in trees growing in semiarid regions having less than 100 cm rain fall per annum. Consumption of *P. pinnata* seeds is not recommended for humans or animals due to the existence of the toxic components karanjin, pongamin and glabrin. The seed of *P. pinnata* has been reported to contain starch. Pongamia seeds and seedcake are used as a fertilizer and are found to be a good source of carbon and nitrogen. The seeds contain 25%–30% oil, which on extraction yields a seedcake that constitutes about 70%–75%

of the seed. *P. pinnata* is known as karanja seed; it is a potential feedstock for bio-diesel production with the productivity of seed more than 200,000 tonnes per annum.

Fully mature Pongamia trees usually yield approximately 900–9,000 kg seeds per hectare after six years. Due to availability, it is the best source for biodiesel production. Production of biodiesel from *P. pinnata* seed oil generates a huge mass of de-oiled seedcake, which is approximately 700–750 g per kg of seeds. Due to the presence of higher amounts of carbohydrates and low cost of seedcake, it is a potential residue for bioethanol production. The proximate analysis carried out using Pongamia seedcake has revealed that it is composed of 29% lignin, 22.3% protein, 42% carbohydrate and 6.8% ash. To obtain sugar from lignocellulosic residues, two types of processes are adopted, namely, chemical and thermochemical hydrolysis with variations in each.

5.2.3.2 Cocoa Pod Shells

Fruit of the cocoa tree is known as cocoa (*Theobroma cacao L.*). Seeds are gener-ally known as cocoa beans and consist of two cotyledons, surrounded by an outer shell and a small germ. The cocoa tree provides by-products such as cocoa pod husk (CPH), cocoa bean meal (CBM) and cocoa pod shell (CPS). CPS is an industrial lignocellulosic waste from cocoa-based chocolate companies. CPSs contain about 43.9%–45.2% carbohydrate. CPS contains hemicelluloses, cellulose, pectin, lignin and crude fibre. Since the cocoa shells contain 12%–20% of the cocoa seed, it can be used as a raw material for hydrolysis. Protein content of the CPS was reported to be 5.9% and crude fibre 21.3%. The outer shell of cocoa is separated along with the germ after or before roasting and the broken fragments of cotyledon, called nibs without the shells, are used in the production of chocolate. Shells which are removed from the cocoa pod are the underutilized by-product of the cocoa industry. Interaction with local cocoa farmers has revealed the problem of disposing CPS directly to the soil, leading to a reduction in fertility and increased acidic condition. Hence, currently CPSs are placed in a jute bag and moisture conditions are maintained to carry out natural fermentation to convert it to fertilizer. In India, cocoa production is reported to be 12,000 metric tonnes annually. About 700–750 kg of CPSs are generated as wastage per tonne of cocoa fruit; i.e. CPS represents between 70% and 75% of the entire weight of the cocoa fruit. During the year 2016–2017 a local chocolate produc-tion company, Campco Ltd., situated in the Dakshina Kannada district, Karnataka state, India, had utilized 3,400 MT of dry cocoa beans. The composition of CPS is quite variable, depending on its origin, geographic location and the way it has been processed [13].

5.2.3.3 Coconut Leaflets

The coconut palm (*Cocos nucifera L.*) belongs to the Palmaceae family. The culti-vation and processing of coconut results in the accumulation of a large quantity of by-products which are rich in lignin and cellulose (PCA 1979) and are available in surplus without any cost attached to them throughout the year. Mature coconut palm grows to a height of approximately 15–30 m. A crown consists of unbranched large 15–20 pinnate leaves on top of the trunk and the leaves originate from the apex of the stem with the tender leaves wrapped and folded together. The number of leaflets present in each leaf is between 200 and 250. The length of the smallest leaflets present at the tip is 25 cm. Coconut leaflets are composed of cellulose (23%) and lignin (38.68%). Coconut palms are cultivated in abundance in the coastal regions of tropical countries and about 100,000 km² has been cultivated in the world. Coconut palm generates surplus residues with only a small quantity used as fuel and the rest is burnt in the fields causing serious environmental problems.

5.3 BIOMATERIAL TO BIOETHANOL: FEASIBLE APPROACH

Biomaterials are processed and pretreated prior to the fermentation process. The pretreatment method should be decided based on the structure of the biomaterial. Starchy materials require mild pretreatment; however, lignocellulosic materials require harsh treatment. The pretreatment method should also be chosen in such a way so as to obtain maximum sugars and minimize the degradation of released sugars. The current section is focused on a feasible approach for the conversion of biomaterials to bioethanol (Figure 5.1).

5.3.1 PROCESSING OF BIOMATERIALS

Cleaning of biomaterials is the initial step to remove any dirt particle associated with the biomaterial. Prior to hydrolysis, the biomaterial has to be processed in order to remove the moisture and size reduction should be carried out to facilitate the interaction of biomaterial with suitable liquid during the pretreatment process. The

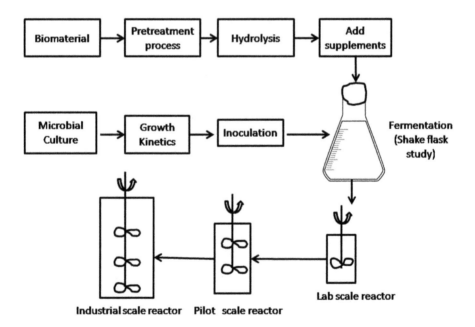

FIGURE 5.1   Conversion of biomaterials to bioethanol.

processing of non-edible seedcake, CPSs and coconut leaflets is discussed in the current section.

5.3.1.1 Non-edible Seedcake

Seedcake has to be powdered using a mixer-grinder. To remove the residual oil distilled water should be mixed to attain the ratio 1:2. The resulting mixture has to be heated for 45 minutes at 70°C and allowed to cool at room temperature. The separated oil accumulated on the top layer can be removed by skimming, followed by decantation and filtration using a muslin cloth. The sample should be dried at 90°C in a hot air oven for a stipulated period, until the moisture is removed. The drying process can be continued until a constant weight is attained. The dried cake should be ground into a fine powder using a suitable mixer-grinder and sieved using Taylor mesh number 10. The powder can be stored in airtight containers and kept in a deep freezer for further use [14].

5.3.1.2 CPSs and Coconut Leaflets

The biomaterial should be thoroughly washed using tap water to remove any sort of particulate matter adhering to the surface. Slice the material into uniform pieces, dry under sunlight and further oven dry at 90°C to remove moisture. The dried material should be ground to a fine powder and the coarse particles can be removed by sieving using Taylor mesh number 10. The powder can be stored in an airtight container and deep freezed for further use [13, 15].

5.3.2 Composition Analysis of Biomaterial

Biomaterials should be assessed for starch, cellulose, hemicellulose, lignin, moisture and ash content. Scanning electron microscopy/transmission electron microscopy with energy dispersive X-ray analysis will provide structural features and elemental composition of the biomaterial.

5.3.3 Pretreatment Process

5.3.3.1 Acid Hydrolysis

Concentrated or dilute acids are used to hydrolyse the hemicellulose and cellulose available in the lignocellulosic residues into sugars. Acid hydrolysis is a commonly used procedure, where a suitable acid can be chosen. During the acid hydrolysis process, lignocellulosic residues are treated with acid in the presence of H_2O to release sugars, i.e. hemicelluloses to pentose and cellulose to hexose. Currently, acid hydrolysis treatment processes consist of two stages, using heterogeneous acids and double acids. Usually, H_2SO_4 attacks biomass in a better way; hence the overall cost is less due to more sugar recovery and dilute acid hydrolysis is used for furfural production. The mode of application of acid is either by agitating the biomass with acid solution and heating the resulting slurry by steam, spraying, developing a bed system for mixing biomass.

For an effective hydrolysis, high pressure and temperature are required. Generation of lower degradation products has been revealed by dilute acid hydrolysis process.

A Bronsted acid characteristic is exhibited by HCl in order to hydrolyse the complex carbohydrates to simpler molecules via solvolytic heterogeneous or homogeneous reactions. Glycosidic linkages are cleaved during the hydrolysis to release the monomeric sugars. The lignocellulosic structure is disrupted by protons (H^+) due to the dissociation in the water. Acid media acting as catalyst causes protonation of O_2 atoms situated in the glycosidic bond. Protonation of hemicellulose event results in carbonium cation formation either at the oxygen atom present in the ring oxygen or at the glycosidic linkage. Further, the charged group leaves the polymer chain, allowing the hydroxyl group to replace the proton. Hence, degradation of polymer takes place to release sugars such as xylose, glucose and fragments of lignin. Acid hydrolysis is comparatively much faster than the enzymatic method. Monosaccharide yield of 90% was achieved using concentrated acid for pretreatment of biomass. Interestingly, organic acids such as oxalic acid, maleic acids and fumaric acid have been found to be effective in the hydrolysis of biomass, due to their greater potential of solution as opposed to mineral acids.

Dilute acid hydrolysis is a frequently adopted technique in biomass pretreatment. Due to the effectiveness in hydrolysis of hemicelluloses and the economic aspect of the process, dilute acid hydrolysis is preferred. Dilute sulphuric acid (H_2SO_4) is the most widely used. Phosphoric acid (H_3PO_4), hydrochloric acid (HCl) and nitric acid (HNO_3) are also used for hydrolysis. H_3PO_4 is a mild acid and is benign to the environment. Even acetyl groups from hemicelluloses contribute to the gradual increase of protons, further increasing the acidity and facilitating digestion. H_2SO_4-based hydrolysis was carried out using *P. pinnata* oil seed residue and from 1 g of the seedcake 0.034 g of glucose was released. Maximum glucose released was 30.7% w/v when 1 M HCl hydrolysis was carried out at 75°C for 4 h. To release glucose, 2%–6% w/w of acid concentration and 80°C–100°C temperature was chosen. Maximum glucose of 173.4 g per kg of seedcake was released with 2% HCl treatment. The concentration used for hydrolysis ranged from 0.25 to 1.25 M at two different temperatures of 75°C and 95°C.

5.3.3.2 Alkali Hydrolysis

During alkali hydrolysis, the residue is treated with ammonium, calcium, potassium and sodium hydroxide at atmospheric pressure and room temperature. The chemicals used in alkali hydrolysis are less caustic compared to acid-based hydrolysis. NaOH solubilized lignin and hemicelluloses drastically under suitable conditions. The acetyl groups in hemicelluloses and uronic acid groups are removed during alkali hydrolysis. Hemicellulose is joined to xylan by ester linkage, which is hydrolysed during alkali treatment process. At ambient temperature and pressure, hydrolysis takes a long time. Partial hemicellulose salvation, alteration of lignin structure, swelling of cellulose and partial cellulose decrystallization are caused due to alkali hydrolysis. The disadvantage of alkali treatment is the formation of irrecoverable salts or the integration of salts into the residues during the pretreatment, making alkali pretreatment a challenging issue. NaOH pretreatment preserves the carbohydrates but substantial reduction of lignin after 24 h of treatment at ambient temperature was revealed. Even though NaOH is expensive, due to its higher alkalinity, it is used widely for fractionating lignocellulosic residues. NaOH is an effective reagent acting on the linkage between hemicellulose and lignin in a dissociated condition such as Na^+ and OH^- ions. As the concentration of OH^- ion increases, the reaction rate of hydrolysis also increases. The alkali hydrolysis process primarily depends on the solubility performance of lignin in the alkali solution. KOH, $NaOH$, $Ca(OH)_2$, NH_4OH and Na_2CO_3 are most suitable for alkaline hydrolysis of lignocellulosic residues. Ester bonds present between the molecules which connect hemicelluloses-xylan and lignin to hemicelluloses undergo a saponification reaction. As the connecting bonds disappear, the pore structure of lignocellulose increases. The process can be carried out at atmospheric pressure and room temperature for seconds to days, and usually releases less sugar compared to acid pretreatment; hence, alkali hydrolysis is suitable for agricultural residues. CPH was hydrolysed using $NaOH$.

5.3.3.3 Microwave-Assisted Hydrolysis

Microwave irradiation exhibits high heating efficiency, by increasing the rate of reaction and in turn reducing the time of reaction. Frequency of the microwaves ranges from 300 MHz to 300 GHz. Microwave irradiation was first carried out using rice straw and this treatment altered the ultrastructure of lignocellulose. Microwave pretreatment is usually carried out at a high temperature (>160°C). Microwave radiation interacts with ions and polar molecules present in the material, resulting in both thermal and non-thermal effects that drive biological, chemical and physical reactions. Microwave-assisted (MA) alkali hydrolysis reduces the biomass structure and cellulose crystallinity and further improves the release of fermentable sugars, leading to degradation of sugars. The MA pretreatment process is an alternative methodology adopted to accelerate the hydrolysis of carbohydrate polymers using acids. The energy saving of microwave heating can be achieved up to 85 fold, in comparison to conventional heating by reducing chemical consumption and reaction time.

Previous reports have revealed that the release of glucose from starch increases with microwave irradiation treatment. The MA hydrolysis process is achieved by mixing biomass in an alkali solution and irradiating the solution for a stipulated time. Combination of MA treatment with either alkali or acid or combined alkali/acid, which is an alternative protocol for pretreatment of lignocellulosic residues, has been explored lately. Since the 1970s, with the arrival of commercial microwave ovens, it has been adopted due to high efficiency for pretreatment of lignocellulosic residues as a non-conventional way of heating. MA treatment of Miscanthus by $NaOH$ and H_2SO_4 yielded 12 times higher sugar compared to conventional processes.

5.3.3.4 Steam Explosion

Steam explosion is a widely used pretreatment method, taking place at a temperature range between 200°C and 280°C. The retention time may vary between 2 and 10 min. Longer residence time and lower temperature are more favourable. The mechanism involves a high-pressure steam, which penetrates through the fibres and is released from the closed pores. Besides, high pressure and temperature enhance the breakage of hydrogen bonds resulting in changes in the structure of cellulose. The hydroxyl generated contributes to lignin hydrolysis and hemicellulose transformation. Steam explosion pretreatment requires low energy and no environmental or recycling cost. For the pretreatment of agricultural residues and hardwoods, steam explosion is found to be a cost-effective method. Addition of H_2SO_4 during steam explosion decreases the reaction time and temperature and also decreases the generation of inhibitory compounds besides improving the rate of hydrolysis; it completely removes hemicelluloses. Biomass size, residence time and steam temperature are the factors that affect the efficiency of the pretreatment. Pretreatment of banana waste was carried out using autoclave-assisted alkali and acid hydrolysis. Autoclave-assisted acid and alkali hydrolysis is suitable for recovering maximum reducing sugars from lignocellulosic residues. Sugarcane bagasse was treated with dilute H_2SO_4 at 120°C for 30 min to release reducing sugar.

5.3.4 Estimation of Sugars

Sugars released during the hydrolysis process can be quantified by the 3,5-dinitrosalicylic acid (DNSA) or phenol sulphuric acid or high-performance liquid chromatography (HPLC) method. The DNSA method is suitable for determining reduced sugars. The phenol sulphuric acid method is adopted to determine the total sugar. The detailed profile of sugars released during the pretreatment process can be determined using HPLC by choosing a suitable column and mobile phase.

5.3.5 Microorganisms

5.3.5.1 Saccharomyces cerevisiae

Using *S. cerevisiae* for ethanol production is an ancient approach that has been around for thousands of years in the wine and beer industries. This species of yeast tolerates a wide range pH, making it less susceptible to contamination. Higher ethanol concentration inhibits the growth of yeast and it also has the ability to ferment only hexose sugar and not pentose. Even though the preferred condition for the growth is anaerobic, oxygen is essential for the synthesis of sterols and fatty acids. The preference of yeast for fermentation is due to its ability to produce ethanol up to 18% (w/v) and tolerance up to 150 g L^{-1} of ethanol.

5.3.5.2 Pichia stipitis

Pichia stipitis has the ability to convert glucose and xylose into ethanol. The hemicellulosic fraction releases xylose as the main sugar during hydrolysis and its conversion through the biological route is an important step to utilize lignocellulosic materials. *P. stipitis* exhibits the conversion of xylose present in the lignocellulosic hydrolysates into ethanol. Excess O_2 supply reduces the yield of ethanol, since *P. stipitis* is a respiration fermentative yeast. As per the studies reported in the literature, under anaerobic conditions, *P. stipitis* produces ethanol and these micro aerobic conditions are found to be optimum for the production of ethanol. *P. stipitis* has exhibited the ability to ferment a variety of sugars available in lignocellulose hydrolysate. It can ferment C5 sugars such as xylose and cellobiose.

5.3.5.3 Maintenance of Culture and Storage

The suitable microorganisms can be procured from the nearest culture bank. The culture needs to be revived prior to the inoculation. The medium components specified by the culture bank should be suspended in distilled water prior to autoclaving at 121°C for 15 minutes at 15 psi. The medium is cooled to approximately 50°C prior to pouring into sterile culture tubes and the slants should be prepared by allowing the molten agar to cool at room temperature for gelation by keeping the tubes in a reclining position. The microorganism should be streaked on the agar slants at the laminar airflow chamber and incubated at 30 ± 0.2°C for 24 hours in the incubator and should be maintained at 4°C. Pure cultures require sub-culturing at regular intervals.

5.3.5.4 Growth Kinetics

The slant culture of the microorganism to be revived to prepare the suspension culture is grown in a suitable liquid medium for 24 hours at 30°C. The side arm of the Erlenmeyer flask containing the media should be inoculated with the revived suspension culture of 2% (v/v) and the optical density of the culture needs to be measured preferably for a 30-min interval at 600 nm till the stationary phase is reached.

5.3.5.5 Inoculum Development

The suspension culture to be developed by growing yeast cells on media should be incubated at the appropriate temperature for 24 h by maintaining suitable pH and rpm. After incubation, 2% v/v of the culture needs to be transferred to the Erlenmeyer flask having a side arm containing the sterilized media. At 600 nm measure the optical density (OD) for a 30-min interval to monitor the log phase with reference to growth kinetics data. As the culture reaches the middle of the log phase, the inoculum should be transferred to hydrolysate of biomaterials incorporated with supplements.

5.3.6 BIOETHANOL PRODUCTION FROM BIOMATERIALS

Metabolic process is involved in converting sugars into alcohol and CO_2 during alcoholic fermentation. As per the stoichiometry, 1 mole of glucose (180 g) is converted into two moles of ethyl alcohol (92 g). The microorganisms involved in the ethanolic fermentation reported were Kluyueromyces yeast species (ferments lactose), Schwanniomycesalluvius (hydrolyses starch), *Pichia stipitis* (ferments xylose) and *Saccharomyces cerevisiae* (ferments mostly hexoses). The Embden-Meyerhof pathway is chosen by yeast to metabolize glucose into ethanol under anaerobic conditions. The hexose and pentose sugar released during hydrolysis from lignocellulosic residue is converted into bioethanol during fermentation.

5.3.6.1 Approach for Bioethanol Production and Mode of Operation

Three types of fermentation are generally adopted in the bioethanol production process. Separate hydrolysis and fermentation (SHF) is preferred for bioethanol production. Biomaterials should be hydrolysed prior to the fermentation step. Optimization of the hydrolysis and fermentation process is recommended to achieve maximum yield. Hydrolysis of biomaterials and fermentation can be carried out in separate reactors. Process optimization is the major advantage of this approach. However, there is a potential risk of contamination during the biomaterials hydrolysis stage. The simultaneous saccharification and fermentation (SSF) approach ensures that both the saccharification and the fermentation step can be carried out in a single reactor. As the sugar is released from biomaterials, it is subsequently converted into ethanol due to the metabolic activity of microorganisms. The concentration of sugar accumulation is reduced to avoid substrate inhibition. As the fermentation progresses, the concentration of ethanol is increased in the broth, preventing the risk of contamination. Thus, it is difficult to optimize the factors considering hydrolysis and fermentation at a time.

Simultaneous saccharification and co-fermentation (SSCF) adopts culturing of two different microorganisms having the ability to metabolize hexose and pentose sugar, respectively. Hexose metabolizing microorganisms grow relatively faster, leading to higher ethanol yield. An alternative option is to culture a single microorganism possessing the ability to metabolize both pentose and hexose sugar to produce a higher ethanol yield.

Fermentation can be carried out in batch, fed-batch and continuous mode of operations. During the batch process, a fixed concentration of substrate is maintained at the beginning of the process without any removal or addition. The batch mode of operation has the advantage of sterilization and improved control over the process as key advantages.

The substrate is fed intermittently to the fermenter during the fed-batch mode of operation on a commercial scale. In continuous mode of operation, the substrate is provided throughout the process.

5.3.6.2 Shake Flask Studies

Preliminary study on the bioprocess development is carried out in a shake flask. Microorganisms should be kept in the suspended state to avoid settling under the gravitational force and the media should be in continuous contact with cells from three dimensions to supply the nutrient. Kinetic study of the shake flask will provide rate, volume, order, yield, conversion of the fermentable sugar into bioethanol data as a foundation for scale-up.

Sugars released during hydrolysis of biomaterials serve as a carbon source for fermentation. Hydrolysate must be supplemented with peptone, yeast extract, ammonium sulphate, K_2HPO_4, manganese and magnesium sulphates or any other suitable supplements in an Erlenmeyer flask as per the requirement of the microorganism chosen for the fermentation. Process parameters such as temperature and pH that are essential for the chosen microorganism must be maintained. Hydrolysates along with supplements must be autoclaved at 121°C for 15 min. Inoculation should be done when OD at 600 nm reaches the log phase based on the growth kinetics data. At different intervals of time, product concentration has to be determined to understand the kinetics of the fermentation process.

5.3.6.3 Lab Scale Reactor

Based on the kinetics data of the shake flask studies, the production of ethanol should be carried out in a lab scale bioreactor. Mass transfer barrier in the process such as mixing, oxygen transfer rate (kLa), flow dynamics should be studied to determine possible non-ideality in the reactor.

Supplements should be incorporated into the hydrolysate and the reactor should be sterilized.

To monitor and control the process parameters, suitable probes should be installed prior to sterilization. The outlet vent should be connected with an autoclavable syringe filter. Operational procedures will depend on the lab scale reactor chosen. However, temperature, pH, agitation speed should be set in correlation with shake flask studies. As the inoculum is ready in the Erlenmeyer side arm flask, it should be shifted to laminar airflow. Preferably 10% v/v of inoculum can be transferred to a sterilized screw cap bottle having two vents. One of the vents is connected with a syringe filter and another with a silicon tube. The silicon tube should be folded at the end and wrapped with aluminium foil. The bottle is later connected to a peristaltic pump to transfer the inoculum into a bioreactor. Aluminium foil is removed from the silicon tube and the tip of the tube is dipped in 70% ethanol. The inlet vent of the bioreactor is to be wiped with 70% ethanol to maintain aseptic conditions. The tip of the silicon tube dipped in 70% ethanol is connected to the inlet port. The feed pump is operated using a control unit and the inoculum from the bottle is transferred to the bioreactor aseptically. Fermentation can be carried out for stipulated duration.

5.3.6.4 Scale-Up Studies

The main aim of scale-up is to transform the laboratory scale process to an industrial scale commercial process by developing a set of criteria. The constraints in a laboratory reactor/shake flask are minimal due to less interference of heat and mass transfer resistance during the reaction. Due to these factors the chemical or biochemical conversion in laboratory reactors is not reproduced when it comes to an industrial scale process. The problem in transforming the lab scale to an industrial scale process can be overcome by optimizing and modelling the process. Optimizing and modelling the bioethanol process requires a deep knowledge of physical and technical parameters in the production process. The impacts of each parameter that influence the bioethanol process should be investigated, for which mathematical solutions are not always available. In such situations, the designer must rely on both model experiments and the principles of model-to-full-scale plant similarity. This theory improves experiment planning and execution, as well as data analysis, to generate sufficient and accurate data on the full-scale plant size and process characteristics.

In the bioethanol production process, pretreatment is the most crucial step. The amount and quality of sugar released in this step determine the overall fermentation yield. Therefore, scale-up of pretreatment process plays a vital role. During scale-up of the pretreatment process, the wettability and mean residence time of biomass should be calculated and optimized. In enzymatic hydrolysis the recovery of enzyme is more vital to make the process economical. Increase in production volume makes recovery of the enzyme more challenging for a designer. In the fermentation process, more emphasis needs to be given to the fluid dynamic and agitation rate. These factors determine the possible non-ideality and prevent the formation of a substrate limiting zone inside the reactor.

The success of bioethanol production is primarily determined by the reactor scale-up process. Laboratory experiments carried out in the shake flasks/batch reactors provide critical information about reaction kinetics, fluid dynamics and transport barriers. This assists designers in avoiding non-ideality and replicating the lab process on a commercial scale.

5.3.7 Recovery and Purification of Bioethanol

Distillation and membrane separation methods can be adopted to recover the bioethanol. The estimation of ethanol content generated during fermentation should be carried out using a gas chromatography system.

5.4 CONCLUSION

Underutilized biomaterial available abundantly in nature has the potential to be channelized into bioethanol for attaining sustainability in the fuel sector. The pretreatment step should be chosen based on the composition and structure of the biomaterial. Releasing maximum sugar without generating fermentation inhibitors is crucial for deciding the best pretreatment method for the biomaterial. The microorganism for fermentation should be selected on the basis of hexose/pentose sugar released during the pretreatment. Shake flask studies set the base for the bioprocess developed in the lab scale reactor. Scale-up studies will drive the bioethanol production to an industrial scale.

REFERENCES

1. Dias De Oliveira, M.E., Vaughan, B.E., Rykiel, E.J.: Ethanol as fuel: Energy, carbon dioxide balances, and ecological footprint. *Bioscience.* 55(7), 593–602, 2005.
2. Nigam, P.S., Singh, A.: Production of liquid biofuels from renewable resources. *Progress in Energy and Combustion Science.* 37(1), 52–68, 2011.
3. Lee, S.K., Chou, H., Ham, T.S., Lee, T.S., Keasling, J.D.: Metabolic engineering of microorganisms for biofuels production: From bugs to synthetic biology to fuels. *Current Opinion in Biotechnology.* 19(6), 556–563, 2008.
4. Demirbas, A.: Combustion characteristics of different biomass fuels. *Progress in energy and Combustion Science.* 30(2), 219–230, 2004.
5. Mood, S.H., Golfeshan, A.H., Tabatabaei, M., Jouzani, G.S., Najafi, G.H., Gholami, M., Ardjmand, M.: Lignocellulosic biomass to bioethanol, a comprehensive review with a focus on pretreatment. *Renewable and Sustainable Energy Reviews.* 27, 77–93, 2013.
6. Achinas, S., Euverink, G.J.W.: Consolidated briefing of biochemical ethanol production from lignocellulosic biomass. *Electronic Journal of Biotechnology.* 23, 44–53, 2016.
7. Zabed, H., Sahu, J.N., Boyce, A.N., Faruq, G.: Fuel ethanol production from lignocellulosic biomass: An overview on feedstocks and technological approaches. *Renewable and Sustainable Energy Reviews.* 66, 751–774. 2016.
8. Araújo, K., Mahajan, D., Kerr, R., Silva, M.D.: Global biofuels at the crossroads: An overview of technical, policy, and investment complexities in the sustainability of biofuel development. *Agriculture.* 7, 32, 2017.
9. Ragauskas, A.J., Beckham, G.T., Biddy, M.J., Chandra, R., Chen, F., Davis, M.F., Davison, B.H., Dixon, R.A., Gilna, P., Keller, M., Langan, P.: Lignin valorization: Improving lignin processing in the biorefinery. *Science.* 344(6185), 1246843, 2014.
10. Reddy, N., Yang, Y.: Biofibers from agricultural byproducts for industrial applications. *Trends Biotechnol.* 23, 22–27, 2005.
11. Gressel, J., Zilberstein, A.: Let them eat (GM) straw. *Trends Biotechnol.* 21, 525–530, 2003.
12. Lim, S.K., Son, T.W., Lee, D.W., Park, B.K., Cho, K.M.: Novel regenerated cellulose fibers from rice straw. *J. Appl. Polym. Sci.* 82, 1705–1708, 2001.

13. Shet, V.B., sanil, N., Bhat, M., Naik, M., Mascarenhas, L.N., Goveas, L.C., Rao, C.V., Ujwal, P., Sandesh, K., Aparna, A.: Acid hydrolysis optimization of cocoa pod shell using response surface methodology approach toward ethanol production. *Agric. Nat. Resour.* 52, 581–587, 2018.

14. Sharmada, N., Punja, A., Shetty, S.S., Shet, V.B., Goveas, L.C., Rao, C.V.: Optimization of pre-treatment of de-oiled oil seedcake for release of reducing sugars by response surface methodology. *Bioethanol.* 2, 2016. https://doi.org/10.1515/bioeth-2016-0006

15. Shet, V.B., Varun, C., Aishwarya, U., Palan, A.M., Rao, S.U., Goveas, L.C., Raja, S., Rao, C.V., Puttur, U.: Optimization of reducing sugars production from agro-residue coconut leaflets using autoclave-assisted HCl hydrolysis with response surface methodology. *Agric. Nat. Resour.* 52, 280–284, 2018.

6 Biosensors
General Introduction, Working, and Their Applications

Shobhana Sharma

6.1 INTRODUCTION

Biosensors are simple measuring devices used to detect the presence of biological moieties [1]. The specificity, reliability, sensitivity, accuracy, and rapid detection factors help in the up-gradation of biosensors [2]. Other factors like a quick response and low cost are the main advantage of using biosensors compared to conventional methods [3]. Biosensors are essential in the field of water treatment [4], as well as for monitoring of water quality parameters like chemical oxygen demand (COD), biochemical oxygen demand (BOD), pH, heavy metals, bacteria, etc., in the field of environmental monitoring [5]; optical and thermal biosensors are generally used for the detection of organic matters [3]. Biosensors also find applications in various vital fields like analysis of food samples [3], investigation of drugs and medicines [3], detection of blood sugar levels [3], contamination of soil and water [3], etc. Biosensors possess different shapes and sizes. They can detect even low concentrations of toxic substances, enzymes, or pathogens and variable pH levels [6]. Some characteristic properties include range, selectivity, sensitivity, reproducibility, response time, etc. [3]. The research and development in biosensor technology are directly related to the field of medicine, the food industry, and environmental monitoring. Illustrating critical developments in biosensors provides a brief evolutionary history of biosensors. The categorization of biosensors into various classes depends on several factors. Recently, the combinatorial approach followed led to advancements in biosensor technology [7]. The merging of biosensor technology with nanotechnology, fabrication, and molecular biology has contributed toward immense developments in this field.

6.2 CONFIGURATION

Biosensors follow an electro-enzymatic approach which involves the biological reorganization of an analyte by the bioreceptor. The biosensor consists of five components [8], as shown in Figure 6.1.

DOI: 10.1201/9781032636368-6

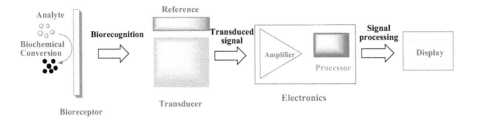

FIGURE 6.1   The five components of a biosensor [8].

6.2.1 ANALYTE

A specific substance or target identified by a biosensor.

6.2.2 BIORECEPTOR

A bioreceptor is a bio-recognizing molecule that detects the target sample and generates a signal.

6.2.3 TRANSDUCER

It transforms the generated signal into a measurable response.

6.2.4 ELECTRONICS

It contains an amplifier and a processor. The amplifier amplifies the electric signals generated by the transducer. A processor or microcontroller converts these analog signals into digital.

6.2.5 DISPLAY

It provides output in an understandable form.

6.3 WORKING

The output of the transducer obtained is current, or voltage, depending on the type of transducer present in the biosensor. If the result is current, it should be converted into voltage first. The signal processing unit or the signal conditioning unit does the amplification and filtration of the signal. The analog signal is converted into a digital signal by the microcontroller. The working of a biosensor is illustrated in Figure 6.2.

6.4 EVOLUTIONARY HISTORY OF BIOSENSORS

Based on the attachment of the components in the biosensor, their evolutionary history can be divided into three generations [9], as seen in Figure 6.3.

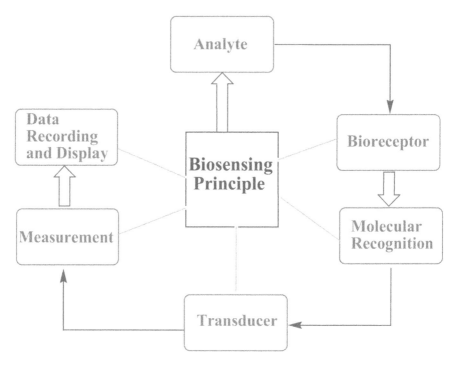

FIGURE 6.2   The illustration of the working of a biosensor.

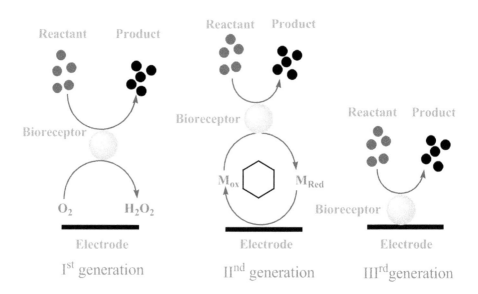

FIGURE 6.3   The three generations of evolutionary history of biosensors [9].

First-generation biosensors: This class includes mediator-less amperometric biosensors in which the product of the reaction directly diffuses into the transducer and generates an electric signal. This type of biosensor depends on oxidase and dehydrogenase enzymes. Cremer et al. observed a proportional relationship between the concentration of an acid in a liquid and the rise of potential between fluids present on opposite sides of a glass membrane [10]. Sorenson founded the concept of pH [11]. In 1916, Griffin et al. reported the immobilization of invertase enzyme on $Al(OH)_3$ and charcoal [12, 13]. The electrodes used for measuring pH have been investigated by Hughes [14]. The first report of Leland Charles Clark Jr. (father of biosensors) discussed the components of biosensors [15]. The report was published in 1956, in which he demonstrated the oxygen electrode. In 1962, he established the working of an amperometric enzyme electrode used for glucose analysis [16]. Updike et al. observed the first functional enzymatic electrode based on glucose oxidase [17]. The first potentiometric enzyme electrode-based biosensor was reported by Guilbault et al. [18]. Bergveld invented the field-effect transistor (FET) sensitized for ions [19]. Guilbault et al. investigated the detection of hydrogen peroxide from a platinum electrode by an enzyme-based biosensor [20]. Mindt et al. reported the first lactate biosensor [21]. This biosensor replaced Cytochrome B2 with the artificial redox mediator ferricyanide. Mosbach et al. prepared the enzyme thermistor [22]. In 1974, biosensors contained a thermal transducer. The fiber-optic biosensor for detecting carbon dioxide and oxygen was introduced by Lubbers et al. [23]. Divie developed a microbial sensor used to measure alcohol [24]. In 1975, Scheller et al. prepared an amperometric glucose sensor. Yellow Spring Instrument (YSI) [25, 26] reported that it was the first commercial biosensor to detect glucose. Suzuki et al. fabricated the first microbe-based immunosensor [27].

Second-generation biosensors: These belong to specific 'mediators' between the reaction and the transducer to generate improved signals. Second-generation biosensors use artificial, partially toxic mediators or nanomaterials as oxidizing agents and electron carriers to transport the electrons to the electrode. Clemens et al. invented the first bedside artificial pancreas [28]. In 1977, Scheller combined enzymatic reactions with sensors. This mediator-free bioelectrocatalysis provides the foundation for the third generation. Goldstein et al. [29] reported the first fiber-optic pH sensor to detect in vivo blood gases. Later on, the fiber-optic biosensor was demonstrated by Schultz et al. [30] to detect glucose. Liedberg et al. [31] reported the surface plasmon resonance (SPR) immunosensor. Roederer et al. [32] developed the first immunosensor based on piezoelectric detection.

Third-generation biosensors depend on bioelectrocatalysis, which involves direct electron transfer from the enzyme to the electrode. The response is caused by only the reaction, and no product or mediator is required for the response.

Estimating glucose by glucose oxidase and ferrocene Cass et al. [33] generated the first mediated amperometric biosensor. In 1990, the first BIAcore was commercially launched based on SPR [34]. In 1992, a lightweight, easy-to-use, portable iSTAT blood analyzer was introduced, which was based on advanced technology. It also provides a point of care testing platform for health professionals [35]. In 1999, Poncharal et al. reported the first nanobiosensor equipped with immobilized bioreceptor probes [36]. Later on, these nanosensors revolutionized the diagnosis of diseases. Biosensor technology has become a research area that connects the elementary idea of science

TABLE 6.1
Evolutionary history of biosensor technology

S.no.	Year	Reported by	Invention made	References
1	1906	Cremer et al.	Study electric potential arising between parts of the fluid	[9]
2	1909	Sorensen	Give the concept of pH and pH scale	[10]
3	1916	Griffin et al.	Immobilization of invertase enzyme on activated charcoal	[11, 12]
4	1922	Hughes	First glass pH electrode	[13]
5	1956	Clark et al.	Invention of the oxygen electrode (Clark)	[14]
6	1962	Clark et al.	Discuss an amperometric enzyme electrode used for glucose detection	[15]
7	1967	Updike et al.	Provide first functional enzyme electrode	[16]
8	1969	Guilbault et al.	Potentiometric enzyme electrode-based sensor used for the detection of urea	[17]
9	1970	Bergveld	Ion-sensitive field-effect transistor (ISFET)	[18]
10	1973	Guilbault et al.	Introduce glucose and a lactate enzyme sensor used for hydrogen peroxide detection	[19]
11	1973	Mindt et al.	Develop first lactate sensor	[20]
12	1974	Mosbach et al.	Enzyme thermistor	[21]
13	1975	Lubbers et al.	Fiber-optic biosensor used for carbon dioxide and oxygen detection	[22]
14	1975	Divie	Microbial sensor for alcohol measurement	[23]
15	1975	YSI	Commercial biosensor for glucose detection	[24, 25]
16	1975	Suzuki et al.	Microbe-based immunosensor	[26]
17	1976	Clemens et al.	Bedside artificial pancreas	[27]
18	1980	Goldstein et al.	Fiber-optic pH sensor used in in vivo blood gases	[28]
19	1982	Schultz	Fiber-optic biosensor used in glucose detection	[29]
20	1983	Liedberg et al.	Observed SPR immunosensor	[30]
21	1983	Roederer et al.	Develop the first immunosensor-based on piezoelectric detection	[31]
22	1984	Cass et al.	First mediated amperometric biosensor	[32]
23	1990	–	Launch of the Pharmacia BIACore SPR-based biosensor	[33]
24	1992	STAT	iSTAT launches hand-held blood analyzer	[34]
25	1999	Poncharal et al.	Demonstrated the first nanobiosensor	[35]

with nanotechnology and micro-technology. Nowadays, biosensors do not need mediators and are free from labels and interference and possess hypersensitivity. The period from 1906 to 1999 provides an evolutionary historical background of biosensors, summarized in Table 6.1.

6.5 CHARACTERISTICS OF BIOSENSORS

Some characteristic properties are incorporated into biosensors for the development of highly effective commercial biosensor systems [3].

6.5.1 OPERATIONAL RANGE

The measurement of concentration over which the biosensor exhibits a response provides the functional capacity of a biosensor. The variation in a biosensor's operating range depends on the biosensor's application.

6.5.2 SELECTIVITY

Selectivity is the most crucial feature of a biosensor. The biosensor's selectivity is the ability to detect a specific target in the sample. Selectivity indicates the binding capacity of the receptor with the particular target from the mixture of other biomolecules.

6.5.3 RESPONSE TIME

The response time varies for different biosensors. A rapid response time of biosensors indicates their high level of sensitivity. Response time is the time needed to obtain the maximum result.

6.5.4 REPEATABILITY OR REPRODUCIBILITY

The biosensor can generate the same results on measuring identical samples many more times. Reproducibility provides high reliability to the biosensor. The transducer and electronic decide the precision and accuracy of the biosensor.

6.5.5 SENSITIVITY

The sensitivity of the biosensor is the correct detection of the presence of a trace amount of analyte in the sample.

6.5.6 LINEARITY

The linearity indicates the resolution power of the biosensor or accuracy of measuring response. Resolution is the minor change in analyte concentration, which requires bringing a change in response. Linearity means a linear change in response to a change in analyte concentration in a biosensor.

6.5.7 STABILITY

Stability is the susceptibility scale for the environmental interruption in and around the biosensor device. Many factors like temperature, the affinity of the bioreceptor, and degradation of bioreceptor influence stability.

6.6 CLASSIFICATION OF BIOSENSORS

It is a diverse and multifaceted field. Several criteria are involved in the category of biosensors [37], as outlined in Scheme 6.1. The type of biosensors depends on

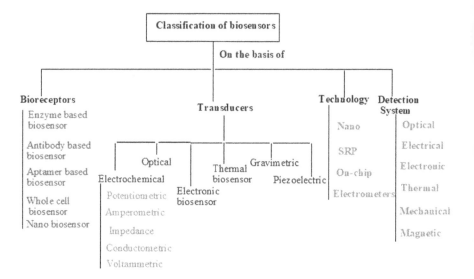

SCHEME 6.1   Classification of biosensors on the basis of various criteria [37].

various factors like bioreceptors, transducers, detection systems, and technology [37].

6.6.1 Classification Based on Bioreceptor

6.6.1.1 Enzyme-Based Biosensor

An enzyme-based biosensor is an analytical device in which enzymes recognize or sense elements. The enzymes used in this type of biosensor are catalytic and specific.

6.6.1.2 Antibody-Based Biosensor

In this type of biosensor, antigen-antibody interaction occurs in the form of a measurable physical signal. These biosensors are employed in bioprocess analysis and other clinical diagnoses.

6.6.1.3 Aptamer-Based Biosensor

This type of biosensor produces an electrochemical signal in the presence of a specific microbial target. The signal is measurable by the variation of current flow through the electrode. These biosensors are highly specific for certain microbial strains and nucleic acids.

6.6.1.4 Microbial or Whole-Cell Biosensor

These analytical devices contain microorganisms that analyze analytes and generate signals for the significant biochemical, physiological, and electrical changes.

6.6.1.5 Nanobiosensor

Nanobiosensors bind bioanalytics like antibodies, nucleic acids, pathogens, etc., resulting in physiological change.

6.6.2 Classification Based on Transducer

The broad classification is as follows:

6.6.2.1 Electrochemical Biosensors

This category of biosensors possesses an electrochemical transducer. These biosensors analyze biological materials like tissues, enzymes, etc., and nonbiological matters. The electrochemical biosensor is further categorized into various types as discussed below:

6.6.2.1.1 Potentiometric Biosensors

This type of biosensor depends on the potential difference between the reference and the working electrode. The biosensor's highly stable and accurate reference electrode provides more sensitivity and selectivity to the biosensor.

6.6.2.1.2 Amperometric Biosensors

These biosensors measure current flow between electrodes during the redox reaction, like in a glucometer.

6.6.2.1.3 Impedimetric Biosensors

This biosensor is applied to detect broad chemical and physical properties.

6.6.2.1.4 Conductometric Biosensors

A conductometric biosensor does not require any reference electrodes. These are operated at low-amplitude alternating voltage. These biosensors use electroactive species for analysis. The working of the biosensor is based on measuring an analyte's preferable and specific conductance.

6.6.2.1.5 Voltammetric Biosensors

Such biosensors measure changes in both current and potential differences. The density of the peak current provides the concentration of the analyte and the value of the peak current is used for the identification of the analyte.

6.6.2.2 Electronic Biosensors

The working of an electronic biosensor based on FETs regulates the flowing current in the biosensor by using an electric field. The surface potential keeps changing due to the binding of a biomolecule with a bioreceptor. This change in surface potential alters the flow of current. FET-based biosensors have high sensitivity and high spatial resolution.

6.6.2.2.1 Thermal Biosensors

A thermal biosensor measures the amount of heat energy absorbed and released during the biochemical reactions by which one can estimate the extent of the response.

6.6.2.2.2 Optical Biosensors

These biosensors follow the principle of electro-chemiluminescence. Such biosensors possess optical transducer systems to generate a signal that provides the analyte concentration.

6.6.2.2.3 Mass-Based or Gravimetric Biosensors

These biosensors respond to changes in mass. They employ quartz crystal microbalance.

6.6.2.2.4 Acoustic or Piezoelectric Biosensors

These work based on the change in acoustic or sound vibrations which the detected substance can cause. The mechanical force applied to the piezoelectric biosensor generates the electric signal.

6.6.3 CLASSIFICATION BASED ON TECHNOLOGY

The following categories broadly divide biosensors:

6.6.3.1 Nanobiosensors

This category of biosensor includes nanomaterial-based biosensors. There are different types of nanomaterial-based biosensors.

- Nanoparticle-based biosensors
- Nanorods-based biosensors
- Nanowire-based biosensors
- Quantum dots-based biosensors
- Carbon nanotube-based biosensors
- Dendrimers-based biosensors

6.6.3.2 Surface Plasmon Resonance

Due to molecular interaction at the metallic surface with surface plasmon waves, a change in refractive index occurs, which is detected by the SPR biosensor. Thus, the SPR biosensor depends on refractive index variation, which varies due to the binding of the analyte with the biorecognition element.

6.6.3.3 Biosensors-on-Chip (Lab-on-Chip)

This biosensor provides qualitative and quantitative information about the target analyte. These biosensors possess such design that physical, chemical, and biological change converts into a measurable signal. Lab-on-chip biosensors are used in the field of diagnosis.

6.6.3.4 Electrometers

This category comprises advanced biosensor technology like hand-held units, disposable strips, etc. It enables rapid blood testing. These biosensors are portable, compact, lightweight, and easy to use.

6.6.4 CLASSIFICATION BASED ON DETECTION SYSTEM

This type of classification is as follows:

6.6.4.1 Optical

Some of the optical-based biosensors used are (a) fluorescence-based optical biosensors, (b) chemiluminescence-based optical biosensors, (c) SPR-based optical biosensors, and (d) optical fiber-based optical biosensors.

6.6.4.2 Electrical

Electrical detection technologies provide advanced biosensor features like sensitivity, miniature size, portability, etc.

6.6.4.3 Electronic

Electronic biosensors are based on FETs. The FET is a three-terminal device that regulates the flow of current with the help of an electric field. The binding of a biomolecule with a bioreceptor surface leads to changes in surface potential, which results in variation in the inflowing current.

6.6.4.4 Thermal

These biosensors are often associated with temperature detection. This method requires no labeling of reactants.

6.6.4.5 Mechanical

Mechanical biosensors follow mass change due to chemical reaction and adsorption. These biosensors contain microcantilevers that convert the molecular identification of biomolecules into a nanomechanical motion.

6.6.4.6 Magnetic

Such biosensors lead to changes in magnetic properties like coil inductance, magneto-optical properties, resistance, etc. The biological interactions are detectable due to changes in these magnetic properties. These biosensors possess paramagnetic or superparamagnetic particles which detect physical interaction with the help of changes in magnetic properties.

6.7 APPLICATIONS OF BIOSENSORS

Biosensors are more sensitive and stable than traditional methods. The biosensor is essential in all those fields which require rapid and reliable analysis. Thus, the biosensor is applied in the environment, for clinical diagnosis, etc. [38]. Some essential biosensor applications are discussed below and are shown in Figure 6.4.

6.7.1 CLINICAL DIAGNOSIS

Biosensors are used by diabetic patients for monitoring blood sugar levels on a routine basis [38]. They provide an easy self-testing monitoring system to detect blood glucose levels in undiluted blood samples. In pathological laboratories and bedside glucose monitors, biosensors that estimate blood glucose levels are applied. The analytical and nucleic acid biosensors provide a powerful diagnostic platform for analyzing various infectious diseases [39]. The biosensor is employed in clinical cancer testing, pregnancy tests, urinary tract infection, etc. [39].

6.7.2 SCREENING OF ENVIRONMENT

Paper-based electrochemical biosensors provide a fast and straightforward method for determining heavy metal ions [40]. Biosensors possess a unique analytical system for monitoring environmental pollutants. Biosensors play an essential role in analyzing the type and concentration of contaminants present in the environment. Microbial biosensors are analytical devices used for sensing substances present in the environment. The different microorganisms for various pollutants, e.g., heavy metals, organic pollutants, pesticides, etc., are used. Biosensors provide a quick evaluation of water contamination [41].

6.7.3 FOOD SAFETY

Biosensors play a vital role in the food industry by monitoring food safety and nutritional quality [6, 43]. The biosensor detects nutrients, antinutrients, natural toxins, etc. Biosensors analyze nutrients, biological toxins, and antinutrients for monitoring food processing. Biosensors are used in the food industry to monitor food processing and analyze genetically modified organisms. Biosensors are portable, easy to use, and do not require any professional skills to operate. Biosensors also detect the level of vitamin B complex, antibiotics, proteins, etc., present in various food materials by enzymatic reactions [42, 44]. Enzymes-based biosensors test food quality to measure amines, amides, cofactors, alcohols, carbohydrates, etc. Biosensors also detect the pathogenic organisms found in meat, poultry, egg, etc.

6.7.4 AGRICULTURAL MONITORING

The extent of pesticides, heavy metal ions, herbicides, and their residual content in soil and water is evaluated by biosensors. Biosensors analyze the possible occurrence of soil diseases [45]. Immunosensors play an essential role in estimating crop diseases based on antigen-antibody interaction. Acoustic-based biosensors help detect plant pathogens that cause diseases in plants. Electrochemical biosensors are used in agronomy, as well as in soil chemistry. Biosensors investigate chemical pollutants, microbes present in soft drinks, food-borne pathogens, etc., in the food industry. Aptamer biosensors help to detect the presence of ochratoxin A in beer and coffee samples [46]. Aflatoxins in agricultural products are detected and quantified by a fluorometric biosensor.

6.7.5 PATHOGEN DETECTION

Electrochemical biosensors contain conducting and semiconducting materials as the transducer or electrode for detecting pathogens in food, soil, and water [47]. Immunosensors detect pathogenic bacteria in various fields like environmental screening, analysis of food, and clinical diagnosis. Amperometric biosensors help to study pathogens like *S. aureus* in food samples like milk, meat, and cheese. Optical biosensors can detect the presence of *E. coli* in the sample [48]. Biosensors are necessary for detecting animal, poultry, and dairy pathogens. They are used for detecting mycotoxins produced by a family of fungi generally present in agricultural products. Biosensors are also helpful in detecting and removing contaminants present in food material.

6.7.6 DRUG DELIVERY

Patients can monitor their health daily with the help of portable biosensors, which can detect specific biomarkers [49]. Various routes of obtaining biomarkers like breath, saliva, urine, etc., indicate various illnesses. Biosensors play an essential role in drug analysis by quantifying active components in drug formulation, deterioration products, and the metabolites present in biological fluids. The presence of a low analyte concentration in saliva requires highly sensitive biosensors. The presence of the particular volatile organic compound in exhaled breath indicates specific pathophysiological conditions [49].

6.7.7 TOXICOLOGY

Biosensors possess a thin DNA film, and pure metabolic enzyme is playing a promising role in predicting genotoxicity. Biosensors indicate the risk of using heavy metal ions present in the recycled materials found in the environment. Biosensors also serve as a potential rapid tool for finding the effectiveness of remediation practices in contaminated soil [50].

6.7.8 FERMENTATION PROCESSES

The simple instrumentation and selective nature of biosensors lead to precise controls in the fermentation industry [51]. Commercial biosensors are pretty helpful for the estimation of biochemical parameters. The factories employ glucose biosensors to control fermentation and saccharification processes.

6.7.9 BIODEFENSE

The sensitive and selective nature of biosensors helps identify biowarfare agents, toxins, bacteria, and viruses. Biosensors are essential for military purposes during biological attacks [52].

6.7.10 BIOSENSORS IN PLANT BIOLOGY

Biosensors act as a selective, sensitive, and rapid tool for disease diagnosis in tissue engineering applications. New technologies like DNA sequencing and molecular imaging are helpful in the advancement of plant science. The biosensors collect information on subcellular and cellular localization and measure metabolite levels and ion tapped [52]. Biosensors measure the dynamic processes under physiological conditions.

6.8 CHALLENGES IN BIOSENSING RESEARCH

Biosensors possess some challenges like detection time, detection limit, and specificity. While designing biosensor systems, detection time poses significant challenges, like predicting a suitable technology while maintaining specificity and sensitivity [52]. Specificity is the ability of a biosensor to differentiate between targeted and

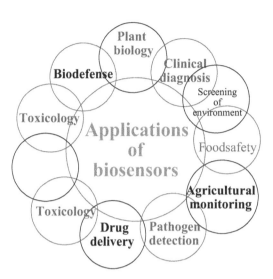

FIGURE 6.4   Various applications of biosensors [38].

non-targeted biological analytes present in a sample. Specificity is considered an essential quality of a biosensor. The detection time becomes necessary when the quality control test needs to be fast, further reducing the cost of storage and minimizing the product's chances of decay [52]. Attempts are being made to minimize biosensors' detection limit while maintaining sensitivity and reliability.

6.9 CONCLUSION

In the present era, biosensor technology is in its infant stage. The critical features of biosensors are sensitivity, stability, reproducibility, and cost. There are endless possibilities to enhance the potential of biosensors with a price reduction. Novel biosensors must monitor and prevent crops from infectious diseases caused by pathogens that control harvest and post-harvest losses, improve productivity, and lead to agriculture sustainability. Quick online search and accurate sensing also open up opportunities for biosensors in environmental monitoring. Diagnostic methods should be simple, sensitive, and easily detectable by low concentrations of multiple biomarkers in biological fluids. Biosensors work as a frontline diagnostic tool in the medicinal field. The commercialization of biosensors in the food industry indicates their massive potential. Salmonella has mainly caused food-borne infectious diseases. Thus, effective identification and monitoring of Salmonella biosensors are essential in the food industry. Biosensors play a crucial role in providing the forewarnings of the transmission of diseases by detecting the presence of toxins and bacteria in samples.

6.10 FUTURE PROSPECTIVE

It is a big challenge to develop more reliable and improved biosensor devices to avoid instrumental drift. Thus, a rigorous study is required to present future trends in the biosensor field. The incorporation of nanomaterials in biosensors leads to advancements in biosensor technology. Dynamic and multifaceted biosensors development can be achieved by combining sensing technology with nanomaterials. By incorporating nanomaterials sensitivity & selectivity, rapid detection, and reproducibility can be enhanced in biosensors. Efforts were made to develop user-friendly and inexpensive nanomaterial-based biosensors. These biosensors furnish fast and precise results. In the case of developing nations, climate-smart agriculture constitutes the central part of their economy. The blended fabrication and biosensor technology approach fulfills future biosensing applications like high performance and real-time monitoring of physiological events. Such biosensors provide diverse applications in climate-smart agriculture required for a sustainable environment. In the future, attempts will have to be made to develop a high-potential biosensor for the medicinal field to improve the nation's health. Incorporating molecular biology with biosensors improves biocomponent stability, which provides highly reproducible results in the medical field.

REFERENCES

1. Rocchitta, G., Spanu, A., Babudieri, S., Latte, G., Madeddu, G., Galleri, G., Nuvoli, S., Bagella, P., Demartis, M. I., Fiore, V., Manetti, R., Serra, P. A. Enzyme Biosensors for Biomedical Applications: Strategies for Safeguarding Analytical Performances in Biological Fluids. *Sensors*, 16(6), 780, 2016.
2. Zucolotto, V. Specialty Grand Challenges in Biosensors. *Front. Sens.*, 2020. https://doi.org/10.3389/fsens.2020.00003settings.
3. Naresh, V., Lee, N. A Review on Biosensors and Recent Development of Nanostructured Materials-Enabled Biosensors. *Sensors*, 21(4), 1109, 2021. https://doi.org/10.3390/s21041109.
4. Mao, K., Zhang, H., Pan, Y., Yang, Z. Biosensors for Wastewater-Based Epidemiology for Monitoring Public Health. *Water Res.*, 191, 116787, 2021.
5. Zhou, T., Han, H., Liu, P., Xiong, J., Tian, F., Li, X. Microbial Fuels Cell-Based Biosensor for Toxicity Detection: A Review. *Sensors*, 17, 2230, 2017. https://doi.org/10.3390/s17102230.
6. Thakur, M. S., Ragavan, K. V. Biosensors in Food Processing. *J. Food. Sci. Technol.*, 50(4), 625, 2013. https://doi.org/10.1007/s13197-012-0783-z.
7. Nagraik, R., Sharma, A., Kumar, D., Mukherjee, S., Sen, F., Kumar, A. P. Amalgamation of Biosensors and Nanotechnology in Disease Diagnosis: Mini-Review. *Sensors Int.*, 2, 100089, 2021. https://doi.org/10.1016/j.sintl.2021.100089.
8. Bhalla, N., Jolly, P., Formisano, N., Estrela, P. Introduction to Biosensors. *Essays Biochem.*, 60(1), 1, 2016. https://doi.org/10.1042/EBC20150001.
9. Fernández, H., Arévalo, F. J., Granero, A., Robledo, S. N., Diaz, C., Riberi, I., Zon, M. A. Electrochemical Biosensors for the Determination of Toxic Substances Related to Food Safety Developed in South America: Mycotoxins and Herbicides. *Chemosens.*, 5(3), 23, 2017. https://doi.org/10.3390/chemosensors5030023.
10. Cremer, M. Über die Ursache der elektromotorischen Eigenschaften der Gewebe, zugleich ein Beitrag zur Lehre von den polyphasischen Elektrolytketten. *Z. Biol.*, 47, 562, 1906. (In German).
11. Sörensen, S. P. L. Enzyme Studies. 2nd Report. On the Measurement and the Importance of Hydrogen Ion Concentration during Enzymatic Processes. *Biochem. Z.*, 21, 131, 1909.

12. Griffin, E. G., Nelson, J. M. The Influence of Certain Substances on the Activity of Invertase. *J. Am. Chem. Soc.*, 38, 722, 1916.

13. Nelson, J. M., Griffin, E. G. Adsorption of Invertase. *J. Am. Chem. Soc.*, 38, 1109, 1916.

14. Hughes, W. S. The Potential Difference between Glass and Electrolytes in Contact with the Glass. *J. Am. Chem. Soc.*, 44, 2860, 1922.

15. Heineman, W. R., Jensen, W. B. Clark Jr., L. C. (1918–2005). *Biosens. Bioelectron.*, 21, 1403, 2006.

16. Clark, L. C., Lyons, C. Electrode Systems for Continuous Monitoring in Cardiovascular Surgery. *Ann. N. Y. Acad. Sci.*, 102, 29, 1962.

17. Updike, S. J., Hicks, G. P. The Enzyme Electrode. *Nature*, 214, 986, 1967.

18. Guilbault, G. G., Montalvo, J. G., Jr. Urea-Specific Enzyme Electrode. *J. Am. Chem. Soc.*, 91, 2164, 1969.

19. Bergveld, P. Development of an Ion-Sensitive Solid-State Device for Neurophysiological Measurements. *IEEE Trans. Biomed. Eng.*, 17, 70, 1970.

20. Guilbault, G. G., Lubrano, G. J. An Enzyme Electrode for the Amperometric Determination of Glucose. *Anal. Chim. Acta*, 64, 439, 1973.

21. Mindt, W., Racine, P., Schlaepfer, P. Sensoren fur Lactat und Glucose. *Ber. Busenges. Phys. Chem.*, 77, 804, 1973.

22. Mosbach, K., Danielsson, B. An Enzyme Thermistor. *Biochim. Biophys. Acta.*, 364, 140, 1974.

23. Lübbers, D. W., Opitz, N. The pCO_2-/pO_2-Optode: A New Probe for Measurement of pCO_2 or pO in Fluids and Gases (authors transl). *Z. Naturforsch C Biosci.*, 30, 532, 1975.

24. Divis C. Notes on Ethanol Oxidation by a Microbial Electrode *Acetobacter zylinum*. *Ann Microbiol.*, 126(A), 175, 1975.

25. Newman, J. D., Turner, A. P. F. Home Blood Glucose Biosensors: A Commercial Perspective. *Biosens. Bioelectron.*, 20, 2435, 2005.

26. D'Orazio, P. Biosensors in Clinical Chemistry. *Clin. Chim. Acta*, 334, 41, 2003.

27. Suzuki, S., Takahashi, F., Satoh, I., Sonobe, N. Ethanol and Lactic Acid Sensors Using Electrodes Coated with Dehydrogenase–Collagen Membranes. *Bull. Chem. Soc. Jap.*, 48, 3246, 1975.

28. Clemens, A. H., Chang, P. H., Myers, R. W. Le développement d'un système automatique d'infusion d'insuline controle par la glycemie, son système de dosage du glucose et ses algorithmes de controle. *J. Ann. Diabétol. Hotel Dieu*, 269, 1976.

29. Goldstein, S. R., Peterson, J. I., Fitzgerald, R. V. A Miniature Fiber Optic pH Sensor for Physiological Use. *J. Biomech. Eng.*, 102, 141, 1980.

30. Schultz, J. S. Oxygen Sensor of Plasma Constituents. U.S. Patent No. 4,344,438A, 17, 1982.

31. Liedberg, W., Nylander, C., Lundstrm, I. Surface Plasmon Resonance for Gas Detection and Biosensing. *Sens. Actuators A Phys.*, 4, 299, 1983.

32. Roederer, J. E., Bastiaans, G. J. Microgravimetric Immunoassay with Piezoelectric Crystals. *Anal. Chem.*, 55, 2333, 1983.

33. Cass, A. E., Davis, G., Francis, G. D., Hill, H. A., Aston, W. J., Higgins, I. J., Plotkin, E. V., Scott, L. D., Turner, A. P. Ferrocene-Mediated Enzyme Electrode for Amperometric Determination of Glucose. *Anal. Chem.*, 56, 667, 1984.

34. Liedberg, B., Nylander, C., Lundstrom, I. Biosensing with Surface Plasmon Resonance - How It All Started. *Biosens. Bioelectron.*, 10, 1995.

35. Vestergaard, M. C., Kerman, K., Hsing, I. M., Tamiya, E., editors. *Nanobiosensors and Nanobioanalyses*. Tokyo: Springer; 2015.

36. Poncharal, P., Wang, Z. L., Ugarte, D., De Heer, W. A. Electrostatic Deflections and Electromechanical Resonances of Carbon Nanotubes. *Science*, 283, 1513, 1999.

37. Curulli, A. Electrochemical Biosensors in Food Safety: Challenges and Perspectives. *Molecules*, 26(10), 2940. https://doi.org/10.3390/molecules26102940.
38. Monošík, R., Streďanský, M., Šturdíka, E. Biosensors — Classification, Characterization and New Trends. *Acta Chimica Slovaca*, 5, 109, 2012. https://doi.org/10.2478/v10188-012-0017-z.
39. Sin, M. L.Y., Mach, K. E., Wong, P. K., Liao, J. C. Advances and Challenges in Biosensor-Based Diagnosis of Infectious Diseases. *Expert Rev. Mol. Diagn.*, 14(2), 225, 2014. 10.1586/14737159.2014.888313.
40. Colozza, N., Caratelli, V., Moscone, D., Arduini, F. Paper-Based Devices as New Smart Analytical Tools for Sustainable Detection of Environmental Pollutants. *Case Stud. Chem. Env. Eng.*, 4, 100167, 2021.
41. Vogrinc, D., Vodovnik, M., Logar, R. M. Microbial Biosensors for Environmental Monitoring. *Acta agriculturae Slovenica*, 106(2), 67–75, 2015. https://doi.org/10.14720/aas.2015.106.2.1.
42. Shams, J., Singh, S., Ashraf, M., Manzoor, A., Dar, H. Application of Biosensors in Food Quality Control. *J. Postharvest Technol.*, 08(1), 53–74, 2020.
43. Shah, A., Chauhan, O. P. Applications of Bio-Sensors in Agri-Food Industry and Mitigation of Bio-Threat Paradigm. *2nd International Symposium on Physics and Technology of Sensors (ISPTS)*, 2015, 10.1109/ISPTS.2015.7220156.
44. Pagkali, V., Petrou, P., Salapatas, A., Makarona, E. Detection of Ochratoxin A in Beer Samples with a Label-Free Monolithically Integrated Optoelectronic Biosensor. *J. Haz. Mater.*, 323, 75–83, 2017. 10.1016/j.jhazmat.2016.03.019.
45. Cesewski, E., Johnson, B. N. Electrochemical Biosensors for Pathogen Detection. *Biosens Bioelectron.*, 1, 159.112214, 2020. 10.1016/j.bios.2020.112214.
46. Massad-Ivanir, N., Shtenberg, G., Sega, E. Optical Detection of *E. coli* Bacteria by Mesoporous Silicon Biosensors. *J. Vis. Exp.*, (81), 50805, 2013. https://doi.org/10.3791/50805.
47. Ngoepe, M., Choonara, Y. E., Tyagi, C., Tomar, L. K., Toit, L. C., Kumar, V. M. K., Ndesendo, P., Pillay, V. Integration of Biosensors and Drug Delivery Technologies for Early Detection and Chronic Management of Illness. *Sensors*, 13(6), 7680, 2013. https://doi.org/10.3390/s130607680.
48. Corbisier, P., Thiry, E., Diels, L. Bacterial Biosensors for the Toxicity Assessment of Solid Wastes. *Environ. Toxicol. Water Qual.*, 11(3), 171, 1996.
49. Yan, C., Dong, F., Chun-yuan, B., Si-rong, Z., Jian-guo, S. Recent Progress of Commercially Available Biosensors in China and Their Applications in Fermentation Processes. *J. Northeast Agric. Univ.*, 21, 73, 2014.
50. Mehrotra, P. Biosensors and Their Applications – A Review. *J. Oral. Biol. Craniofac. Res.*, 6(2), 153, 2016.
51. Okumoto, S. Quantitative Imaging Using Genetically Encoded Sensors for Small Molecules in Plants. *Plant J.*, 70, 108, 2012.
52. Varshney, M., Mallikarjunan, K. Challenges in Biosensor Development: Detection Limit, Detection Time, and Specificity. *Resour.: Eng. Technol. Sustain. World.*, 16(7), 18, 2009.

7 Biomaterials in Electrochemical Applications

Jose A. Lara-Ramos, Fiderman Machuca-Martinez, Jennyfer Diaz-Angulo, Edgar Mosquera-Vargas and Jesús E. Diosa

7.1 INTRODUCTION

Biomaterials are compounds that can be obtained directly from nature or synthesized in laboratories, the latter employing a considerable number of chemical methods that use metallic components, polymers, ceramics, or composite materials (Prado da Silva 2016; Zore et al. 2021; Li et al. 2022). They are often used and adapted in medical applications and, for this reason, augment or replace a natural function by being all or part of a structure of a living being or a biomedical device that performs a biological function of an organism (Zore et al. 2021; Song et al. 2022). In the last century, the study of biomaterials focused only on physically replacing the soft or hard tissue of a recipient host that had been damaged or destroyed through disease or accidental processes (Wojcik et al. 2022). However, currently, there are different areas where biomaterials are of interest including chemistry, biology, nanoscience, engineering, environment, medicine (one of the most prominent areas in applications), and electrochemistry (Chae et al. 2018; Wang, Xu, and Lei 2023).

A number of investigations and documents in the literature present very important advances in recent years (Hedayat, Du, and Ilkhani 2017; Chae et al. 2018; Zore et al. 2021; Nik Md Noordin Kahar et al. 2022). This progress and growth have been motivated by scientific advances in new materials and their analysis techniques (Jeyaraman and Slaughter 2021; Li et al. 2022). Many of the analysis tools feature electronic and electrical components that must interact with the morphology and composition of new materials. In addition to the above, the global concern for more environmentally friendly processes, devices, and procedures means that electrochemistry plays a leading role in this regard (Rusling et al. 2014; Pandey and Jain 2021). In the specific case of electrochemistry, there are different fields of expansion where biomaterials are very important in transforming processes, making them simpler, more viable, stable, and economically sound for various sectors of industry and commerce (Rusling et al. 2014; Jeyaraman and Slaughter 2021).

Electrochemical processes are often the fundamental components for operating different devices such as sensors, capacitors, batteries, dialysis equipment, solar

DOI: 10.1201/9781032636368-7

panels, and waste treatment reactors, among others (Prado da Silva 2016; Chae et al. 2018; Kiran and Ramakrishna 2021; Taghvaie Nakhjiri et al. 2022). For all the above, this chapter focuses on the main applications of biomaterials in electrochemistry as this document's area of interest.

7.1.1 FUNDAMENTALS

7.1.1.1 Biomaterials

Biomaterials are elements that are used to replace, support, or complement a biological system or tissue (Nik Md Noordin Kahar et al. 2022). The study of biomaterials in science is approximately 50 years old. The branch of science that studies biomaterials is called biomaterials science or biomaterial engineering (Song et al. 2022). It is currently experiencing constant and strong growth in the development of new products (Azmi et al. 2018; Pandey and Jain 2021) (see Figure 7.1). Biomaterials science encompasses some applications in medicine, biology, chemistry, engineering, electrochemistry, metallurgy, and materials science (Czerwińska-Główka and Krukiewicz 2020; Li et al. 2022; Taghvaie Nakhjiri et al. 2022).

The use of devices or pieces of biomaterial in living beings can be called implants; the most common examples of biomaterials are pacemakers, heart valves, encapsulating drugs, bone grafts, or dental pieces (Pandey and Jain 2021; Li et al. 2022). These biomaterials can be natural or made from polymers, ceramics, plastics, metals, and composites (Salgado et al. 2011; Wojcik et al. 2022). In the scheme of Figure 7.1, a synthesis of the type of biomaterials and their functions is shown. It is important to note that a biomaterial differs from a biological material, such as bone; a biological organism generates that. The most commonly used materials are made of metals (Pandey and Jain 2021). These materials must be biocompatible and must not affect the biological system. Furthermore, care must be taken when defining a biomaterial as biocompatible, as it is specific to its application. A biocompatible biomaterial in one application may not be biocompatible in another application (Song et al. 2022).

Meanwhile, biomaterials are being used in different applications, such as materials science (Rusling et al. 2014), electrochemistry (Taghvaie Nakhjiri et al. 2022), the environmental sciences (Taghvaie Nakhjiri et al. 2022), sensors (Pandey and Jain 2021), piezoelectricity (Chae et al. 2018), fuel cells (Pandey and Jain 2021), among others (Rusling et al. 2014); each of them is associated with some type of particular application. It is important to mention that a fundamental property of biomaterials is that they are biocompatible and are accepted by the body of a living being, because not all substances or materials can be considered a biomaterial but with the property of being biocompatible (Song et al. 2022).

7.1.1.2 Desired Properties of Biomaterials

Biomaterials cover different knowledge areas, as seen in Figure 7.3. In the specific case of the design of a certain device, for example, a screw for a bone requires knowledge from multiple disciplines (Salgado et al. 2011). On the other hand, the design of a simple biomaterial requires the harmonic integration of materials, medicine, biology, chemistry, and mechanical sciences (Prado da Silva 2016; Song et al. 2022). The World Health Organization reports an estimated 1.5 million medical devices in

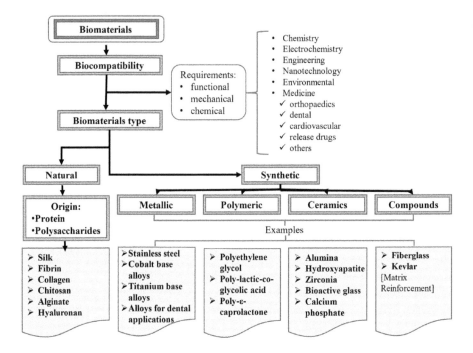

FIGURE 7.1   Types and functional roles of biomaterials for applications.

humans, with the availability of 10,000 standardized devices worldwide (Kiran and Ramakrishna 2021). It considers this type of data and the importance of biomaterials in medicine to improve, treat, and replace organ tissues or body functions (Salgado et al. 2011; Pandey and Jain 2021). It is necessary to consider the properties of the nature of the system to be replaced, device material, and physiological specifications (Li et al. 2022; Wojcik et al. 2022). For all the above, Table 7.1 presents the most important properties that must be considered in medical applications of biomaterials.

7.1.1.3 Electrochemistry

Electrochemical processes basically consist of a chemical reaction generated by applying an electric current (Kapałka, Fóti, and Comninellis 2010). Therefore, the electric current is produced as a result of electron transfer from the atoms or molecules involved in the reaction (oxidation-reduction reactions) (Sandoval, Calzadilla, and Salazar 2022). Various fields of application of electrochemical processes include energy storage, batteries, electrolysis, electrochemical deposition, and water treatment (McBeath et al. 2021). In an electrochemical system, a cathode and anode (electrodes) are used in the specific case of the use of electrochemical processes in water treatment (Kapałka, Fóti, and Comninellis 2010). The molecules or atoms in the solution that approach the anode are oxidized in other species (direct oxidation) or can be oxidized in bulk (indirect oxidation). For example, a hydroxyl radical is an oxidizing species that is generated at the anode through electrical discharges in a water molecule (Rabaaoui et al. 2013). This is why the electrochemical processes

TABLE 7.1

Desirable properties in the medical applications of biomaterials (Prado da Silva 2016; Kiran and Ramakrishna 2021)

Property	Key elements
Biocompatibility	Biocompatibility is the ability of a living organism to accept a synthetic or natural material in its body (Li et al. 2022).
Host response	Host response is defined as the "response of the host organism to the implanted material or device" (Kiran and Ramakrishna 2021). The success of an implant in a living being depends on the reaction of the tissue with the implant (they are rarely inert). In most cases, the reaction of the biomaterial and the host tissue depends on the implantation duration, purpose, and site.
Non-toxicity	A designed biomaterial must fulfill its purpose without negatively influencing the organism of the living being. Therefore, a biomaterial should not release or generate anything foreign. Finally, the toxicity of biomaterials allows the detection of adverse elements in the receiving host organism (Kiran and Ramakrishna 2021; Song et al. 2022).
Mechanical properties	The mechanical properties are part of a biomaterial design before the device implementation and distribution. Biomaterials are subjected to several tests, including deformation, tension, shear, and a combination of the latter. Furthermore, corrosion, tensile strength, elastic modulus, and hardness are among the most important properties of biomaterials, which must be carefully analyzed before use. In the case of hard tissue applications, mechanical properties are a priority.
Corrosion, wear, and fatigue properties	One of the main causes of implant failure is wear. In addition, wear can also accelerate the corrosion process in biomedical implants. In the case of fatigue resistance, the biomaterial must withstand repeated cyclic loads.
Design and manufacturability	Design is a critical factor in biomaterials. For example, in the last century, plates of bone material were used to repair broken long bones (by their primitive design, they broke). On the other hand, manufacturability is manufacturing an item under the requirement of intended use with high reliability and low cost. Regarding biomaterials, it is expected that the material will not deteriorate with sterilization techniques, dry heat, radiation, etc. In short, a biomaterial must: • Be biocompatible, that is, non-toxic, non-allergenic, among others. • Be easy to process and stable in durability over the stipulated period of use. • Be sterilizable, cost-effective, and accessible. • Have the required/appropriate physical, chemical, and mechanical properties.

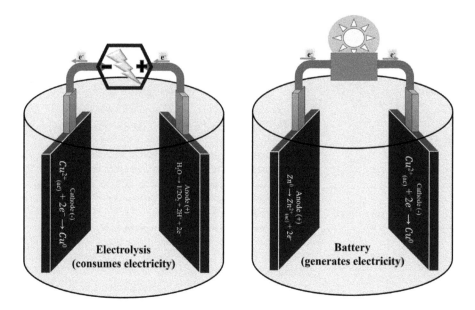

FIGURE 7.2   Scheme of an electrolysis cell (left) and a galvanostatic cell (right).

used in water treatment are part of the advanced oxidation processes (Comninellis and Chen 2010; Kapałka, Fóti, and Comninellis 2010).

The redox reactions at the electrode surfaces are spontaneous, i.e., without an external energy supply, energy is produced (electric current), and the system is called a battery (also called a galvanic or voltaic cell). On the other hand, if external electricity must be applied to induce the redox reaction, we have an electrolysis process (Figure 7.2). The oxidation-reduction reactions occur spontaneously on the surface of the electrodes, meaning an external energy supply is unnecessary. Instead, energy (electric current) is produced, and the system is also called a galvanic or voltaic cell. If an external electrical current has to be applied to induce the oxidation-reduction reaction, then we have an electrolysis process (Figure 7.2).

The electrode material is critical to achieving adequate performance and selectivity in electrochemical processes. Although the type of material has an important influence on the kinetics of reaction, migration, and transfer of electrons, these parameters often define the feasibility of an electrochemical transformation. Previously, the selection of electrodes used to be empirical since the mechanisms and processes in the electrodes were unknown. This chapter is intended to highlight the recent advances and studies related to biomaterials in which usage of the materials is justified and the prospects of future developments in electrodes and applications of biomaterials in electrochemistry are discussed. Some important aspects of the applications of biomaterials are:

- Electrochemistry is widely used to measure various analytes (neurotransmitters, large biomolecules, and biological gases) in many different biological areas.

- Due to the ability to use a wide range of techniques and sensors, measurements can be performed in vivo, ex vivo, or in vitro.

7.1.1.3.1 Electrochemical Cells

At the laboratory or industrial level, it is usual to work with a collection of interfaces that involve an electrochemical cell. These cells usually consist of two electrodes (anode and cathode) separated by at least one electrolytic phase (Comninellis and Chen 2010; Barhoumi et al. 2017). Electrochemical cells in which an oxidation-reduction reaction occurs and an electrolysis current intervenes can be classified as galvanic or electrolytic (Comninellis and Chen 2010), as illustrated in Figure 7.2.

- In a galvanic cell, the electrochemical reaction occurs spontaneously, and the chemical energy is transformed into electrical energy (Comninellis and Chen 2010). This energy transformation is because one of the analytes is a reducer (an electron donor), while the other analyte is an oxidant (an electron acceptor), as can be seen in Figure 7.2.
- In an electrolytic cell, the electrochemical reaction occurs by the external imposition of a potential difference greater than the reversible potential of the cell (Bruguera-Casamada et al. 2017). Additionally, it is possible to impose an electric electrolysis current that allows the electrochemical transformation of the substances in the medium (see Figure 7.2) (Comninellis and Chen 2010; Abdessamad et al. 2013).

7.2 APPLICATION AREA AND NUMBER OF PUBLICATIONS

Biomaterials and their applications are of great interest to researchers in different areas of science. Therefore, publications on biomaterials have increased considerably since 2002 (Figure 7.3a). Works on biomaterials are continually being published; due to the large number of publications, it is necessary to synthesize the research achievements, focusing on the key aspects of the application of biomaterials. For example, the heart repair app found in the Li et al. (2022) application focused on the recovery of resources, and environmental sustainability through the recovery of precious metals from industrial wastewater can be found in the article by Taghvaie Nakhjiri et al. (2022). Rusling et al. (2014) present a detailed description of biomaterials in electrochemical arrays for protein detection.

On the other hand, Pandey and Jain (2021) focus on introducing electrochemical sensors from polymer-based biomaterials. The fields of application of the most current biomaterials and the largest number of publications discuss material sciences, nanotechnology, chemical engineering, electrochemistry, chemistry, among others. That is, the application of biomaterials in electrochemistry is highly relevant, with 1,541 publications to date, being the fourth field of interest based on the Scopus database (Figure 7.3b). Finally, readers interested in the detailed aspects of biomaterials and their applications should review the literature as mentioned in earlier reports.

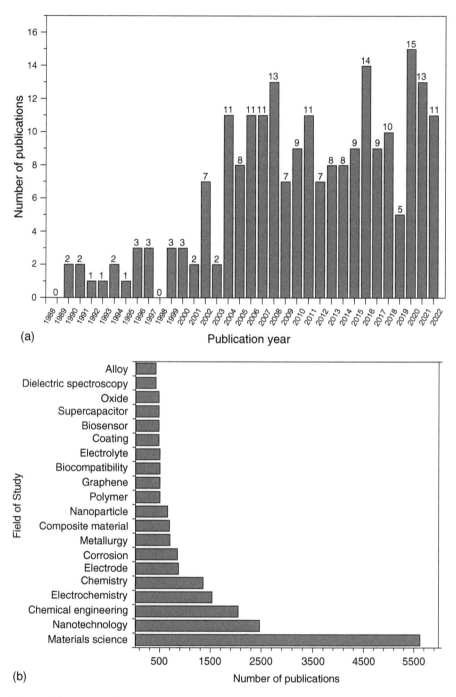

Figure 7.3 (a) Number of publications containing the phrase "Biomaterials in electrochemical applications" published in the literature between 1988 and 2022, (b) the fields of study in which biomaterials research has been addressed (* until 7th August 2022) according to the Scopus database.

7.3 BIOMATERIALS USED IN ELECTROCHEMISTRY

7.3.1 Medical Field

Biomaterials are substances that can perform a biological function, including treating, improving, or replacing damaged tissues, organs, or bodily functions. Ceramics, alloys, conductive polymers, and composites are examples of biomaterials used in different medical applications such as implants. On the other hand, electrochemistry can be applied to biomaterials in two differentiated ways: (i) synthesis and improvement of biomaterials and (ii) biosensors.

7.3.1.1 Synthesis and Improvements of Biomaterials

Biomaterials are used to replace and repair diseased or damaged parts such as joints, bones, and teeth, whence they are continuously in contact with tissues and must exhibit biocompatible, mechanical, and non-toxic characteristics (Salgado et al. 2011). Biomaterials based on metallic compounds and alloys such as stainless steel, titanium, and magnesium alloys are widely used as implants. However, these must be coated to improve the surfaces' bioactive character and prevent corrosion (Singh, Singh, and Bala 2018). Polymeric, ceramic, and composite biomaterials are also susceptible to the attack of colonies of microorganisms that trigger the material's corrosion and possible infection of the user, so the manufacture of antimicrobial biomaterials is necessary. In addition, electrochemical methods are used to improve these materials' biocompatibility for medical applications.

Myocardial infarction is one of the fatal diseases in humans. Its incidence is steadily increasing every year around the world. The problem is accompanied by the limited regenerative capacity of cardiomyocytes, leading to fibrous scar tissue formation (Phung et al. 2016; Li et al. 2022). The propagation of electrical impulses in such tissue is severely hampered, negatively influencing the heart's normal pumping function. Therefore, reconstruction of the internal cardiac electrical connection is currently one of the main concerns of myocardial repair. Conductive biomaterials with or without cellular charge have been extensively investigated to address this issue (Li et al. 2022). In Table 7.2, as we can see, conductive biomaterial is used for cardiac repair.

7.3.1.2 Synthesis of Conductive Polymers

Electrochemical synthesis to obtain conductive polymers dates back to 1968. During an experiment, an aqueous pyrrole and sulfuric acid solution was exposed to an oxidative potential, forming a "pyrrole black" precipitate on a platinum electrode (Guimard, Gomez, and Schmidt 2007). Electrochemical synthesis is currently carried out using a three-electrode system (working, counter, and reference electrodes) immersed in a solution of the monomer, the appropriate solvent, and an electrolyte (dopant) (Filimon 2016).

Some advantages of this method are the possible thin film synthesis, ease of synthesis, entrapment of molecules in conductive polymers, and simultaneous doping. However, at the same time, it presents disadvantages such as difficulty in removing the film from the electrode surface; also, post-covalent modification of bulk conductive polymers is difficult (Pandey and Jain 2021).

TABLE 7.2
Conductive biomaterials used in cardiac repair[a]

Form	Compositions	Methods	Cell type	Improvements	Conductivity
Conductive film	PPy/PCL	Polymerization	HL-1	Adhesion, calcium wave propagation speed	1.2×10^{-3} S cm^{-1}
	PPy/PGS/collagen	Mold casting	H9c2	Adhesion, proliferation	6.0×10^{-2} S cm^{-1}
	PPy/acid-modified silk	Deposition	hiPSC-CMs	Sarcomere and gap junctions, maturation	~1.0 S cm^{-1}
	PPy/CS	Polymerization	Neonatal Rat CardioMyocytes (NRCMs)	Maturation, synchronization, vascularization	~1.0×10^{-2} S/cm
	PANi/PLA	Electrospinning	H9c2	Cell viability, maturation, synchronization	3.1×10^{-5} S cm^{-1}
	PANi/PLGA	Electrospinning	NRCMs	Maturation, synchronization	3.1×10^{-3} S cm^{-1}
	GO/CS	Lyophilization	H9c2	Adhesion, proliferation	1.34×10^{-3} S cm^{-1}
	AuNPs/PU	Deposition	H9c2	Adhesion, proliferation, cell morphology and alignment	—
Conductive hydrogel	PPy/ECM-derived pre-cardioge	Lyophilization	NMCMs	Adhesion, migration, maturation, synchronization	2.3×10^{-2} S cm^{-1}
	poly(3,4-ethylene dioxythiophene) (PEDOT):PSS/collagen-alginate	Self-cross-linking by blending	hiPSC-CMs	Maturation, beating properties	3.5×10^{-3} S cm^{-1}
	PAMB/gelatin	Polymerization	NRCMs	Synchronization, electrical impulse propagation	1.9×10^{-4} S cm^{-1}

	CNTs/pericardial matrix	Lyophilization	hiPSC-CMs	Maturation, proliferation contraction amplitude	1.5×10^{-2} S cm^{-1}
	AuNPs/CS	Self-cross-linking by blending	hMSCs	Metabolism, migration, proliferation	1.3×10^{-3} S cm^{-1}
Other conductive construct	PPy/PGS-nanofibrillated cellulose	3D printing	H9c2	Proliferation, adhesion, cell viability	3.4×10^{-2} S cm^{-1}
	CNTf	Wet spinning	In vivo epicardial scar	Conduction velocity	5.8×10^{4} S cm^{-1}
	CNTs/PCL/silk fibroin	Weaving	NRCMs	Maturation, orientation, anisotropy	8.1×10^{-7} S cm^{-1}

DOPA: dopamine, PAMB: poly-3-amino-4-methoxybenzoic acid, hiPSC-CMs: human induced pluripotent stem cell-derived cardiomyocytes, NMCMs: neonatal mouse cardiomyocytes.

[a] Taken and edited (Li et al. 2022).

7.3.1.3 Electrochemical Deposition

Electrochemical deposition is a versatile and commercial technique for coating biomedical implants and offers several advantages such as (i) the need for a simple apparatus, (ii) short processing time, (iii) few restrictions on substrate shape, and (iv) room temperature processing. Electrochemical deposition consists of two stages: electrophoretic and electrolytic (Harun et al. 2018; Singh, Singh, and Bala 2018).

The electrophoresis process can produce impregnated ceramic particles toward a porous substrate and composite consolidations, while the electrolytic process gives rise to the formation of metal salts from solutions. Electrochemical deposition has been extensively employed for coating a titanium substrate since it can form a homogeneous coating layer, enhancing the adhesion strength between the coating layer and the implant surface (Sankar et al. 2017).

7.3.1.4 Electrodischarge Coating Process

Electrical discharge coating is a method used to produce coatings from tool electrodes in a dielectric fluid onto a selected workpiece. This method utilizes the principle of sparking by applying electrical discharge machining (Gordin et al. 2005). The performance of coatings produced by electrical discharge depends on the fine-scale microstructure and elemental distribution within the layer. The electrical discharge coating process involves is a complex mechanism of localized melting, on the magnitude of an isolated discharge, of the workpiece together with substance removal, substance set down from the tool electrode, and intermingling within the near-surface of the work object (Kumar, Roy, and Dey 2021).

7.3.1.5 Anodizing

The anodizing process is based on applying a current between the material to be treated (anode), for example, titanium, and a non-reactive/inert material (cathode), which can be platinum or graphite, which must previously be immersed in a conductive solution. The anodizing process is complex and depends on several parameters such as applied current density, concentration, and composition of the electrolyte, the electrolyte temperature, duration of the process, the agitation speed of the solution during treatment, and the cathode-to-anode surface area ratios. In addition, an electron flux between both materials is generated in the anodizing process, and the oxide film is grown on the anode (titanium surface) because of oxidation reactions (Murray et al. 2017).

7.3.1.6 Biosensors

Biosensors are devices that use a recognition system based on the coupling of a biological entity (bionanomaterials) and a physical transducer (electrode substrate) to measure the interaction between the analyte of interest and the device that can measure the electrical signal (Soloducho and Cabaj 2015). Biosensors are of great interest and a promising technology due to several advantages such as low cost, high sensitivity, the possibility of simultaneous analysis of single and multiple biomarkers, high reusability, and compactness. Bio-devices have applications in clinical diagnostics, bioprocess monitoring, environmental monitoring, etc. (Filimon 2016). Figure 7.4

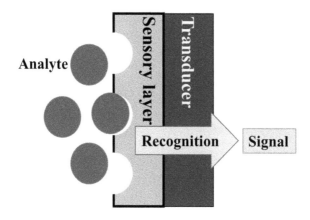

FIGURE 7.4 Schematic of an electrochemical biosensor.

shows a schematic of a biosensor, a device composed of two elements: a bioreceptor that recognizes the analyte and a transducer used to convert a biochemical signal into an electronic one.

Biosensors can be classified according to the biological molecules used and according to the signal transmitter. They can also be divided into clinical and non-clinical. Depending on the natural molecules used, they can be catalytic-based or affinity-based. In the catalytic-based biosensor, the analyte is transformed, while the affinity-based one binds directly to the analyte. Depending on the signal transmitter, the biosensor can also be defined as electrochemical, optical, or piezoelectric. Clinical biosensors can be in vivo (long-term implantable or short-term invasive) or in vitro (single-shot or multi-analysis), while non-clinical biosensors can be single-assay, reactive monitoring, or bioagent detection (Soloducho and Cabaj 2015).

The electrochemical reactions that give rise to the mechanisms of action of organic compounds such as drugs and phenomena such as stereochemistry, diffusion, solubility, permeability, and electron transfer allow information on biological data to be obtained from correlations (Chae et al. 2018).

7.3.1.7 Electrochemical Nucleic Acid Biosensors

Electrochemical nucleic acid biosensors are among the most studied and applied due to their high portability, biocompatibility, easy access, and use (Tan et al. 2016). They have been used to detect several pathogens, such as *Escherichia coli*, *Cryptosporidium parvum*, *Salmonella choleraesuis*, and *Staphylococcus saprophyticus* (Soloducho and Cabaj 2015). In addition, nucleic acid-based biosensors have the advantage that they can measure metal ions, organic molecules such as drugs, pesticides, or glucose, and even proteins thanks to their multiple negative charges (Onur Uygun, Deniz Ertuğrul Uygun, and Girgin Sağin 2021).

In electrochemical nucleic acid biosensors, a biochemical reaction on the transducer generates an electrical signal due to the interaction between the biorecognition agent and the analyte of interest (Tan et al. 2016; Sharma et al. 2017). Thus, biosensors can be classified according to the working principle of the biomolecule or the working principle of the transducer. Different techniques can be used depending on the biochemical reaction. Amperometry measurements can be performed when an electron is generated after a biochemical reaction, while if a molecule is generated, potentiometric measurements can detect it. On the other hand, if there is a direct affinity, it can be measured by impedimetric/capacitive. Nucleic acids are considered ideal transistors due to their structure. They show conductive nanocable characteristics and their use is very efficient in electrochemical biosensors (Onur Uygun, Deniz Ertuğrul Uygun, and Girgin Sağin 2021).

7.3.1.8 Electrochemical Biosensors in the Pharmaceutical Industry

Biosensors can be used to analyze pharmaceutical compounds quantitatively, and their main application is in the analysis of drugs and the interaction and kinetics of the drug and the biocomponent. An adequate pharmacological analysis of compounds is fundamental to controlling social health. A careful watch must be maintained over some drugs because of their secondary effects. Therefore, many techniques such as chromatography and fluorometric and spectrophotometric methods are used. However, these are usually labored. So, electrochemical methods have been used to determine pharmaceutical compounds (Gil and de Melo 2010).

Acyclovir is an example of a drug with apparent absence of secondary effects, so it becomes an interesting drug. Therefore, various electrochemical methods have been employed to determine its properties, including electrocatalytic oxidation on copper nanoparticles-modified carbon paste electrodes (Soloducho and Cabaj 2015), electrochemical technique coupled with carbon nanotube (CNT) modified electrodes (Shah, Lafleur, and Chen 2013), and fullerene-C60-modified glassy carbon by cyclic and differential pulse voltammetry. Furthermore, the study of neurotransmitters and neuronal medicines have also benefited from electrochemistry because the coupling of voltammetry and microelectrodes altered with neurotransmitter receptors serves to qualify the physiopathological process and physiological aims (Shetti, Malode, and Nandibewoor 2012).

7.3.2 Environmental Field

7.3.2.1 Online Monitoring of Wastewater Quality

Biomaterials in wastewater quality monitoring offer a great opportunity to take advantage of biological resources (Zhang and Angelidaki 2012; Kim and Han 2013; Do et al. 2020). Their design is simple, cost-effective, and compact. Therefore, the application of biosensors based on microbial fuel cells for real-time wastewater monitoring is very innovative (Do et al. 2020). This section presents the central axes of the applications of biomaterials in electrochemicals for wastewater quality management. The fuel cell design has two presentations: single and double cells. The single-chamber cell is composed of an anode chamber and an air cathode (Zore et al. 2021). In contrast, the double-chamber cell is composed of two chambers, one for the anode and the other for the cathode, which is separated by a proton exchange membrane. At the anode, microbes oxidize the substrates that are added, generating electrons and protons. H^+ and other cations pass to the cathode through migration and combine with an electron acceptor, such as oxygen, to generate water (Do et al.

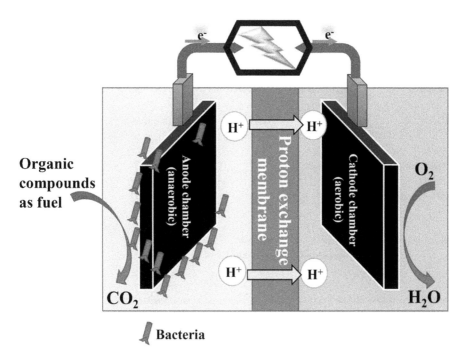

FIGURE 7.5 Diagram of the mechanism for a basic microbial fuel cell system.

2020; Tanvir et al. 2021). Figure 7.5 describes a biosensor's design and basic components based on a fuel cell.

Biochemical Oxygen Demand (BOD) is one of the most important parameters when evaluating water quality (Do et al. 2020). Conventional methods for measuring BOD are universally accepted and inexpensive. However, the response time requires five to seven days to get a result, which complicates online monitoring (Peixoto et al. 2011). Therefore, fuel cell-based biosensors are a promising alternative for online and real-time measurement of BOD. Meanwhile, the detection of toxicity is very important for the quality inspection of drinking and wastewater. Now, toxic substances have an inhibitory effect on an organism's metabolism (especially electron transfer). This means that toxic substances can decrease the electrical current compared to water without toxic substances. Therefore, biomaterials such as microbial fuel cell-based biosensors can be used to detect toxic compounds in water. On the other hand, it is important to mention other biosensor applications based on microbial fuel cells (Do et al. 2020), which are:

- Microbial fuel cells-based biosensor for monitoring volatile fatty acids
- Monitoring of microbial activities
- Monitoring of corrosive biofilms

Finally, microbial fuel cells as biosensors have outperformed traditional biosensors since the latter requires an external power supply and expensive devices for their

TABLE 7.3

Latest publications on biosensors based on microbial fuel cell

Parameter measured	Type of microbial fuel cell	Type of wastewater	Measuring range	Ref.
BOD	Submersible microbial fuel cell	Domestic wastewater	17–183 mg O_2L^{-1}	Peixoto et al. (2011)
Quantification of *E. coli*	One chamber		–	Kim and Han (2013)
Microbial activity			0–13 nmol L^{-1}	Zhang and Angelidaki (2012)
Dissolved oxygen	Submersible Two chamber	Domestic wastewater	0–8.8 mg L^{-1}	Vishwanathan, Rao, and Siva Sankara Sai (2013)
Volatile fatty acid	H type		0–40 mg L^{-1}	Kaur et al. (2013)
Anaerobic digestion process	Wall-jet type	Artificial wastewater	–	Liu et al. (2011)
Assimilable organic carbon	Two chambers	Sea water	0–75 mg L^{-1}	Quek, Cheng, and Cord-Ruwisch (2015)

operation. This means that biosensors from microbial fuel cells are cheap, simple, and capable of meeting your requirements remotely (Do et al. 2020). Table 7.3 shows the comparative advances of microbial fuel cells as biosensors in recent years.

7.3.2.2 Metals Recovery

Plants have antioxidant properties and reducing capacities that have been reported in the literature (Taghvaie Nakhjiri et al. 2022). In addition, some of the natural phytochemicals that plants have are phenolic compounds that can act as protective and reducing agents, capable of establishing metal particles and reducing metal ions into zero valence metals (Azmi et al. 2018; Taghvaie Nakhjiri et al. 2022).

The structure of phenolic compounds extracted from plants is composed of hydroxyl and carboxyl groups, which can capture free radicals, donate electrons or hydrogen atoms, and chelate metal ions. Metal chelation is related to the nucleophilic character in the aromatic rings of phenolic compounds (Azmi et al. 2018; Taghvaie Nakhjiri et al. 2022); the biorecovery of heavy metals is aided by plant extracts whose mechanism consists of three phases: in the first phase, activation occurs in which the metal ions are reduced; in the second phase, metal atoms agglomerate into larger particles accompanied by an increase in the thermodynamic stability of the particle agglomerates; in the third phase, metal particles are coated (Azmi et al. 2018) (see Figure 7.6).

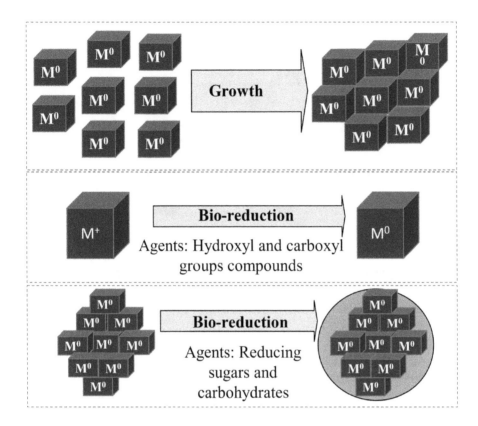

FIGURE 7.6 Scheme of metal biorecovery mechanisms.

The hydrogen ions belonging to phenolic compounds are released in the activation phase through the hydroxyl groups and are oxidized into carbonyl groups and, simultaneously, metal ions (Azmi et al. 2018; Taghvaie Nakhjiri et al. 2022). On the other hand, negatively charged oxygen in the hydroxyl groups of macromolecules interacts with positively charged metal particles through electrostatic surface interactions and binds due to the presence of van der Waals forces. Ultimately, the metal particles are completely clogged on the surface and they stabilize in the solution (Azmi et al. 2018).

7.3.3 ENERGY FIELD

7.3.3.1 Supercapacitors

Supercapacitors are devices that store energy through the assimilation of ions in the electrical double layer or redox reactions on the surface (Catañeda et al. 2019; Zore et al. 2021). The mechanism that allows the charging and discharging of supercapacitors is due to the interaction of ions on the surface (charged double layer) (Krishnamoorthy et al. 2022). This process allows the ions to accumulate reversible

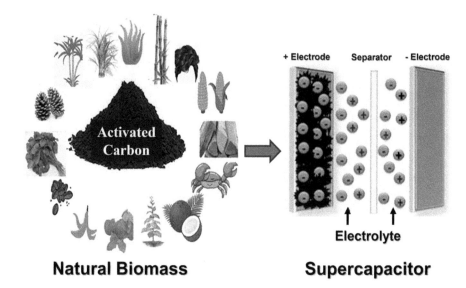

+ Electrode   Separator   - Electrode

Electrolyte

Natural Biomass   **Supercapacitor**

FIGURE 7.7   Schematic of biomass-derived supercapacitor. Taken and edited from Manasa, Sambasivam, and Ran (2022).

electrostatic charge. When the system is polarized, positive ions are released from the cathode, while negative ions are released from the anode (Mehta, Jha, and Liang 2020). Diffusion of ions through the electrolyte causes a condensed layer to lie parallel to the electrode surface.

The electrode material of a supercapacitor is very important, and, in general, most of the electrodes are based on material derived from carbon (due to its mechanical, physical, and mechanical stability) (Manasa, Sambasivam, and Ran 2022). The amount of carbon material that is produced by carbonization, activation, and precursors such as activated carbon or even graphene and carbon nanotubes that are very sophisticated for the construction of supercapacitors is enormous (Krishnamoorthy et al. 2022; Li et al. 2022). Additionally, the use of biomaterials to obtain carbon material is an economically and environmentally viable strategy that has aroused much interest in research (Mehta, Jha, and Liang 2020; Manasa, Sambasivam, and Ran 2022). Some images of different new types of biomass-derived carbon electrode biomaterials for supercapacitor applications are shown in Figure 7.7.

7.3.3.2 Batteries

A battery is a device that contains a cathode and an anode, with the ability to convert stored chemical energy into electrical energy like a supercapacitor. The electrolyte is responsible for transferring the electrons, which allows the electric current to flow from the battery to the outside. Current research on new alternative energy sources has promoted research on the production of new systems that improve energy storage and are more economical. Lithium-ion, nickel-cadmium, and sodium-sulfur batteries are the focus of solid-state battery research. The supporting electrolyte plays an important role in electrochemical energy storage systems. This allows ionic

TABLE 7.4

Electrodes are derived from used biomaterials and electrolytes

Electrolyte	Type	Specific capacitance (F g^{-1})
H_2SO_4	Acidic	33–400
KOH	Basic	89–311
Na_2SO_4	Neutral	66–125
$NaNO_3$	Neutral	47–165
[EMIM]BF$_4$	Ionic liquid	138
MeEt$_3$NBF$_4$/CH$_3$CN	Organic	120–251
MeEt$_3$NBF$_4$/AN	Organic	100
$(C_2H_5)_4$NBF$_4$/CH$_3$CN	Organic	69–120

Extracted from Mehta, Jha, and Liang (2020).

conductivity, a reversible redox process, to store the charge in the system. Also, it determines the capacitive performance of the battery (Mehta, Jha, and Liang 2020).

Ion concentration (including size and type), electrode-electrolyte interface, and the interaction between ions, solvents, and cell power window are the most significant parameters of the supporting electrolyte that determine the capacitance current density, power, and cycling of the cell (Comninellis and Chen 2010; Mehta, Jha, and Liang 2020). Table 7.4 shows some of the supporting electrolytes used with electrodes derived from biomaterials, and this includes the range of specific capacitance obtained for the battery.

7.4 CONCLUSIONS

In this chapter, we review the progress of the different applications of biomaterials in the area of electrochemical phenomena and processes. We provide an overview of the fundamentals of biomaterials (relevant characteristics and main properties) and the key aspects of electrochemistry (redox reactions and electrochemical cells, etc.) that are of interest during the application of a biomaterial. Also, the current interest of the scientific community and the number of documents that are related to biomaterials are reviewed. Finally, the main contributions are presented in different interdisciplinary fields where the fundamentals of electrochemistry are the central focus, such as:

- Applications in the field of medicine are the methods of synthesis of conductive polymers and biosensors.
- Applications in the field of the environment include the use of biomaterials that allow real-time monitoring of BOD, toxic compounds, among others. The recovery and mechanisms of biomaterials are also of interest in the environment.
- In the field of energy, the use of biomaterials for the manufacture of supercapacitors from biomass stands out. Biomaterial-derived electrolytes are the other applications of interest in the energy field.

ACKNOWLEDGMENTS

The author thanks Tecnoparque Nodo Cali and Universidad del Valle. Diaz-Angulo and Lara-Ramos thank the Minciencias for Ph.D. Scholarship 647, 727. The authors also thank Minciencias (CTeI-SGR BPIN 2020000100377). This work is supported by the Universidad del Valle. Fiderman Machuca and Lara-Ramos thank Dismares-Biohidroingeniería (C.I. 0521177).

REFERENCES

Abdessamad, NourElHouda, Hanene Akrout, Ghaith Hamdaoui, Kais Elghniji, Mohamed Ksibi, and Latifa Bousselmi. 2013. "Evaluation of the Efficiency of Monopolar and Bipolar BDD Electrodes for Electrochemical Oxidation of Anthraquinone Textile Synthetic Effluent for Reuse." *Chemosphere* 93 (7): 1309–16. doi:10.1016/j.chemosphere.2013.07.011.

Azmi, A. A., J. Jai, N. A. Zamanhuri, and A. Yahya. 2018. "Precious Metals Recovery from Electroplating Wastewater: A Review." *IOP Conference Series: Materials Science and Engineering* 358 (1): 012024. doi:10.1088/1757-899X/358/1/012024.

Barhoumi, Natija, Hugo Olvera-vargas, Nihal Oturan, and David Huguenot. 2017. "Environmental Kinetics of Oxidative Degradation/Mineralization Pathways of the Antibiotic Tetracycline by the Novel Heterogeneous Electro-Fenton Process with Solid Catalyst Chalcopyrite." *Applied Catalysis B, Environmental* 209: 637–47. doi:10.1016/j. apcatb.2017.03.034.

Bruguera-Casamada, Carmina, Ignasi Sirés, Enric Brillas, and Rosa M. Araujo. 2017. "Effect of Electrogenerated Hydroxyl Radicals, Active Chlorine and Organic Matter on the Electrochemical Inactivation of Pseudomonas Aeruginosa Using BDD and Dimensionally Stable Anodes." *Separation and Purification Technology* 178. doi:10.1016/j.seppur.2017. 01.042.

Catañeda, Locksley F., Fernando F. Rivera, Tzayam Pérez, and José L. Nava. 2019. "Mathematical Modeling and Simulation of the Reaction Environment in Electrochemical Reactors." *Current Opinion in Electrochemistry* 16 (August): 75–82. doi:10.1016/j. coelec.2019.04.025.

Chae, Inseok, Chang Kyu Jeong, Zoubeida Ounaies, and Seong H. Kim. 2018. "Review on Electromechanical Coupling Properties of Biomaterials." *ACS Applied Bio Materials* 1 (4): 936–53. doi:10.1021/acsabm.8b00309.

Comninellis, Christos, and Guohua Chen, eds. 2010. *Electrochemistry for the Environment*, 1–563. New York: Springer. doi:10.1007/978-0-387-68318-8.

Czerwińska-Główka, Dominika, and Katarzyna Krukiewicz. 2020. "A Journey in the Complex Interactions between Electrochemistry and Bacteriology: From Electroactivity to Electromodulation of Bacterial Biofilms." *Bioelectrochemistry* 131. doi:10.1016/j. bioelechem.2019.107401.

Do, Minh Hang, Huu Hao Ngo, Wenshan Guo, Soon Woong Chang, Dinh Duc Nguyen, Yiwen Liu, Sunita Varjani, and Mathava Kumar. 2020. "Microbial Fuel Cell-Based Biosensor for Online Monitoring Wastewater Quality: A Critical Review." *Science of The Total Environment* 712 (April): 135612. doi:10.1016/j.scitotenv.2019.135612.

Filimon, Anca. 2016. "Perspectives of Conductive Polymers toward Smart Biomaterials for Tissue Engineering." *Conducting Polymers*. doi:10.5772/63555.

Gil, Eric de Souza, and Giselle Rodrigues de Melo. 2010. "Electrochemical Biosensors in Pharmaceutical Analysis." *Brazilian Journal of Pharmaceutical Sciences* 46 (3): 375–91. doi:10.1590/S1984-82502010000300002.

Gordin, D. M., T. Gloriant, G. Nemtoi, R. Chelariu, N. Aelenei, A. Guillou, and D. Ansel. 2005. "Synthesis, Structure and Electrochemical Behavior of a Beta Ti-12Mo-5Ta Alloy as New Biomaterial." *Materials Letters* 59 (23): 2936–41. doi:10.1016/j.matlet.2004.09.063.

Guimard, Nathalie K., Natalia Gomez, and Christine E. Schmidt. 2007. "Conducting Polymers in Biomedical Engineering." *Progress in Polymer Science* 32 (8–9): 876–921. doi:10.1016/j.progpolymsci.2007.05.012.

Harun, Wan Sharuzi Wan, Rahil Izzati Mohd Asri, Abu Bakar Sulong, Saiful Anwar Che Ghani, and Zakri Ghazalli. 2018. "Hydroxyapatite-Based Coating on Biomedical Implant." In *Hydroxyapatite - Advances in Composite Nanomaterials, Biomedical Applications and Its Technological Facets.* InTech. doi:10.5772/intechopen.71063.

Hedayat, Nader, Yanhai Du, and Hoda Ilkhani. 2017. "Review on Fabrication Techniques for Porous Electrodes of Solid Oxide Fuel Cells by Sacrificial Template Methods." *Renewable and Sustainable Energy Reviews* 77 (April): 1221–39. doi:10.1016/j. rser.2017.03.095.

Jeyaraman, Shankara Narayanan, and Gymama Slaughter. 2021. "Membranes, Immobilization, and Protective Strategies for Enzyme Fuel Cell Stability." *Current Opinion in Electrochemistry* 29: 100753. doi:10.1016/j.coelec.2021.100753.

Kapałka, Agnieszka, György Fóti, and Christos Comninellis. 2010. "Basic Principles of the Electrochemical Mineralization of Organic Pollutants for Wastewater Treatment." *Electrochemistry for the Environment.* doi:10.1007/978-0-387-68318-8_1.

Kaur, Amandeep, Jung Rae Kim, Iain Michie, Richard M. Dinsdale, Alan J. Guwy, and Giuliano C. Premier. 2013. "Microbial Fuel Cell Type Biosensor for Specific Volatile Fatty Acids Using Acclimated Bacterial Communities." *Biosensors and Bioelectronics* 47 (September): 50–55. doi:10.1016/j.bios.2013.02.033.

Kim, Taegyu, and Jong-In Han. 2013. "Fast Detection and Quantification of Escherichia Coli Using the Base Principle of the Microbial Fuel Cell." *Journal of Environmental Management* 130 (November): 267–75. doi:10.1016/j.jenvman.2013.08.051.

Kiran, A. Sandeep Kranthi, and Seeram Ramakrishna. 2021. "Biomaterials: Basic Principles." In *An Introduction to Biomaterials Science and Engineering*, 82–93. World Scientific. doi:10.1142/9789811228186_0004.

Krishnamoorthy, Hemalatha, R. Ramyea, Ayyadurai Maruthu, Kannan Kandasamy, Monika Michalska, and Senthil Kumar. 2022. "Bioresource Technology Reports Synthesis Methods of Carbonaceous Materials from Different Bio-Wastes as Electrodes for Supercapacitor and Its Electrochemistry - A Review." *Bioresource Technology Reports* 19 (August): 101187. doi:10.1016/j.biteb.2022.101187.

Kumar, Sanjay, Dijendra Nath Roy, and Vidyut Dey. 2021. "A Comprehensive Review on Techniques to Create the Anti-Microbial Surface of Biomaterials to Intervene in Biofouling." *Colloids and Interface Science Communications* 43 (April): 100464. doi:10.1016/j.colcom.2021.100464.

Li, Yimeng, Leqian Wei, Lizhen Lan, Yaya Gao, Qian Zhang, Hewan Dawit, Jifu Mao, Lamei Guo, Li Shen, and Lu Wang. 2022. "Conductive Biomaterials for Cardiac Repair: A Review." *Acta Biomaterialia* 139 (February): 157–78. doi:10.1016/j.actbio. 2021.04.018.

Liu, Zhidan, Jing Liu, Songping Zhang, Xin Hui Xing, and Zhiguo Su. 2011. "Microbial Fuel Cell Based Biosensor for in Situ Monitoring of Anaerobic Digestion Process." *Bioresource Technology* 102 (22): 10221–29. doi:10.1016/j.biortech.2011.08.053.

Manasa, Pantrangi, Sangaraju Sambasivam, and Fen Ran. 2022. "Recent Progress on Biomass Waste Derived Activated Carbon Electrode Materials for Supercapacitors Applications—A Review." *Journal of Energy Storage* 54 (May): 105290. doi:10.1016/j. est.2022.105290.

McBeath, Sean T., Adrián Serrano Mora, Fatemeh Asadi Zeidabadi, Brooke K. Mayer, Patrick McNamara, Madjid Mohseni, Michael R. Hoffmann, and Nigel J. D. Graham. 2021. "Progress and Prospect of Anodic Oxidation for the Remediation of Perfluoroalkyl and Polyfluoroalkyl Substances in Water and Wastewater Using Diamond Electrodes." *Current Opinion in Electrochemistry* 30 (December): 100865. doi:10.1016/j. coelec.2021.100865.

Mehta, Siddhi, Swarn Jha, and Hong Liang. 2020. "Lignocellulose Materials for Supercapacitor and Battery Electrodes: A Review." *Renewable and Sustainable Energy Reviews* 134 (September): 110345. doi:10.1016/j.rser.2020.110345.

Murray, J. W., S. J. Algodi, M. W. Fay, P. D. Brown, and A. T. Clare. 2017. "Formation Mechanism of Electrical Discharge TiC-Fe Composite Coatings." *Journal of Materials Processing Technology* 243: 143–51. doi:10.1016/j.jmatprotec.2016.12.011.

Nik Md Noordin Kahar, Nik Nur Farisha, Mariatti Jaafar, Nurazreena Ahmad, Abdul Razak Sulaiman, Badrul Hisham Yahaya, and Zuratul Ain Abdul Hamid. 2022. "Biomechanical Evaluation of Biomaterial Implants in Large Animal Model: A Review." *Materials Today: Proceedings*. doi:10.1016/j.matpr.2022.07.332.

Onur Uygun, Zihni, Hilmiye Deniz Ertuğrul Uygun, and Ferhan Girgin Sağin. 2021. "Nucleic Acids for Electrochemical Biosensor Technology." In *Biosensors - Current and Novel Strategies for Biosensing*. IntechOpen. doi:10.5772/intechopen.93968.

Pandey, Annu, and Rajeev Jain. 2021. "Polymer-Based Biomaterials: An Emerging Electrochemical Sensor." In *Handbook of Polymer and Ceramic Nanotechnology*, 1309–27. Cham: Springer International Publishing. doi:10.1007/978-3-030-40513-7_60.

Peixoto, Luciana, Booki Min, Gilberto Martins, António G. Brito, Pablo Kroff, Pier Parpot, Irini Angelidaki, and Regina Nogueira. 2011. "In Situ Microbial Fuel Cell-Based Biosensor for Organic Carbon." *Bioelectrochemistry* 81 (2): 99–103. doi:10.1016/j. bioelechem.2011.02.002.

Phung, Dung, Phong K. Thai, Yuming Guo, Lidia Morawska, Shannon Rutherford, and Cordia Chu. 2016. "Ambient Temperature and Risk of Cardiovascular Hospitalization: An Updated Systematic Review and Meta-Analysis." *Science of the Total Environment* 550: 1084–102. doi:10.1016/j.scitotenv.2016.01.154.

Prado da Silva, M. H. 2016. "Biomaterials Concepts." In *Reference Module in Materials Science and Materials Engineering*. Elsevier. doi:10.1016/B978-0-12-803581-8.04086-8.

Quek, Soon Bee, Liang Cheng, and Ralf Cord-Ruwisch. 2015. "Microbial Fuel Cell Biosensor for Rapid Assessment of Assimilable Organic Carbon under Marine Conditions." *Water Research* 77: 64–71. doi:10.1016/j.watres.2015.03.012.

Rabaaoui, Nejmeddine, Younes Moussaoui, Mohamed Salah Allagui, Bedoui Ahmed, and Elimame Elaloui. 2013. "Anodic Oxidation of Nitrobenzene on BDD Electrode: Variable Effects and Mechanisms of Degradation." *Separation and Purification Technology* 107: 318–23. doi:10.1016/j.seppur.2013.01.047.

Rusling, James F., Gregory W. Bishop, Nhi M. Doan, and Fotios Papadimitrakopoulos. 2014. "Nanomaterials and Biomaterials in Electrochemical Arrays for Protein Detection." *Journal of Materials Chemistry B* 2 (1): 12–30. doi:10.1039/C3TB21323D.

Salgado, Patrícia C., Plínio C. Sathler, Helena C. Castro, Gutemberg G. Alves, Aline M. de Oliveira, Rodrigo C. de Oliveira, Mônica D. C. Maia, et al. 2011. "Bone Remodeling, Biomaterials and Technological Applications: Revisiting Basic Concepts." *Journal of Biomaterials and Nanobiotechnology* 02 (03): 318–28. doi:10.4236/jbnb.2011.23039.

Sandoval, Miguel A., Wendy Calzadilla, and Ricardo Salazar. 2022. "Influence of Reactor Design on the Electrochemical Oxidation and Disinfection of Wastewaters Using Boron-Doped Diamond Electrodes." *Current Opinion in Electrochemistry* 33 (June): 100939. doi:10.1016/j.coelec.2022.100939.

Sankar, Magesh, Satyam Suwas, Subramanian Balasubramanian, and Geetha Manivasagam. 2017. *Comparison of Electrochemical Behavior of Hydroxyapatite Coated onto WE43 Mg Alloy by Electrophoretic and Pulsed Laser Deposition. Surface and Coatings Technology*, Vol. 309. Elsevier B.V. doi:10.1016/j.surfcoat.2016.10.077.

Shah, Badal, Todd Lafleur, and Aicheng Chen. 2013. "Carbon Nanotube Based Electrochemical Sensor for the Sensitive Detection of Valacyclovir." *Faraday Discussions* 164: 135–46. doi:10.1039/c3fd00023k.

Sharma, Taruna, Neeraj Dohare, Meena Kumari, Upendra Kumar Singh, Abbul Bashar Khan, Mahendra S. Borse, and Rajan Patel. 2017. "Comparative Effect of Cationic Gemini Surfactant and Its Monomeric Counterpart on the Conformational Stability and Activity of Lysozyme." *RSC Advances* 7 (27): 16763–76. doi:10.1039/C7RA00172J.

Shetti, Nagaraj P., Shweta J. Malode, and Sharanappa T. Nandibewoor. 2012. "Electrochemical Behavior of an Antiviral Drug Acyclovir at Fullerene-C60-Modified Glassy Carbon Electrode." *Bioelectrochemistry* 88: 76–83. doi:10.1016/j.bioelechem.2012.06.004.

Singh, Sandeep, Gurpreet Singh, and Niraj Bala. 2018. "Electrophoretic Deposition of Bioactive Glass Composite Coating on Biomaterials and Electrochemical Behavior Study: A Review." *Materials Today: Proceedings* 5 (9): 20160–69. doi:10.1016/j.matpr.2018.06.385.

Soloducho, Jadwiga, and Joanna Cabaj. 2015. "Electrochemical and Optical Biosensors in Medical Applications." In *Biosensors - Micro and Nanoscale Applications*, edited by Toonika Rinken. InTech. doi:10.5772/60967.

Song, Xu, Zhonglan Tang, Wenbo Liu, Kuan Chen, Jie Liang, Bo Yuan, Hai Lin, et al. 2022. "Biomaterials and Regulatory Science." *Journal of Materials Science & Technology* 128 (November): 221–27. doi:10.1016/j.jmst.2022.04.018.

Taghvaie Nakhjiri, Ali, Hamidreza Sanaeepur, Abtin Ebadi Amooghin, and Mohammad Mahdi A. Shirazi. 2022. "Recovery of Precious Metals from Industrial Wastewater towards Resource Recovery and Environmental Sustainability: A Critical Review." *Desalination* 527 (January): 115510. doi:10.1016/j.desal.2021.115510.

Tan, Aaron, Candy Lim, Shui Zou, Qian Ma, and Zhiqiang Gao. 2016. "Electrochemical Nucleic Acid Biosensors: From Fabrication to Application." *Analytical Methods* 8 (26): 5169–89. doi:10.1039/C6AY01221C.

Tanvir, Rahamat Ullah, Jianying Zhang, Timothy Canter, Dick Chen, Jingrang Lu, and Zhiqiang Hu. 2021. "Harnessing Solar Energy Using Phototrophic Microorganisms: A Sustainable Pathway to Bioenergy, Biomaterials, and Environmental Solutions." *Renewable and Sustainable Energy Reviews* 146 (August): 111181. doi:10.1016/j.rser.2021.111181.

Vishwanathan, A. S., Govind Rao, and S. Siva Sankara Sai. 2013. "A Novel Minimally Invasive Method for Monitoring Oxygen in Microbial Fuel Cells." *Biotechnology Letters* 35 (4): 553–58. doi:10.1007/s10529-012-1109-y.

Wang, Min, Peng Xu, and Bo Lei. 2023. "Engineering Multifunctional Bioactive Citrate-Based Biomaterials for Tissue Engineering." *Bioactive Materials* 19 (May 2022): 511–37. doi:10.1016/j.bioactmat.2022.04.027.

Wojcik, Michal, Paulina Kazimierczak, Anna Belcarz, Anna Wilczynska, Vladyslav Vivcharenko, Lukasz Pajchel, Lukasz Adaszek, and Agata Przekora. 2022. "Biomaterials Advances Biocompatible Curdlan-Based Biomaterials Loaded with Gentamicin and Zn-Doped Nano-Hydroxyapatite as Promising Dressing Materials for the Treatment of Infected Wounds and Prevention of Surgical Site Infections." *Biomaterials Advances* 139 (March): 213006. doi:10.1016/j.bioadv.2022.213006.

Zhang, Yifeng, and Irini Angelidaki. 2012. "Self-Stacked Submersible Microbial Fuel Cell (SSMFC) for Improved Remote Power Generation from Lake Sediments." *Biosensors and Bioelectronics* 35 (1): 265–70. doi:10.1016/j.bios.2012.02.059.

Zore, Ujwal Kishor, Sripadh Guptha Yedire, Narasimha Pandi, Sivakumar Manickam, and Shirish H. Sonawane. 2021. "A Review on Recent Advances in Hydrogen Energy, Fuel Cell, Biofuel and Fuel Refining via Ultrasound Process Intensification." *Ultrasonics Sonochemistry* 73 (May): 105536. doi:10.1016/j.ultsonch.2021.105536.

8 A Comprehensive Review on Electrochemical Sequestration of Carbon Dioxide (CO$_2$) into Value-Added Chemicals
A Materials Perspective Approach

*A. Karthik Chandan, C. Uday Kumar,
P. Uma Maheshwari, R. Kiran Kumar
Reddy and Dilip Kumar Behara*

8.1 INTRODUCTION

In today's world, the earth's temperature is rising at an alarming rate, which is both dangerous and upsetting. The terms climate change and global warming are interchangeable, although the word climate change encompasses all aspects of global warming and its consequences, such as changes in precipitation and harmful effects that vary by place. Many scientific findings have systematically demonstrated the impact of climate change. The existing literature misjudges global damages by failing to account for adaptation and climate change. Climate change poses a very substantial global threat, requiring immediate global action, according to technological evidence. Environmental pollution is a major cause for climate change and high population is one of the contributors to the pollution which impacts not only human life but also damages the ecological balance [1–3]. Water pollution is the most common kind of pollution which occurs due to man-made and natural calamities. The water pollution causing contaminants can cause stomach pains, vomiting, diarrhea, and typhoid fever in humans. Furthermore, soil pollution can also be one of the major contributors causing streams to get contaminated from industrial and residential wastes. The most significant sources of soil contamination are open-pit dumping, trash burning, and insufficient landfills.

DOI: 10.1201/9781032636368-8

Apart from water and soil, air pollution is also one of the major contributors of environmental pollution [4]. Because of global warming and ozone layer depletion, air pollution has a detrimental influence on the quality of life on earth. Several factors like industrialization, mining, deforestation, agriculture, fossil fuels, urbanization, particulate matter, etc., contribute to environmental pollution. Due to rapid industrialization and to meet the needs of the growing population, plastic consumption increased. Industrial waste includes polluted water, harmful gases, aerosol ashes, smokes, and chemical precipitates, among others. Untreated waste, such as chlorine, sulfate, magnesium nitrate bicarbonate, sodium, phosphate, through sewage effluents from industries is responsible for deaths of aquatic ecosystems causing ecological imbalance. The combustion of hydrocarbon fuels (coal/petroleum) maximizes the concentration of CO_2, which causes a change in radiation and heat balance, as CO_2 intensifies the greenhouse gases which allow the solar radiation to pass. Mining and exploration produce various degrees of pollution, affecting the quality of the land, air, and water. Lead (Pb) may erupt during the extraction of valuable materials, such as gold ore, causing land and water contamination. Forests act as natural carbon dioxide sinks because they consume carbon dioxide to produce their food during photosynthesis. Conversion of forest area into agricultural land, overgrazing, forest fires, logging, multifunctional river projects, among others are important factors of deforestation locally and globally. The combustion of waste materials or burning of fossil fuels result in the release of a variety of air pollutants like NO, CO_2, and CH_4 that not only pollute the environment but also damage the ecological cycle. Oil, coal, and gas that mainly fulfill the global energy demands contribute majorly to global warming. The rise in the temperatures across the globe results in lower agricultural yield as the productivity of the land decreases. Furthermore, the ecosystem and soil are degraded, while biodiversity is lost. In addition to this freshwater resources are diminished, oceans are acidified, and atmospheric ozone is disturbed and depleted. The impact of CO_2 on global warming is around 65%. It's important to comprehend that around 95% of the energy for transportation across the globe is derived from combustion of fossil fuels, mostly gasoline and diesel. Road emissions account for nearly half of all transportation emissions, accounting for 7% of CO_2 emissions globally. Even though this is comparatively modest in quantity, significant emphasis is given on reduction of these emissions [5]. Carbon sequestration is crucial in the fight against the greenhouse effect, which is generated by an excess of carbon dioxide in the atmosphere as a result of combustion of fossil fuels and decrease in forest area, which are some of the other detrimental man-made activities. Carbon sequestration is the long-term retention of carbon in vegetation, sediments, rocky outcrops, and the ocean [6]. Carbon sequestration occurs naturally as well as due to anthropogenic activities. It generally refers to the storage of carbon that can potentially produce CO_2 gas. In response to concerns about climate change because of high carbon dioxide concentration in the atmosphere, there was significant interest in accelerating the quantity of carbon sequestered by variations in land use and forestry, as well as geo-engineering techniques like carbon capture and storage [7]. So far, geosequestration approaches have mostly relied on knowledge and experience obtained from oil and gas production, underground natural gas storage, and coal-bed methane storage. Although these approaches provide viable CO_2 sequestration alternatives, improved technology for CO_2 sequestration in geologic formations may considerably cut prices, expand capacity, and improve safety [8]. The goal is to store carbon both in solid and in other diffused forms, thus preventing it from overheating the atmosphere. Carbon dioxide emissions from energy generation are increasing, while industrial emissions are alarming. Approximately 45% of greenhouse gases remain in the atmosphere, with the rest becoming naturally absorbed by the ecosystem. More CO_2 injected into the atmosphere generates a blanket-like effect, thereby entrapping large amounts of heat, leading to global warming. Global warming is the cause of more dry seasons, fewer rainy seasons, melting ice caps, rising ocean levels, more flooding, less agricultural output, and, eventually, more famine, among many other unfavorable crisis for both terrestrial and aquatic life worldwide.

Oceans naturally absorb around 25% of the carbon dioxide generated by human activities each year. The warmer, less nutrient-rich parts of the ocean absorb more CO_2 than nutrient and colder parts. Plant-rich habitats such as forests, pastureland, and grasslands absorb around 25% of global carbon emissions. Fortunately, soil can also retain carbon as carbonates, which form carbon dioxide over countless generations by dissolving in water and percolating through the soil. Capturing of carbon from CO_2 released by industrial sources such as steel and cement factories and its storage permit the usage of fossil fuels until a large-scale alternative energy source is provided. Technological sequestration is a relatively novel method of collecting and storing carbon dioxide that scientists are now investigating. Because the approach employs novel technology, scientists are investigating it further, to use CO_2 as a resource instead of eliminating it from the environment and redirecting it elsewhere. Direct air capture (DAC) is a method of removing CO_2 from the atmosphere utilizing sophisticated technological facilities in the same way as the natural plants do to collect carbon dioxide from the air. It is a viable technical approach for carbon sequestration. Industrial sequestration is not a well-known approach, although it can be useful in some sectors. They collect carbon from power plants in three ways: pre-, post-combustion, and oxy-fuel. Carbon is collected in power stations before the fuel is burned in a process known as pre-combustion. The post-combustion process removes carbon from a power plant's output after the fuel has been burnt. This way toxic gases are collected and washed free of carbon dioxide before they ascend the smokestacks. The goal of oxyfuel or oxy-combustion is to burn fossil fuels in more oxygen and store all of the gases generated. The basic advantage of oxy-combustion method is that it captures and stores the full output from the smokestacks instead of tediously extracting carbon dioxide from other flue gases. Once the steam is drained by cooling and condensation and converted to water, the carbon dioxide may be kept securely.

In recent years, the electrochemical methods for the reduction of CO_2 in the segregated emissions (e.g., ocean and atmosphere) and gases evolving from the industrial regions are gaining attention due to their flexibility and sustainability compared to conventional approaches. Processes like pH swing, redox-active carriers, molten carbonate, hybrid methods are commercially viable ones in CO_2 reduction using electrochemical methods [9]. The electrochemical reduction process of CO_2 requires an electrical energy source supply that creates a potential difference between two electrodes that allows CO_2 to be available in the reduced form, which majorly depends upon factors like the material nature of the electrodes (anode, cathode) chosen in the reaction, and other parameters like pressure and temperature maintained in the overall reaction of the reduction process. The cathode electrode made with copper metal yields a mixture of hydrocarbons (methane and ethylene) and alcohols. CO is the by-product when using metal electrodes like Ag, Au, and Zn [10]. Other possibilities for the reduction of CO_2 using the electrochemical approach are based on products that are used in the electrolysis, the change like the cathode electrodes, and also the type of the solvents used in the supporting electrodes in the reaction. Various reduction techniques such as radiochemical, chemical, thermo-chemical, photochemical, and biochemical processes were introduced for the removal of CO_2 and also to replace fossil fuels with by-products like methanol, and other liquid energy resources into valuable chemicals [11]. The heterogeneous conversion of carbon dioxide electrochemically can yield a broad range of hydrocarbon groups with high current efficiency, such as formic acid, carbon monoxide, methane, and ethylene, allowing the generation of renewable fuels. These materials can be used as chemical synthesis substitutes or transformed into hydrocarbon fuels for energy generation [12]. The possible by-products for the electrochemical reduction of CO_2 using the above approaches tend to the synthesize the value-added reactants for process materials like solid oxide devices and also in the liquid electrolyte devices, for the formation of the methanol, ethanol, propanol, formaldehyde, etc. [13].

The electrochemical parameters for the homogeneous and heterogeneous catalysts that were used in the reduction process of CO_2 depend on the comparison between the turnover frequency and overpotential of the catalyst used in the reaction [14]. Various research groups have developed unique Cu-based catalysts with higher activity parameters including selectivity and overpotential for CO_2 reduction via electrochemical methods. The developed catalyst activity was tested to identify the structure of the catalyst and the conditions that were used for the catalytic activity toward CO_2 reduction and product selectivity using electrolytes like $KHCO_3$ and $CsHCO_3$ [15]. It was observed that the addition of ionic liquid to the electrolyte enhances CO_2 reduction in the electrochemical process by utilizing different types of materials [16], where metal nanoparticles' size and shape (i.e., structural modification) play a key role. The high conductivity of the catalyst used in electrodes, and in electrolytes, allows for appropriate product formation and reactant mass transfer, including formic acid and carbon monoxide [17]. Organic molecules like tetra alkyl ammonium salts, aroma compounds and nitriles, ionic liquids, and pyridinium derivatives act as mediators and catalysts in both photo-electrochemical and electrochemical CO_2 reduction to produce value-added by-products like CO, formic acid, oxalate, and methanol [18].

The electrochemical process that converts renewable energy to useful fuelssparked an interest as a means of closing the anthropogenic carbon cycle and producing high-energy-density hydrocarbon fuels including methane and ethanol. Materials play an important role in converting CO_2 into useful chemicals. However, the material properties need to be harnessed by proper tuning of parameters like size, bandgap, defect states, electronic states, etc. For example, material properties for copper, gold, and silver were tuned to achieve good strength to withstand reactions in the electrodes [19, 20] by tailoring the size. Further, the alloys of Cu-Au and also Cu layers over platinum have shown promising results in the production of CO and hydrocarbons from CO_2. Furthermore, the selectivity and activity for production of CO were adjusted by minimizing the geometric as well as electronic properties of the material of Pt, which also showed the differences in strain allowed by the tuning of the selectivity between ethylene and methane production [21]. There are various types of electronic structure methods for tuning materials, such as metal alloying, ligan coating, molecular metal organics, and metal organic frameworks, among others, where oxygen-derived metal catalysts have gained enormous importance in the electrochemical reduction of CO_2. Various characteristics like lattice defects, distortions, dislocations, local strain, or grain boundaries play a significant role in harnessing important properties from materials [22]. The development of CuO_x-Vo was partially reduced copper oxide nanodendrites with many surface oxygen vacancies, which seem to be good Lewis' base sites for the electrochemical reduction of CO_2. These oxygen vacancies were abundant on CuO_x surfaces, resulting in a high binding affinity for CO and COH intermediates; however, it has a low affinity for CH_2, resulting in more efficient C_2H_4 formation. The oxygen vacancy frequency in CuO_x, which can be regenerated for multiple cycles, has also shown a massive effect on the C_2H_4 Faradaic efficiency in the electrochemical stability test, indicating that oxygen vacancies play a vital role in C_2 product selectivity [23]. Covalent organic frameworks (COFs) are a type of substance framework with finely adjusted parameters made up of covalently bonded organic molecules bound together in precise geometric and spatial arrangements of the molecularly activated sites that are embedded within the reticular super structure reactivity that is involved in the reduction of the CO_2 into sustainable fuels [24]. Overall, materials and surfaces play a dominant role in the electrochemical CO_2

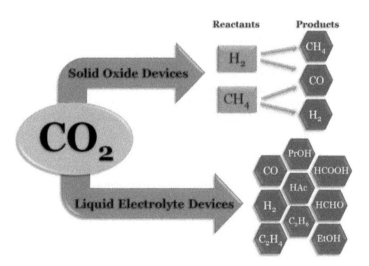

FIGURE 8.1 Summary of electrochemical conversion pathways for CO_2 EtOH: ethanol; PrOH: propanol; HAc: acetic acid; HCHO: formaldehyde. Reprinted from reference [12]. Copyrights 2012 Catalysis Science and Technology.

sequestration and the present review focuses on electrochemical CO_2 sequestration techniques emphasizing materials in particular.

The goal of this review is to provide a summary of all possible CO_2 reduction pathways via electrochemical approaches. Several prominent approaches such as biological, geological, technical, and electrochemical methods for CO_2 conversion to useful products are shown in Figure 8.1 in lowering the greenhouse gases, but the efficiency and efficacy of electrochemical approaches occupy frontline in CO_2 conversion methods.

This review encompasses various materials such as metals, metal oxides, composites, doped metal oxides, and surfaces which play an important role as a catalyst in CO_2 reduction, as shown in Figure 8.2. The electrocatalytic conversion of CO_2 into feasible liquid fuels is the most significant challenge that has a positive impact on the global problem of CO_2 conversion into usable fuels, maintaining the carbon balance. In this process, the efficiency of the particular method that produces the most beneficial goods comes into picture, as well as the necessity to be familiar with its usage for the actual implementation of the process in locations where CO_2 constituents are extremely high.

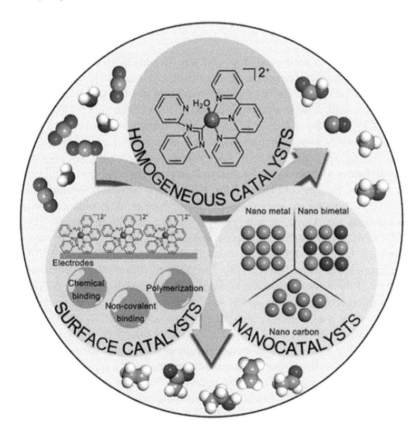

FIGURE 8.2 Catalysts used in the reduction process of CO_2. Reprinted with permissions from reference [37]. Copyrights 2020 pubs.acs.org.

8.2 SIGNIFICANCE/ROLE OF MATERIALS IN CO_2 SEQUESTRATION

8.2.1 SIGNIFICANCE/ROLE OF METALS AND METAL OXIDES IN CO_2 SEQUESTRATION

Electrochemical reduction of CO_2 by chemical process results in CO and hydrocarbons. Transition metals possess good catalytic activity as they can easily donate and accept electrons from other molecules, and have the ability to change the oxidation state by adsorbing other metals onto their surface, thus activating them. Various metals find numerous applications because of the unique properties they possess, the freely available electronic cloud on the metal surface makes them suitable for CO_2 sequestration (specific characteristics) applications. Some of these metals were utilized for the reduction of CO_2 using the electrochemical approach. In line with metals, metal oxides, doped structures, hetero structures or composites are favorable for CO_2 sequestration due to abnormal changes in material properties. For example, doping shifts the absorption band edge toward the visible range that finds prominent applications in the by-product analysis of CO_2 sequestration. However, research investigations on metals encouraged doping of metals for better results. The CO_2 reduction by the electrochemical approach using various materials and optimization strategies is shown in Figure 8.3.

8.2.1.1 Silver (Ag)

Silver is a noble metal with good shape and size, as well as used as an LSPR (localized surface plasmon resonance) at the nanoscale level. Silver sulfide nanowire was used as a catalyst for the electrochemical reduction of CO_2, Liu et al. synthesized it by using a modified facile one-step method. The nanowire showed good catalytic reaction, selectivity but lower efficiency in CO formation in the ionic liquid [18]. However, Samah A. Mahyoub et al. developed nano silver electrodes and explained the effect of various parameters like size, morphology, porosity, alloying, support, and the surface modification on the electrode performance in the reduction of CO_2. It was revealed that morphology of the Ag-based mixture alloy catalyst can be regulated by adjusting the compositions. The CO by-product obtained from CO_2 reduction from the Ag-based nanocatalyst showed maximum Faradic efficiency (FE) of 99.3%, which was very high in comparison to the traditional silver-foil and various silver (Ag)-based materials and their impact in the reduction of CO_2 using electrochemical reduction (ECR]) [19]. The catalytic effectiveness attributes to the active sites of the silver ions in the silver sulfide that are doped with N-S-rGO, as N breaks the electro neutrality of the carbon materials, S helps in enhancing the catalytic activity, and rGO has high electro conductibility. CO_2 reduction to CO with CO-doped Ag_2S/N-S-rGO showed the greatest selectivity of 87.4% and a low overpotential of 0.23 V [20]. Cheonghee Kim et al. discovered the immobilized Ag nanoparticles by facile one-pot synthesis supported over the carbon that enhanced the FE and low overpotentials for the reduced CO_2 to the required by-products. Here, three different forms of electrocatalysts had been created with unique sizes. The Ag catalyst of 5 nm showed a greater CO_2 reduction reaction activity utilizing the low overpotential, high CO Faradaic efficiency, high mass activity, and also high current exchange density with increased durability [21]. Jonathan Rosen

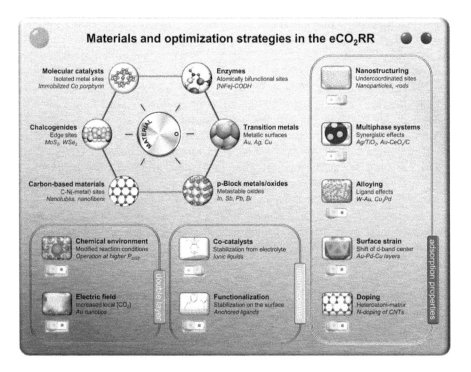

FIGURE 8.3 Description of different classes of materials and optimization strategies as selectable parameters for the construction of eCO$_2$RR catalysts. Materials families are classified according to the different architecture of active sites, whereas optimization strategies refer to modifications of relevant surface properties and/or the surrounding electrochemical environment. The images representing enzymes, chalcogenides, chemical environment and the electric field, co-catalysts, and functionalization are reprinted with permission from reference [17], respectively. Copyrights 2017 American Chemical Society.

et al. reported a prominent way of CO$_2$ reduction using silver as an electrocatalyst. A significant overpotential was necessary for the polycrystalline silver electrocatalyst, whereas the electrocatalyst of nanoporous silver under moderate overpotential rates of less than 0.50 V could reduce CO$_2$ to CO by electrochemical approach with approximately 92% selectivity, that is, 3,000 times higher than the polycrystalline counterpart [22].

8.2.1.2 Gold (Au)

R. Kortlever et al. proposed a unique way for the reduction of CO$_2$ by palladium and gold composite, i.e., a Pd-Au electrocatalyst that could decrease carbon dioxide into a combination of C$_1$ to C$_5$, such as 2-methyl-butane, pentane, 1-butene, isobutane, butane, and pentene, which are almost all forms of hydrocarbons. In addition to soluble compounds like formic acid, methanol, ethanol, and acetic acid, there are a variety of insoluble compounds. It was reported for the first time that higher hydrocarbons (C$_3$-C$_5$) are very useful as electrocatalysts for the future of CO$_2$

reduction to produce valuable added hydrocarbons to increase the ability of CO to attach to the catalyst surface to be fine-tuned to create a unique pattern [23].

Likewise, Monika M. Bienerand's team has shown an appropriate method for the reduction process using a nanoporous gold catalyst [25]. Catalyst loading impact and integration in the reduction performance of CO_2 were explored using a one-step de-alloying method to integrate 100-nm-thick np-Au leaf catalysts into 25 cm^2 for a cutting-edge zero-gap gas diffusion electrode for the electrochemical membrane reactor (ECMR). The zero-gap ECMR turned out to be an excellent contender for combining high current and low voltage due to the lack of liquid electrolytes, and it has a high energy efficiency [24]. Shuo Zhao et al. mentioned that gold and silver are distinct metals with unique properties for the creation of CO, with high selectivity. Furthermore, syn-gas (a compound of CO and H_2) was easily attainable in carbon dioxide electroreduction adjusting silver- and gold-based catalysts, under electrocatalytic conditions, and this syn-gas can then be used as a feedstock for Fischer-Tropsch synthesis, which converts it to CH_3OH, gasoline, and diesel. Materials like gold and silver, along with other metals, are being widely examined as suitable catalysts for electrocatalytic reduction of CO_2 to CO [26]. Eduardus Budi Nursanto et al. used nanostructured gold as a catalyst for CO_2 reduction using the electron beam (e-beam) deposition procedure to deposit Au nanoparticles onto carbon paper. As the amount of Au increased, the FE of the by-product CO and output proportion increased. At samples thicker than Au$_4$, CO Faradaic efficiency reached saturation, which is equivalent to a commercial Au foil [27]. Mariana C. O. Monteiro et al. utilized a gold electrode for gas diffusion converting CO_2 into value-added products. The actual electrode geometries and the possibility of acidic CO_2 electrolysis that performed CO_2 reduction in sulfate electrolytes at varied current densities (10–200 mA cm^2) using a 10-cm^2 gold gas diffusion electrode. The experimental results revealed that significant CO selectivity (90%) may be attained at 100–200 mA cm^2, as long as the weakly hydrated cations (Cs$^+$, K$^+$) are present in moderately acidic conditions. Furthermore, it was discovered that CO_2 reduction can be carried out at significantly lower cell potentials in 1 M Cs_2SO_4 electrolyte than in neutral media (1 M $KHCO_3$), resulting in a 30% reduction in process energy expenditures [28].

8.2.1.3 Copper (Cu)

Cu catalysts generated from copper oxide possess higher energy efficiencies and selectivities in carbon dioxide conversion to commercially important by-products like ethylene (C_2H_4). Yun Huang et al. showed a different approach of reducing CO_2 using copper as an electrocatalyst on Cu_2O/Cu catalysts and found that the onset potentials for C_2H_4 and CO are all the same [29]. At ambient temperature and pressure, Reyhaneh Fazel Zarandi et al. showed electrochemical CO_2 reduction using a copper nanofoam electrochemical system using 1-butyl-3-methyl-imidazolium bromide as the uniform co-catalyst. To benefit from the synergistic effect both the catalysts using copper nanofoam or imidazolium-based catalyst for CO_2 reduction can be analyzed [30].

Methane is produced electrochemically from carbon dioxide on copper electrodes, as depicted in Figure 8.4, using two separate analysis i.e. thermodynamic, combined thermodynamic and kinetic analysis the products, reaction conditions or

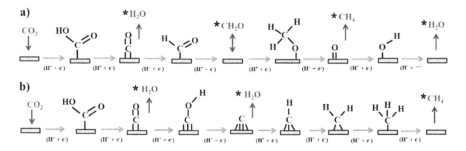

FIGURE 8.4 Pathways for the electrochemical production of methane from CO_2 on Cu electrodes.

by-products are shown by ★. Both methods are similar until *CO is generated by a second proton-electron transfer, thus resulting in *CHO and *COH.

Similarly, various paths for carbon dioxide conversion to carbon monoxide, CH_4 CH_3OH, and HCOO, respectively, are shown in Figure 8.5 with different pathways like section: (a) symbolizes the pathways from carbon dioxide to carbon monoxide (CH_4 is shown by the arrow pointed by 2, CH_3OH with arrow pointed by #, and HCOO with arrow pointed by 1); (b) corresponds to pathways from carbon dioxide to ethylene (arrows pointed by 4 and 6). Adsorbates are shown before the arrows pointed by 3, whereas reactants and products in solution are shown after the arrows pointed by 3. Potentials are provided as a function of the reversible hydrogen electrode (RHE), with Drain-Source on Resistance (RDS) denoting rate-determining steps and (H^+ + e^-) denoting steps involving either continuous or proton-electron transfer that's also separated [23].

S. Wang et al. reported a method for synthesizing Cu catalyst and explained all the reactions involved. Furthermore, the impact of the mixed oxidation states of Cu, local pH, subsurface oxygen, and electrolyte were all addressed [31]. Qinggong Zhu et al. successfully fabricated a copper composite for the reduction of CO_2 by a more direct and simple approach, for electro-depositing a Cu-complex layer with a large surface area onto conductive substrates. Cu–Cu_2O/Cu dendritic across the three-dimensional Cu-Cu_2O composite catalyst is made by growing and decomposing a Cu-Cu_2O complex film in situ with high Faraday energy of 80% and a current density of 11.5 mA cm^2 and the applied potential was as low as 0.4 V compared to RHE in KCl aqueous electrolyte with an overpotential of only 0.53 V and 0.48 V for acetic acid and ethanol [32]. Chunsong Li et al. mentioned that copper, the monometallic electrocatalyst, is really the only one capable of converting CO_2 into valuable compounds such as hydrocarbons and oxygenates. Although it has low selectivity and activity, CO_2 co-electrolysis with low O_2 levels results in the formation of surface hydroxyl species that greatly boost the activity of copper-catalyzed CO_2 electroreduction, using surface-enhanced Raman spectroscopy, and computational analyses [33].

8.2.1.4 Platinum (Pt)

Platinum received less attention from researchers compared to other transition metals like Au, Cu, etc., as Pt is an inert catalyst. Minoru Umeda et al. reported that CO_2

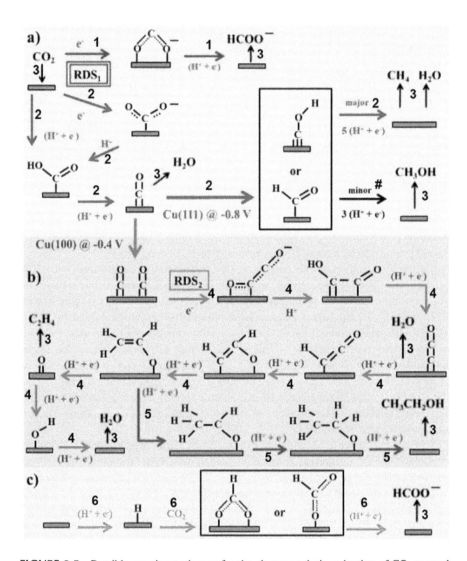

FIGURE 8.5 Possible reaction pathways for the electrocatalytic reduction of CO_2 to products on transition metals and molecular catalysts: (a) pathways from CO_2 to CO, CH_4 (2), CH_3OH (#), and HCOO− (1); (b) pathways from CO_2 to ethylene (4) and ethanol (5); (c) pathway of CO_2 insertion into a metal−H bond yielding formate (6). Species shown by 3 are reactants or products in solution. Potentials are reported versus RHE, while RDS indicates rate-determining steps and ($H^+ + e^-$) indicates steps in which either concerted or separated proton-electron transfer takes place; reprinted from reference [22]. Copyrights 2015 *Journal of Physical Chemistry Letters*.

is reduced dramatically to CO because of the high CO adsorption on the Pt surface. The studies revealed that the CO adsorbed is selectively reduced to methane at low CO_2 partial pressures on carbon-supported Pt catalysts, at potentials near to reaction thermodynamic equilibrium potential (0.16 V versus RHE) [34].

8.2.1.5 Iron (Fe)

Ruggero Bonetto et al. explained the mechanistic aspects of molecular iron catalysts on CO_2 reduction routes. Both iron hydrides and iron-CO_2 adduct use Bronsted acids, nucleophiles, and other co-additives for carbon dioxide capturing moieties to control the process selectivity by incorporating cobalt phthalocyanine in a flow cell; 95% selectivity for CO was attained at a current density of 150 mA [35]. Iqbal Bhugun and team reported that adding moderate Brownsted acids such as 1-propanol, 2-pyrrolidone, and CF_3CH_2OH improves the catalysis of CO_2 reduction by iron. CO is the main product, followed by formic acid, and both the catalytic currents and the catalyst lifetime increase without any substantial hydrogen generation. As the acidity increases, the generation of formic acid decreases synergistically, until it becomes insignificant with CF_3CH_2OH. A chemical route with weaker iron-CO_2 interactions produces formic acid with higher overpotentials [36].

8.2.1.6 Cobalt (Co)

Co-based materials offer distinct properties with mechanical, chemical, and thermal stability, electrical conductivity, high catalytic performance, and distinctive electronic properties, making them potential materials for CO_2RR applications. Cobalt is an excellent substitute for noble metals like platinum, iridium, and ruthenium because of its intriguing characteristics such as d-electrons, which are loosely arranged and hence have a huge number of oxidation states. Co has been used as a significant noble metal-free source electrocatalyst for a wide range of applications in CO_2RR.

Abdalaziz Aljabour and his team reported a unique way of producing CO from CO_2 utilizing Co_3O_4 nanofiber electrodes with a Faradaic efficiency of 65%, without using other metal additives. The electrolyte solution composition aids in monitoring the residual by-product production toward formate (27%) [37]. Muhammad Usman et al. briefly reviewed the various cobalt-based materials like Co complexes, Co multimetals, Co-based metal organic frameworks (MOFs), Co-single atom, or Co-based covalent organic frameworks (COFs). Co oxides and co nitrides are examples of COFs that exhibited good performance [38].

Kristian Torbensen and team overviewed cobalt phthalocyanine with trimethyl ammonium group attached to the phthalocyanine macrocycle over a pH range of 4–14 that reduced CO_2 to CO in water by achieving high activity. CO generation occurred with great selectivity (about 95%) and good stability in a flow cell configuration with highest partial current density of 165 mA cm^2 with basic operational parameters [39].

8.2.2 Metal Oxides and Doped Metal Oxides

The valence band is created by superpositioning the completely filled 2p orbitals of oxide (O_2) ions, but the vacant conduction band is produced by superpositioning the orbitals of metal cations; hence, doping nitrogen into metal oxides enhances the number of active sites. Due to the energy gap between atomic valence orbitals of N and O, the presence of nitrogen in the anionic sub-lattice of the oxide promotes the development of new electronic properties that generally fall within the band gap, not far from the valence band edge.

Doped nitrogen (N) carbon catalysts, such as nitrogen-doped pyrolyzed carbon and multiwall carbon nanotubes (CNTs), a non-precious metal-free alternative electrochemical catalyst for CO_2 reduction, were extremely selective for CO generation over H_2 in an electrochemical flow cell (around 98% CO and 2% H_2) and had high activity [40]. CO had a current density 3.5 times that of silver nanoparticle catalysts at cathode potentials (V = 1.46 V vs. Ag/AgCl), with a max current density of 90 mA cm^2. Also, the energy efficiency and mass activity of the silver nanoparticles were both 48% higher [41]. Liquid Chemical Vapor Deposition (CVD) was used to create nitrogen-doped CNTs that operate as a catalyst with high activity in an electrochemical CO_2 reduction strategy. In comparison to noble metals like silver and gold, these nitrogen-doped CNTs have low overpotentials to obtain similar selectivity for CO generation. At an overpotential of 0.26 V, the maximal FE of CO was about 80%. In contrast, pure CNTs have low activity and show good selectivity in order to reduce CO_2 electrochemically. The precise binding energy of the critical intermediates allows for significant COOH adsorption as well as the potential of CO absorption, which adds to the high selectivity toward CO production. The studies to understand the behavior of Pt- or Fe-doped tetra phenylethylene-conjugated microporous polymer (TPE-CMP) as carbon dioxide reduction electrocatalysts into liquid fuels using a specially designed Gas diffusion Membrane (GDM) assembly revealed that the Pt-doped polymer showed high productivity when used as an electrocatalyst [35]. Moreover, to increase the conductivity and quality, a small amount of CNT was mixed. Furthermore, Pt was replaced with Fe, the variation in morphology and the high pore size resulted in more efficiency with CNTs as the nanoparticles of Fe were well fixed inside CNTs. The electrocatalytic activities of copper nanoparticles of different sizes have been synthesized and studied [36]. The selectivity of carbon-supported Cu nanoparticles was also studied and compared with the carbon-doped Cu nanoparticles for the generation of C_2H_4. Carbon-doped Cu nanoparticles at a potential of −1.6 V generated a high amount of CH_4 compared to others. Faradic efficiency percentage of C_2H_4/CH_4 was observed to reduce with a reduction in the size of the Cu nanoparticles. Using rotating cobalt oxide/graphene composites, single- or multiple-layer graphene sheets were synthesized by low-temperature hydrothermal step. Using N-doped graphene as support, a higher catalytic activity of a metal oxide was recorded when compared to graphene without doping, demonstrating the importance of doping that helps in the reduction of CO_2. The NG-Co_3O_4 catalyst exhibited, a current density of 10.50 mA cm^{-2} and a selectivity of 83%, the highest onset potential of −0.82 V. The mechanism of carbon dioxide reduction using N-doped Sp^2 nanocarbon catalytic agent and $NaHCO_3$ as an electrolyte was explained to provide a clear and concise picture of the situation of carbonaceous products of CO and HCO_2. The CO was obtained from the dissolved CO_2, whereas HCO_2 was produced by reducing HCO_3^-. The production of various other carbonaceous products (C_xH_y or $C_xH_yO_z$) on the sp^{2-} nanocarbon catalyst was possible by balancing the potential between the CO reduction energy and desorption, which can be achieved by the modification of Nitrogen doped sp^{2-} nanocarbon (NDSN) catalysts. A simple and easy approach to build HNCM/CNT nanoporous membranes was technologically advanced [42]. The nanoporous HNCM/CNT electrocatalytic membranes possessed large surface area, high electrical conductivity, higher population of active areas, hierarchical pore architecture, and membrane mechanical integrity. They became preferable as diffusion electrodes thus showing impressive performance and selectivity along with durability. The numerous options available for converting CO_2 into valuable chemicals and fuels is due to the metal functionalized N-doped carbons.

Electrocatalytic reduction studies were conducted using Zn and Cu electrodes individually and in combination with Zn-modified copper electrodes as working electrodes for lowering the CO_2. While copper enhances CO_2 reduction, zinc reduces hydrogen growth due to a porous and ordered structure because of which selectivity can be synergistically enhanced at active sites of Zn-Cu, resulting in high methane formation. The Faradic efficiency improved on Zn-Cu with 58% when compared to undoped copper (28%), which also reduced the hydrogen efficiency when compared to a bare Cu electrode [43]. For electrochemical reduction of CO_2, the carbon-carbon doping is very complex, and so in order to simplify nanostructured iron oxy-hydroxide is deposited on N-doped carbon, which permits high FE of 97.4% and a selectivity of 61% [44]. To study the effect of nitrogen-doped carbon catalysts on the selectivity and activation sites in CO_2 reduction, different structures of CNTs with curvature and graphene without any curvature were used. It was observed that the range of the curvature effect can be used to adjust selectivity. The weak bonds of graphene favor the formation of CO/HCOOH, whereas the strong bonds of CNTs favor HCHO and CH_3OH. The potentials can also be varied using curvature to develop highly selective and cost-effective graphene/carbon catalysts for CO_2 reduction.

8.2.3 Significance of Composites in CO_2 Sequestration

A composite is a material composed of two or more components having varying chemical and physical properties. They are merged to create a material that is adapted to support the main idea whenever they are mixed, such as becoming stronger, lighter, or more electrical resistant. Composites are usually classified based on the type of matrix material utilized.

8.2.3.1 Ag/TiO$_2$

Young Eun Kim et al. worked on the composite of silver/titanium dioxide and examined Ag/TiO$_2$ catalysts for syn-gas and were able to produce a steady H$_2$ and CO ratio with higher FE (93%–100%) and achieved partial current densities (164 mA cm^2). Catalyst characteristics were changed by oxygen vacancies and TiO$_2$ phase differences to manipulate the H$_2$ and CO ratio. By increasing the adsorbed oxygen defects in the bulk and on the surface of Ag/TiO$_2$ catalyst anatase TiO$_2$ was produced. Even though the rutile phase had fewer oxygen vacancies, conversion anatase to rutile TiO$_2$ phases enhanced the H$_2$/CO ratio. Given the fact that Ag/TiO$_2$ rutile air had a higher FE of H$_2$ than Ag/TiO$_2$ anatase-H$_2$/Ar, rutile air had less surface oxygen vacancies [45].

8.2.3.2 Ni$_3$N/MCNT

Zhuo Wang and colleagues reported the electrochemical CO_2 reduction using nickel nitrate and multiwalled carbon nanotubes as a catalyst with CO of FE (89%) and current density of 6.5 mA cm^2 at 0.73 V addressing the reversible hydrogen electrode

(RHE). Ammonolysis-generated Ni_3N/MCNT (multiwalled carbon nanotube) nano-composites showed high CO_2 electrochemical reduction reactivity. The carbon monoxide FE remained at 50.1% even at pH 2.5, showing good selectivity for CO_2 reduction. Increased CO_2 adsorption on the surface of Ni_3N, under an acidic environment, which can compete with evolution of hydrogen, could account for the high catalytic selectivity [41].

8.2.3.3 SnO_2-CuO

Cheng yu Ma and his team developed a catalyst by using a nanocomposite of SnO_2/CuO to reduce CO_2, by mixing nanostructured CuO with SnO_2 nanocatalyst. It was observed that during electrochemical reactions, the selection and activities of SnO_2 nanocatalysts were effectively controlled. CuO combination can enhance the activity of SnO_2-CuO composite dramatically. The SnO_2-CuO (50% SnO_2–50% CuO) nanocomposite with an onset potential of −0.75 V and a current density of roughly − 24 mA cm^{-2} at −1.25 V exhibited the best catalytic performance [46].

Andrews Nirmala Grace and coworkers reported a facile approach for the electrochemical reduction (ECR) of CO_2 under ambient settings, using a Cu_2O nanoparticle-patterned polyaniline matrix (PANI/Cu_2O) in 0.1 M tetrabutylammonium perchlorate (TBAP) and methanol electrolyte. The study was performed in an H-type cell divided into two compartments with Nafion segregating the cathodic and anodic compartments, thus acting as a diaphragm. The catalyst was synthesized electrochemically as a thin layer. At different potentials of polarization, CO_2 was reduced, and formic and acetic acid were the primary products, with Faradaic efficiencies of 30.4% and 63.0%, respectively [47].

8.2.3.4 ZnO-Fe-MXene

Karthik Kannan et al. worked on the reduction of CO_2 using the nanocomposite of ZnO-Fe-MXene as a catalyst. The structural, morphological, and electrochemical properties of a nanocomposite of ZnO, Fe, and MXene were synthesized through a hydrothermal process. The ZnO-Fe-MXene nanocomposite showed best features for the electron-proton coupling transport, during CO_2 reduction reactions. The ZnO-Fe-MXene nanocomposite was a well-organized material prepared by hydro-thermal process and was also used in direct methanol fuel cells to perform the oxida-tion of methanol to formic acid [48].

8.2.3.5 Copper/Reduced Graphene Oxide Thin Film for CO_2 Reduction

M. Nur Hossain and his team gave a basic approach where copper and graphene oxide were reduced directly and the precursors yielded a Cu nanoparticle (NP) with reduced graphene oxide (rGO) nanostructured thin sheet on glassy carbon electrode (GCE). Because the physicochemical interaction between rGO and Cu nanoparticles have different functional groups, Cu nanoparticles and rGO sheets were combined at 0.6 V, with a good current efficiency of 69.4% against the revers-ible hydrogen electrode; for the electrochemical reduction of CO_2 in an aqueous solution, the nanostructured Cu/rGO thin film exhibited outstanding stability and catalytic activity, indicating that it could be used to efficiently convert CO_2 to valu-able products [49].

8.2.3.6 Carbon Nanotubes/Copper Sheets

Young Koo and coworkers reported an innovative way for CO_2 reduction using the electrochemical approach. Cu nanostructures on aligned CNT sheets had various morphologies, which increased the efficiency of CO_2 reduction. The pulsed electrode-position of copper in functionalized CNT sheets increased in terms of electrochemical activity. During the electrochemical reduction of CO_2 to generate hydrocarbons, CO, ethylene, and methane gases at room temperature and atmospheric pressure, the CNT/Cu sheet electrocatalyst performed better [40].

8.2.3.7 Chrysanthemum-Like Cu Nanoflower

Most of the investigations reported that the Cu-based electrocatalysts that are used in electrochemical reactions for the CO_2 reduction into chemical commodities possess high reaction overpotentials and are unstable. Jia-Fang Xie with his peer team reported a unique Cu nanoflower (NF) catalyst with a three-dimensional structure chrysanthemums-like assembly. Cu NFs that are derived from CuO NFs efficaciously catalyzed CO_2 reduction with a 400 mV lesser overpotential than polycrystalline Cu. In addition, in a wide potential window, the production of hydrogen was subdued to under 25% in terms of FE, implying high catalytic selectivity [50].

8.3 EFFECT OF REACTION CONDITIONS IN PRODUCT SELECTIVITY DURING CO₂ REDUCTION

As discussed, electrocatalysts play a vital role in product selectivity during the CO_2 reduction process. Various value-added products like CO, HCHO, HCOOH/HCOO$^-$, hydrocarbons, and alcohols were produced with different electrocatalysts where the reaction conditions and the pathways dictate the product selectivity. Various electrocatalysts such as Ag, Au, Cu, Cu oxides, metal complexes, metal carbides, and carbon materials exhibited high activities and Faradaic efficiencies, thus producing value-added hydrocarbon products as shown in Table 8.1.

8.4 MECHANISM OF COMPOSITE AND METAL-DOPED ASSEMBLES IN OUTREACHING THE PERFORMANCE

Basic steps in electrochemistry include proton-linked transfer of electrons and electron mobility between an electrode and a substrate. Catalytic materials are solids that help chemical processes run more efficiently and cost-effectively. Oxides, mixed and complicated oxides and salts, halides, sulfides, carbides, unassisted and assisted metals are all considered catalytic materials. Electrochemical processes are used in redox-active photochemistry, such as photosynthesis, and other natural processes requiring electron transport chains and other categories of homogeneous and heterogeneous electron transfer. To evaluate the reactions involved in the CO_2 reduction process, cyclic voltammetry (CV), chronoamperometry, and bulk electrolysis are widely used in addition to more complex experiments containing rotating plate electrodes and rotating ring-disk electrode and an electrochemical approach.

TABLE 8.1

Examples of electrochemical reduction of CO_2 with different electrocatalysts

Electrocatalyst	Electrolyte	Major products [Faradaic efficiency, %]	Current density/mass activity	Ref.
AgNPs	0.5 M $KHCO_3$	CO (79.2%)	1 mA cm^{-2}	[21]
CuNPs	0.1 M $KHCO_3$	CH_4 (57%)	23 mA cm^{-2}	[51]
		C_2H_4 (<20%)		
		CO (<5%)		
		HCOOH (<5%)		
Nanoporous Ag	0.5 M $KHCO_3$	CO ($\approx$92%)	$\approx$18 mA cm^{-2}	[22]
Cu_2O	0.1 M $KHCO_3$	C_2H_4 (34%–39%)	30–35 mA cm^{-2}	[52]
		C_2H_5OH (9%–16%)		
Cu_2O	0.5 M $KHCO_3$	C_2H_5OH (96.2%)	4.5 mA cm^{-2}	[53]
Boron-doped graphene	0.1 M $KHCO_3$	$HCOO^-$ (13.6%) CH_3COO^- (77.6%)	2 mA cm^{-2}	[54]
N-doped carbon	[bmim]BF_4/H_2O	CH_4 (93.5%)	1.42 mA cm^{-2}	[55]
Pd_xPt_{100-x}/C	0.1 M KH_2PO_4/0.1 M K_2HPO_4	HCOOH (88%)	$\approx$5 mA cm^{-2}	[23]
Cu nanowires	0.1 M $KClO_4$ 0.1 M $KHCO_3$ 0.1 M K_2HPO_4	C_2H_6 (20.3%) C_2H_6 (17.4%) C_2H_6 (10%) C_2H_4 (<20%) CO (<5%) HCOOH (<5%)	4–5 mA cm^{-2}	[13]
Cu_2O/Cu	0.1 M $KHCO_3$	C_2H_4 (42.6%) C_2H_5OH (11.8%) C_3H_7OH (5.4%)	13.3 mA cm^{-2} 3.7 mA cm^{-2} 1.7 mA cm^{-2}	[56]

8.5 A BETTER UNDERSTANDING OF CO_2 REDUCTION MECHANISM

The reduction of CO_2 molecules on the surface of an electrocatalyst is an enigma. Why the electrochemical reduction of bicarbonate and carbonate can't be achieved at a faster rate is also a mystery. A thorough knowledge of the electrochemical reaction mechanism of CO_2 might give solutions to these questions, enabling the coherent design of next-generation electrocatalysts for CO_2 reduction and innovative CO_2 capture technologies. New operando spectroscopic methods are extremely useful for studying catalysts at the electrode/electrolyte interface; therefore, any efforts to manufacture them are welcome. Other characterization technologies, along with surface-sensitive and time-resolved approaches , may give greater insight into the reaction process.

Arpita Ghosh and colleagues presented a brief overview of an environmentally acceptable approach for production of nitrogen-doped porous carbon metal oxide that self-assembles composites for environmental remediation. In self-assembly of nitrogen-doped carbon into the structure of mesopores, which has been produced in an inexpensive and ecologically acceptable manner, the pore-producing process based on the hydrogen bonding interaction has been discovered. CO_2 adsorption and lithium storage capacity were significantly enhanced by adding metal oxide nanoparticles to the porous carbon network, and the catalytic performance in the oxygen reduction reaction (ORR) was improved. The pore creation mechanism for the synthesized doped material was described in a step-by-step manner [57].

Alessandra D'Epifanio et al. have simply summarized that developing effective electrocatalytic activity to carry out the reaction process resulted in desired products, but required a deep understanding of catalytic processes. The most recent advancement in the development of nanostructured metal-based catalysts for ECO_2RR were discussed. The synthesis processes and the performance of metal-nitrogen-carbon catalysts were highlighted outlining the catalytic mechanisms. The electron transport process was also discussed based on homogeneous, heterogeneous catalysts with immobilized metal complexes.

Iron porphyrins or phthalocyanines are among the metals and composite materials investigated for homogenous CO_2 electroreduction. Iron porphyrins and iron phthalocyanines exhibited better catalytic activity than Ni- and Co-based stages in the process because Fe(0) may provide electrons for CO_2 reduction, decreasing the CO_2RR overpotential. Neng Li et al. developed a method for capturing CO_2 and converting it into hydrocarbon fuels in transition metal carbides (MXenes), as well as the mechanics behind the reactions. Using well-resolved density, functional theory calculations for two-dimensional (2D) transition metal (groups IV, V, VI) carbides (MXenes) with the formula M_3C_2 were evaluated as CO_2 conversion catalysts. MXenes in classes IV to VI showed active behavior in CO_2 collection, and Cr_3C_2 and Mo_3C_2 MXenes were quite promising for selective conversion of CO_2 to CH_4. The results revealed that during the early phases of hydrogenation, both OCHO and HOCO radical species would arise spontaneously. This gives atomic-level information of the essential elements of computer-aided screening for a high-performance catalyst for CO_2 fixation [58]. Qi Lu et al. briefly reviewed the reaction mechanism, the electrocatalyst for the reduction of CO_2, process engineering, and examined the recent progress in electrochemical CO_2 reduction. The new electrocatalyst development, process engineering work, and a mechanistic insight into the CO_2 electroreduction reaction in CO_2 electrolyis were also focused. Furthermore, significant problems with state-of-the-art systems were uncovered and research directions were discussed. CO_2 reduction to CO on polycrystalline Au and oxide-derived Au by one of the proposed mechanisms is explained in Figure 8.6.

8.6 CHALLENGES AND UNADDRESSED ISSUES IN THE PRESENT CONTEXT

Carbon usage is an expensive way of producing chemicals and fuels apart from mitigating climatic changes. It's a reduced, limited "renewable" resource. One of the most important impediments in the energy-intensive conversion to fuels makes

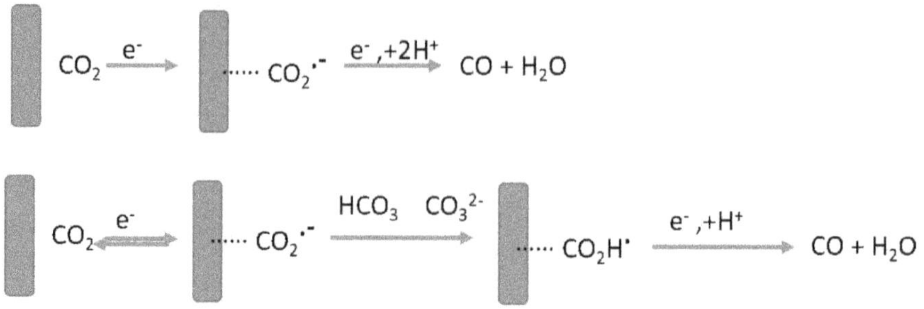

FIGURE 8.6 Proposed mechanism for the electrochemical reduction of CO_2 to methane, including the rate-limiting step (RLS).

them unfeasible. Using CO_2 to produce fuel simply delays rather than eliminates CO_2 emissions. As a consequence, the impacts must be addressed based on the process of the life cycle [59]. Carbon capture and Storage (CCS) technologies involve capturing CO_2, concentrating, transporting, and storing it in a reservoir which might be geological or oceanic. Diverse CCS techniques are now being deployed in energy-producing facilities. Pre-combustion, post-combustion, and oxy-combustion are the three stages of combustion. Transportation may be a particularly costly step in the CCS scheme.

To increase the cost benefits, it is critical to reduce the distance between carbon capture sites, compress the gas, utilize bulk pipelines, and share transportation networks [60].

8.6.1 FACTORS AND CHALLENGES THAT ARE PART OF THE CO_2 REDUCTION PROCESS ARE AS FOLLOWS

1. Transportation of CO_2 to the required unit for the process.
2. Cost is the biggest challenge for handling the equipment.
3. The reactor chambers used for CO_2 processing were accurate with regard to efficiency.
4. The materials that were used for CO_2 reduction must have very good catalytic properties.
5. The electrodes used in the reaction chamber must be stable for longer duration in both acidic and basic medium without corrosion.
6. In electrocatalytic reduction when zinc is used as a working electrode it reduces the growth of hydrogen due to the presence of porous structure. Although zinc has a greater number of active sites it's not stable in pure form; hence it can be utilized more efficiently by doping.
7. Gold exhibits some complications in acidic medium like formation of oxides at lower potentials and it dissolves easily at higher potentials.

8.7 CONCLUSIONS

This chapter emphasizes on the materials and their significance in CO_2 sequestration, as the efficiency and efficacy of the methods is influenced by the nature of the catalyst surface, the electrode potentials, membranes, electrochemical cell design, choice of electrolyte ions, pH, etc. Furthermore, various materials like metals, metal oxides, doped metals and metal oxides, composites, their significance, and the mechanism involved in CO_2 reduction using the electrochemical approach are discussed. Furthermore, the challenges and the unaddressed issues are also addressed. It can be concluded that the doping of metals, metal oxides along with composite resulted in greater Faradic efficiency in CO_2 sequestration.

REFERENCES

[1] R. R. Appannagari, "Environmental pollution causes and consequences: a study," *North Asian Int. Res. J.*, vol. 3, no.8, pp. 151–161, 2017.

[2] P. O. Ukaogo, U. Ewuzie, and C. V. Onwuka, *Environmental pollution: causes, effects, and the remedies.* INC, 2020.

[3] M. A. Khan and A. M. Ghouri, "Environmental pollution: its effects on life and its remedies," *Res. World J. Arts, Sci. Commer.*, vol. 2, no. 2, pp. 276–285, 2011.

[4] R. Hannappel, "The impact of global warming on the automotive industry," *AIP Conf. Proc.*, vol. 1871, no. August, pp. 1–7, 2017, doi: 10.1063/1.4996530.

[5] R. Sharifian, R. M. Wagterveld, I. A. Digdaya, C. Xiang, and D. A. Vermaas, "Electrochemical carbon dioxide capture to close the carbon cycle," *Energy Environ. Sci.*, vol. 14, no. 2, pp. 781–814, 2021, doi: 10.1039/d0ee03382k.

[6] M. Alvarez-Guerra, S. Quintanilla, and A. Irabien, "Conversion of carbon dioxide into formate using a continuous electrochemical reduction process in a lead cathode," *Chem. Eng. J.*, vol. 207–208, pp. 278–284, 2012, doi: 10.1016/j.cej.2012.06.099.

[7] M. Jitaru, D. A. Lowy, M. Toma, B. C. Toma, and L. Oniciu, "Electrochemical reduction of carbon dioxide on flat metallic cathodes," *J. Appl. Electrochem.*, vol. 27, no. 8, pp. 875–889, 1997, doi: 10.1023/A:1018441316386.

[8] D. T. Whipple and P. J. A. Kenis, "Prospects of CO_2 utilization via direct heterogeneous electrochemical reduction," *J. Phys. Chem. Lett.*, vol. 1, no. 24, pp. 3451–3458, 2010, doi: 10.1021/jz1012627.

[9] N. S. Spinner, J. A. Vega, and W. E. Mustain, "Recent progress in the electrochemical conversion and utilization of CO_2," *Catal. Sci. Technol.*, vol. 2, no. 1, pp. 19–28, 2012, doi: 10.1039/c1cy00314c.

[10] C. Costentin, M. Robert, and J. M. Savéant, "Catalysis of the electrochemical reduction of carbon dioxide," *Chem. Soc. Rev.*, vol. 42, no. 6, pp. 2423–2436, 2013, doi: 10.1039/c2cs35360a.

[11] Y. Lum, B. Yue, P. Lobaccaro, A. T. Bell, and J. W. Ager, "Optimizing C-C coupling on oxide-derived copper catalysts for electrochemical CO_2 reduction," *J. Phys. Chem. C*, vol. 121, no. 26, pp. 14191–14203, 2017, doi: 10.1021/acs.jpcc.7b03673.

[12] G. O. Larrazábal, A. J. Martín, and J. Pérez-Ramírez, "Building blocks for high performance in electrocatalytic CO_2 reduction: materials, optimization strategies, and device engineering," *J. Phys. Chem. Lett.*, vol. 8, no. 16, pp. 3933–3944, 2017, doi: 10.1021/acs.jpclett.7b01380.

[13] M. Ma, K. Djanashvili, and W. A. Smith, "Controllable hydrocarbon formation from the electrochemical reduction of CO_2 over Cu nanowire arrays," *Angew. Chem. - Int. Ed.*, vol. 55, no. 23, pp. 6680–6684, 2016, doi: 10.1002/anie.201601282.

[14] Z. Cao et al., "A molecular surface functionalization approach to tuning nanoparticle electrocatalysts for carbon dioxide reduction," *J. Am. Chem. Soc.*, vol. 138, no. 26, pp. 8120–8125, 2016, doi: 10.1021/jacs.6b02878.

[15] K. Jiang, H. Wang, W. Bin Cai, and H. Wang, "Li electrochemical tuning of metal oxide for highly selective CO_2 reduction," *ACS Nano*, vol. 11, no. 6, pp. 6451–6458, 2017, doi: 10.1021/acsnano.7b03029.

[16] Z. Gu *et al.*, "Oxygen vacancy tuning toward efficient electrocatalytic CO_2 reduction to C_2H_4," *Small Methods*, vol. 3, no. 2, pp. 1–8, 2019, doi: 10.1002/smtd.201800449.

[17] C. S. Diercks *et al.*, "Reticular electronic tuning of porphyrin active sites in covalent organic frameworks for electrocatalytic carbon dioxide reduction," *J. Am. Chem. Soc.*, vol. 140, no. 3, pp. 1116–1122, 2018, doi: 10.1021/jacs.7b11940.

[18] S. Liu, H. Tao, Q. Liu, Z. Xu, Q. Liu, and J. L. Luo, "Rational design of silver sulfide nanowires for efficient CO_2 electroreduction in ionic liquid," *ACS Catal.*, vol. 8, no. 2, pp. 1469–1475, 2018, doi: 10.1021/acscatal.7b03619.

[19] S. A. Mahyoub, F. A. Qaraah, C. Chen, F. Zhang, S. Yan, and Z. Cheng, "An overview on the recent developments of Ag-based electrodes in the electrochemical reduction of CO_2 to CO," *Sustain. Energy Fuels*, vol. 4, no. 1, pp. 50–67, 2019, doi: 10.1039/c9se00594c.

[20] L. Zeng, J. Shi, J. Luo, and H. Chen, "Silver sulfide anchored on reduced graphene oxide as a high-performance catalyst for CO_2 electroreduction," *J. Power Sources*, vol. 398, no. July, pp. 83–90, 2018, doi: 10.1016/j.jpowsour.2018.07.049.

[21] C. Kim *et al.*, "Achieving selective and efficient electrocatalytic activity for CO_2 reduction using immobilized silver nanoparticles," *J. Am. Chem. Soc.*, vol. 137, no. 43, pp. 13844–13850, 2015, doi: 10.1021/jacs.5b06568.

[22] Q. Lu *et al.*, "A selective and efficient electrocatalyst for carbon dioxide reduction," *Nat. Commun.*, vol. 5, pp. 1–6, 2014, doi: 10.1038/ncomms4242.

[23] R. Kortlever *et al.*, "Palladium-gold catalyst for the electrochemical reduction of CO_2 to C1-C5 hydrocarbons," *Chem. Commun.*, vol. 52, no. 67, pp. 10229–10232, 2016, doi: 10.1039/c6cc03717h.

[24] Z. Qi *et al.*, "Electrochemical CO_2 to CO reduction at high current densities using a nanoporous gold catalyst," *Mater. Res. Lett.*, vol. 9, no. 2, pp. 99–104, 2021, doi: 10.1080/21663831.2020.1842534.

[25] Biener, Juergen, Monika M. Biener, Robert J. Madix, and Cynthia M. Friend. "Nanoporous gold: understanding the origin of the reactivity of a 21st century catalyst made by pre-columbian technology." *Acs Catal.*, vol. 5, no. 11, pp. 6263–6270, 2015.

[26] S. Zhao, R. Jin, and R. Jin, "Opportunities and challenges in CO_2 reduction by gold- and silver-based electrocatalysts: from bulk metals to nanoparticles and atomically precise nanoclusters," *ACS Energy Lett.*, vol. 3, no. 2, pp. 452–462, 2018, doi: 10.1021/acsenergylett.7b01104.

[27] E. B. Nursanto *et al.*, "Gold catalyst reactivity for CO_2 electro-reduction: from nano particle to layer," *Catal. Today*, vol. 260, pp. 107–111, 2016, doi: 10.1016/j.cattod.2015.05.017.

[28] M. C. O. Monteiro, M. F. Philips, K. J. P. Schouten, and M. T. M. Koper, "Efficiency and selectivity of CO_2 reduction to CO on gold gas diffusion electrodes in acidic media," *Nat. Commun.*, vol. 12, no. 1, pp. 1–7, 2021, doi: 10.1038/s41467-021-24936-6.

[29] Y. Huang, A. D. Handoko, P. Hirunsit, and B. S. Yeo, "Electrochemical reduction of CO_2 using copper single-crystal surfaces: effects of CO* coverage on the selective formation of ethylene," *ACS Catal.*, vol. 7, no. 3, pp. 1749–1756, 2017, doi: 10.1021/acscatal.6b03147.

[30] R. F. Zarandi, B. Rezaei, H. S. Ghaziaskar, and A. A. Ensafi, "Electrochemical reduction of CO_2 to ethanol using copper nanofoam electrode and 1-butyl-3-methyl-imidazolium bromide as the homogeneous co-catalyst," *J. Environ. Chem. Eng.*, vol. 7, no. 3, 2019, doi: 10.1016/j.jece.2019.103141.

[31] S. Wang, T. Kou, S. E. Baker, E. B. Duoss, and Y. Li, "Recent progress in electrochemical reduction of CO_2 by oxide-derived copper catalysts," *Mater. Today Nano*, vol. 12, 2020, doi: 10.1016/j.mtnano.2020.100096.

[32] Q. Zhu *et al.*, "Carbon dioxide electroreduction to C_2 products over copper-cuprous oxide derived from electrosynthesized copper complex," *Nat. Commun.*, vol. 10, no. 1, 2019, doi: 10.1038/s41467-019-11599-7.

[33] M. He *et al.*, "Oxygen induced promotion of electrochemical reduction of CO_2 via co-electrolysis," *Nat. Commun.*, vol. 11, no. 1, pp. 1–10, 2020, doi: 10.1038/s41467-020-17690-8.

[34] M. Umeda, Y. Niitsuma, T. Horikawa, S. Matsuda, and M. Osawa, "Electrochemical reduction of CO_2 to methane on platinum catalysts without overpotentials: strategies for improving conversion efficiency," *ACS Appl. Energy Mater.*, vol. 3, no. 1, pp. 1119–1127, 2020, doi: 10.1021/acsaem.9b02178.

[35] C. Ampelli *et al.*, "CO_2 capture and reduction to liquid fuels in a novel electrochemical setup by using metal-doped conjugated microporous polymers," *J. Appl. Electrochem.*, vol. 45, no. 7, pp. 701–713, 2015, doi: 10.1007/s10800-015-0847-7.

[36] O. A. Baturina *et al.*, "CO_2 electroreduction to hydrocarbons on carbon-supported Cu nanoparticles," *ACS Catal.*, vol. 4, no. 10, pp. 3682–3695, 2014, doi: 10.1021/cs500537y.

[37] A. Aljabour *et al.*, "Nanofibrous cobalt oxide for electrocatalysis of CO_2 reduction to carbon monoxide and formate in an acetonitrile-water electrolyte solution," *Appl. Catal. B Environ.*, vol. 229, pp. 163–170, 2018, doi: 10.1016/j.apcatb.2018.02.017.

[38] M. Usman *et al.*, "Electrochemical reduction of CO_2: a review of cobalt based catalysts for carbon dioxide conversion to fuels," *Nanomaterials*, vol. 11, no. 8, pp. 1–27, 2021, doi: 10.3390/nano11082029.

[39] M. Wang *et al.*, "CO_2 electrochemical catalytic reduction with a highly active cobalt phthalocyanine," *Nat. Commun.*, vol. 10, no. 1, 2019, doi: 10.1038/s41467-019-11542-w.

[40] Y. Koo *et al.*, "Aligned carbon nanotube/copper sheets: a new electrocatalyst for CO_2 reduction to hydrocarbons," *RSC Adv.*, vol. 4, no. 31, pp. 16362–16367, 2014, doi: 10.1039/c4ra00618f.

[41] Z. Wang, P. Hou, Y. Wang, X. Xiang, and P. Kang, "Acidic electrochemical reduction of CO_2 using nickel nitride on multiwalled carbon nanotube as selective catalyst," *ACS Sustain. Chem. Eng.*, vol. 7, no. 6, pp. 6106–6112, 2019, doi: 10.1021/acssuschemeng.8b06278.

[42] H. Wang *et al.*, "Efficient electrocatalytic reduction of CO_2 by nitrogen-doped nanoporous carbon/carbon nanotube membranes: a step towards the electrochemical CO_2 refinery," *Angew. Chem. - Int. Ed.*, vol. 56, no. 27, pp. 7847–7852, 2017, doi: 10.1002/anie.201703720.

[43] C. Genovese *et al.*, "Operando spectroscopy study of the carbon dioxide electroreduction by iron species on nitrogen-doped carbon," *Nat. Commun.*, vol. 9, no. 1, pp. 1–12, 2018, doi: 10.1038/s41467-018-03138-7.

[44] G. L. Chai and Z. X. Guo, "Highly effective sites and selectivity of nitrogen-doped graphene/CNT catalysts for CO_2 electrochemical reduction," *Chem. Sci.*, vol. 7, no. 2, pp. 1268–1275, 2016, doi: 10.1039/c5sc03695j.

[45] Y. E. Kim *et al.*, "Highly tunable syngas production by electrocatalytic reduction of CO_2 using Ag/TiO_2 catalysts," *Chem. Eng. J.*, vol. 413, p. 127448, 2021, doi: 10.1016/j.cej.2020.127448.

[46] M. Fan, C. Ma, T. Lei, J. Jung, D. Guay, and J. Qiao, "Aqueous-phase electrochemical reduction of CO_2 based on SnO_2–CuO nanocomposites with improved catalytic activity and selectivity," *Catal. Today*, vol. 318, pp. 2–9, 2018, doi: 10.1016/j.cattod.2017.09.018.

[47] A. N. Grace *et al.*, "Electrochemical reduction of carbon dioxide at low overpotential on a polyaniline/Cu_2O nanocomposite based electrode," *Appl. Energy*, vol. 120, pp. 85–94, 2014, doi: 10.1016/j.apenergy.2014.01.022.

[48] K. Kannan, M. H. Sliem, A. M. Abdullah, K. K. Sadasivuni, and B. Kumar, "Fabrication of zno-fe-mxene based nanocomposites for efficient CO_2 reduction," *Catalysts*, vol. 10, no. 5, pp. 1–15, 2020, doi: 10.3390/catal10050549.

[49] M. N. Hossain, J. Wen, S. K. Konda, M. Govindhan, and A. Chen, "Electrochemical and FTIR spectroscopic study of CO_2 reduction at a nanostructured Cu/reduced graphene oxide thin film," *Electrochem. Commun.*, vol. 82, no. July, pp. 16–20, 2017, doi: 10.1016/j.elecom.2017.07.006.

[50] J. F. Xie, Y. X. Huang, W. W. Li, X. N. Song, L. Xiong, and H. Q. Yu, "Efficient electrochemical CO_2 reduction on a unique chrysanthemum-like Cu nanoflower electrode and direct observation of carbon deposite," *Electrochim. Acta*, vol. 139, pp. 137–144, 2014, doi: 10.1016/j.electacta.2014.06.034.

[51] R. Reske, H. Mistry, F. Behafarid, B. Roldan Cuenya, and P. Strasser, "Particle size effects in the catalytic electroreduction of CO_2 on Cu nanoparticles". *ACS*, vol. 136, no. 19, pp. 6978–6986. 2014, doi: 10.1021/ja500328k.

[52] D. Ren, Y. Deng, A. D. Handoko, C. S. Chen, S. Malkhandi, and B. S. Yeo, "Selective electrochemical reduction of carbon dioxide to ethylene and ethanol on copper(I) oxide catalysts," *ACS Catal.*, vol. 5, no. 5, pp. 2814–2821, 2015. doi: 10.1021/cs502128q.

[53] V. S. K. Yadav and M. K. Purkait, "Electrochemical studies for CO_2 reduction using synthesized Co_3O_4 (anode) and Cu_2O (cathode) as electrocatalysts," *Energy Fuels*, vol. 29, no. 10, pp. 6670–6677. 2015, doi: 10.1021/acs.energyfuels.5b01656.

[54] Y. Liu, S. Chen, X. Quan, and H. Yu, "Efficient electrochemical reduction of carbon dioxide to acetate on nitrogen-doped nanodiamond. *J. ACS*, vol. 137, no. 36, pp. 11631–11636, 2015. doi: 10.1021/jacs.5b02975.

[55] X. F. Sun, X. C. Kang, Q. G. Zhu, J. Ma, G. Y. Yang, Z. M. Liu, and B. X. Han, "Very highly efficient reduction of CO_2 to CH_4 using metal-free N-doped carbon electrodes," *Chem. Sci.*, vol. 7, no. 4, pp. 2883–2887, 2016. doi: 10.1039/c5sc04158a.

[56] A. D. Handoko, C. W. Ong, Y. Huang, Z. G. Lee, L. Lin, G. B. Panetti, and B. S. Yeo, "Mechanistic insights into the selective electroreduction of carbon dioxide to ethylene on Cu_2O-derived copper catalysts," *Phys. Chem. C*, vol. 120, no. 36, pp. 20058–20067, 2016. doi: 10.1021/acs.jpcc.6b07128.

[57] A. Ghosh, S. Ghosh, G. M. Seshadhri, and S. Ramaprabhu, "Green synthesis of nitrogen-doped self-assembled porous carbon-metal oxide composite towards energy and environmental applications," *Sci. Rep.*, vol. 9, no. 1, pp. 1–13, 2019, doi: 10.1038/s41598-019-41700-5.

[58] N. Li *et al.*, "Understanding of electrochemical mechanisms for CO_2 capture and conversion into hydrocarbon fuels in transition-metal carbides (MXenes)," *ACS Nano*, vol. 11, no. 11, pp. 10825–10833, 2017, doi: 10.1021/acsnano.7b03738.

[59] B. Kanjilal, M. Nabavinia, A. Masoumi, M. Savelski, and I. Noshadi, *Challenges on CO_2 capture, utilization, and conversion.* Elsevier Inc., 2020.

[60] M. Bertolini and F. Conti, "Capture, storage and utilization of carbon dioxide by microalgae and production of biomaterials," *Environ. Clim. Technol.*, vol. 25, no. 1, pp. 574–586, 2021, doi: 10.2478/rtuect-2021-0042.

9 Materials for Drug Delivery

Kamalika Sen and Priti Sengupta

9.1 INTRODUCTION

The unfathomable wealth of natural resources is a vast area of chemical research and allied fields that has provided us with benign alternatives in the development of modern pharmaceutical technology. Better health and extended lives have been achieved by the use of active pharmaceutical ingredients, i.e., drugs. However, adverse health effects are observed in plenty due to the uncontrolled dosage of several drugs even after their controlled administration. This is mostly because of the lack of a suitable sustained delivery mechanism that leads to over-accumulation of the chemical in the body over a short period. Synthesized drugs have specific physical, chemical, and biological activities such as bioavailability, in vivo pharmacokinetics, and stabilities [1]. These parameters sometimes pose unavoidable problems and thereby constitute their critical limitations. To encounter these limitations, ideal drug delivery systems (DDSs) are introduced which are known for their cost-effectiveness compared to the development of a new drug. Issues related to conventional drug molecules can be manipulated using delivery systems as per the specific requirements. The design of DDSs largely depends on base materials derived from natural resources. Before going into details, we need to understand some key aspects of DDSs.

9.2 WHY DO WE NEED TO DESIGN MATERIALS FOR DRUG DELIVERY?

The development of a drug is a time-consuming process and at the same time it involves significant financial support. Additionally, many drug molecules show undesirable side effects due to non-specific interactions with other body parts apart from the area of consideration. Repeated administration of doses increases the drug concentration in the blood plasma, which is also a reason for side effects that lead to unwanted complications in patients. Therefore, low drug doses are preferable, which is practically achieved by incorporating methods for their controlled release. Most of the drugs used for cancer treatment suffer from low specificity. The toxicity of those chemotherapeutic agents can destroy normal healthy cells, which leads to unwanted side effects for the patients. Similarly, various therapeutic agents for neurodegenerative diseases and infections of the brain cannot penetrate the central nervous system (CNS) due to the blood-brain barrier (BBB) [2]. Thus, to overcome BBB, proper delivery systems are required to reach the CNS for the desired curative effect. Besides these, the protection of drug molecules in external environmental parameters is one of the most challenging criteria for getting maximum efficacy [3]. Effective delivery of drugs also depends on their *in vivo* pharmacokinetic parameters such as absorption, distribution, metabolism, and excretion. However, the pharmacokinetics of a drug directly depends on its half-life ($t_{1/2}$), which can be defined as the time required by a drug to reach half of its initial concentration in plasma/blood. A major drawback of several drugs includes their short $t_{1/2}$ and thus researches are focused on the extended $t_{1/2}$ of drug molecules. Several strategies have been undertaken to prolong $t_{1/2}$ of drugs among which DDSs with sustained-release properties are one of the best available options [4]. Apparently, sustained release can reduce dosing frequency by prolonging $t_{1/2}$ (≥ 8 hours) of the drug.

In this context, the development of materials for drug delivery purposes should have the following properties to deliver maximum impact on the treatment of various diseases (Figure 9.1):

- Water solubility for ready access in physiological medium
- Stability in physiological conditions
- Site specificity towards affected tissues/cells
- Reduced toxicity and side effects
- Controlled release for prolonged availability of drugs (long $t_{1/2}$) at the affected sites by incorporating stimuli-responsive nature in DDSs
- Use of DDSs with high drug-loading capacity to increase the doses as per requirements
- Cost-effectiveness as compared to conventional drug molecules
- To overcome the multi-drug-resistant effect once the drug is delivered
- Enhanced cellular uptake capacity

FIGURE 9.1 Schematic presentation of advantages of using DDSs.

All the mentioned features can enhance the chance of better treatment for any particular disease using DDSs from both an economic and a biological point of view.

9.3 WHICH DISEASES/DRUGS NEED DRUG DELIVERY MATERIALS?

Many of the diseases need drug delivery vehicles for advanced therapeutic strategies and the longevity of millions of patients. Recent researches on the development of DDSs are focused on the treatment of various complicated diseases such as diabetes, cancer, cardiovascular diseases, retinal complications, brain diseases, etc. Besides, delivery systems have been utilized for the transport of vaccines, proteins, genes, etc., for immunotherapy [5]. Collectively, all kinds of chronic diseases need DDSs for better therapeutic activity with minimal side effects. From the therapeutic point of view, various types of diseases that need DDSs are mentioned in Figure 9.2.

Different kinds of drugs require drug delivery materials to enhance their efficacy in targeted delivery by overcoming various obstacles present in physiological systems. We can classify the drugs according to their physiological impacts and way of action in the case of a particular disease. We can categorize the requirements for a DDS into two major types as per the situation under consideration.

9.3.1 REQUIREMENT FOR TARGETED DRUG DELIVERY

In some diseases, targeted delivery is important to minimize the side effects and maximize the impact of the drug in the affected cells, tissues, or organs. Drugs used in cancer treatment are highly toxic as these can adversely affect normal cells and also have enormous side effects on patients' bodies. Due to the non-specific action of anticancer drugs, the therapeutic activity towards cancer also decreases. Similarly, drugs used for HIV, tuberculosis, malaria, diabetes, ischaemic heart disease, brain diseases, etc., require specific targeted delivery systems for prolonged and efficient treatment [6]. Targeted delivery of drugs can be achieved by using proper DDSs.

FIGURE 9.2 Different types of diseases need DDSs.

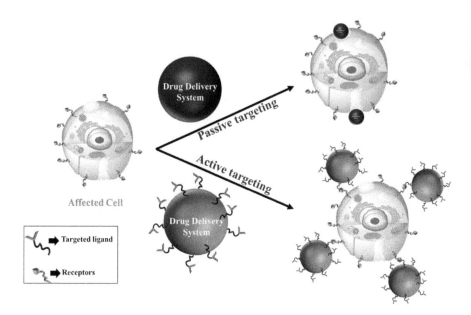

FIGURE 9.3 Schematic presentation of targeted drug delivery.

Targeted delivery methods are based on two types of mechanisms. The first one is passive targeting drug delivery and the other is active targeting drug delivery. In the case of passive targeting, drug efficacy is fully dependent on circulation time and the ability to overcome several unfavourable parameters inside the body. In this process, DDSs accumulate in the affected area based on the enhanced permeability and retention (EPR) effect of the cell which further depends on the size, shape, and composition of the delivery system [7]. For example, polyethylene glycol (PEG)-coated (hydrophilic coating) nanodelivery systems can act as an antiphagocytic substance and thus survive for a long time. But the passive targeting approach has its limitation in the case of solid tumour treatments as in these cases the EPR effect is minimized because of lower permeability. In the case of the active targeting process, DDS surfaces are modified by several ligands which specifically bind with receptors present in the affected cells. Due to the presence of such a specific binding ligand, active targeting-based delivery systems show greater efficiency than passive targeting DDSs. A schematic presentation of the above-mentioned targeting processes is depicted in Figure 9.3 for better understanding.

9.3.2 REQUIREMENT TO ENHANCE THE SOLUBILITY OF DRUGS

Oral administration is the most efficient, cost-effective, and widely utilized drug delivery process for the majority of drugs. But most of these drugs show low bioavailability due to poor water solubility. Water solubility is the most important parameter for a drug to get its desired concentration in the body for maximum pharmacological activity [8]. In this regard, the enhanced water solubility of drugs can render the required efficacy. The incorporation of a proper DDS that enhances the water

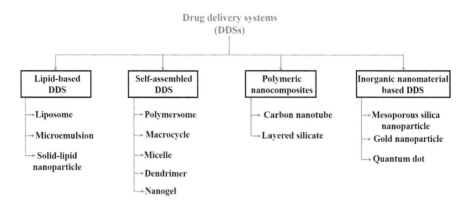

FIGURE 9.4 Classification of different types of DDSs.

solubility and permeability of insoluble drugs can produce increased pharmacological activity for better treatment of deadly diseases.

Depending upon the requirements, a few other kinds of DDSs are also in demand. For example, certain diseases require administration of several injections per day that causes impairment of normal working lives of human beings. Subcutaneous injections of insulin to diabetic patients on a daily basis cause a lot of uneasiness in the normal lifestyle. Oral administration of the drug is not possible as the protein will get denatured in the gastric medium. Therefore, suitable oral DDSs need to be designed that can deliver the drug to the body system bypassing the gastric environment. Similarly, certain drugs have very low metabolic rates which may be improvised using DDSs coupled with stimuli-responsive mechanisms (discussed later).

9.4 TYPES OF DDSs

At present, several DDSs are available to the scientific community. A majority of the components used in DDSs are derived from several natural resources. A basic classification of DDSs can be done as lipid-based systems, self-assembled systems, polymeric nanocomposites, and inorganic nanomaterials. Further, several sub-classes of these DDSs are possible (Figure 9.4). However, some of them which are found most frequently in recent literature are discussed in the coming sections.

A brief listing of the components of DDSs and their corresponding natural sources have been discussed below (cf. Table 9.1).

9.4.1 LIPID-BASED DDSs

9.4.1.1 Liposome
Liposomes (40–300 nm) are excellent candidates with low toxic effects for the delivery of hydrophobic as well as hydrophilic drugs [9]. Liposomes are spherical structures containing natural phospholipids or synthetic lipids along with cholesterol molecules. Advantages of using liposomes as DDSs include their biocompatibility, unique

TABLE 9.1

List of natural sources of different DDSs

Type of DDS	Component of natural resources
Liposomes	Phospholipids extracted from soya, rapeseed, sunflower, chicken eggs, flax seed, wheat germ
Microemulsions	Mineral and vegetable oils
Solid-lipid nanoparticles	Meat, poultry, fish, soya, sunflower, chicken eggs
Macrocycles	Corn, potato, cucumber
Carbon nanotubes	Biomass resources
Layered silicates	Montmorillonite, magnesium aluminium silicate, magadiite

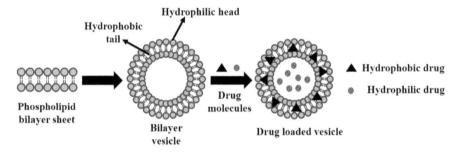

FIGURE 9.5 Pictorial presentation of liposome formation and encapsulation of drugs within the liposome.

structural features, stability, controllable release kinetics, etc. [10]. Phospholipids are obtained from natural resources like soya, rapeseed, sunflower, chicken eggs, etc., and are constituted of a polar head and nonpolar tail. During liposome formation the phospholipids in an aqueous medium associate to form a bilayer sheet where the hydrophilic heads are oriented towards water and hydrophobic tails remain in between the polar heads. In the presence of external energies (e.g., sonication, stirring, lyophilization, etc.) the phospholipid sheets aggregate to form an organized spherical core-shell structure which is conventionally called liposomes (Figure 9.5). As evident from Figure 9.5, hydrophilic drugs are encapsulated within the inner aqueous core and hydrophobic drugs remain within the bilayer of liposomes.

The effect of internal (pH, redox effect, enzyme action, protein interaction, etc.) or external (temperature, magnetic fields, ultrasounds, light, ionic strength, etc.) stimuli triggers the drug to get released at the desired site. At this point, the topic related to 'stimuli' needs a brief introduction. Stimuli-responsive DDSs are generally designed for cell-specific treatment and thus can reduce the side effects of any drug. These smart systems can be developed by grafting a DDS with suitable tools that readily respond to different stimuli which trigger the release of drug molecules at the specific site (Figure 9.6) [11]. Numerous biological factors which are dominant in the cellular environment can be used as potential stimuli for the delivery of drugs from stimuli-responsive delivery systems. Let us now discuss a few examples of DDSs where

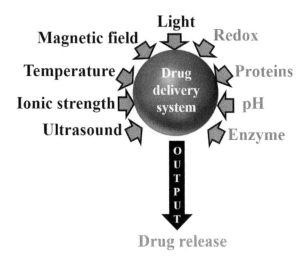

FIGURE 9.6 Stimuli-responsive systems with internal (blue) and external (black) stimuli.

researchers have successfully utilized the stimuli-responsive nature to deliver various drugs at specific sites. Recently, a huge number of DDSs have been reported where acidic pH is used as the stimulus for the delivery of anticancer drugs such as doxorubicin, chlorambucil, etc. In some of the cases, ester linkers and hydrazone linkers that are highly labile towards low pH were used for the preparation of DDSs. The low pH microenvironment of cancer cells triggers the hydrolysis of the above-mentioned linkers which helps to deliver anticancer drugs [12, 13, 40]. Another such example commonly used in the stimuli-responsive drug delivery field is a redox responsive linker. A commonly reported linker comprises disulphide bonds that are known for their redox responsive self-immolative properties. Researches have been carried out using disulphide linker-based DDSs, where reducing agents such as glutathione (GSH) act as the stimulus for the release of the drug to the affected tissues [14–16].

9.4.1.2 Microemulsion

Microemulsions (10–100 nm) are well-utilized liquid vehicles in the field of drug delivery for their thermodynamic stability, transparency, spontaneous formation, lower viscosity, long *in vitro* stability, delivery of both lipophilic and hydrophilic drugs, etc. [17, 18]. It contains oil and water phases along with surfactant (cationic, anionic, zwitterionic, and non-ionic) and co-surfactant (short- to medium-chain alcohol, amines, acids, etc.). The presence of a surfactant along with a co-surfactant helps to reduce the oil-water interfacial tension, which expedites the dispersion process. Generally, two types of microemulsions are available, viz., water-in-oil (W/O) and oil-in-water (O/W). Several mineral and vegetable oils are used for the purpose. Dispersion of W/O continuous phase and dispersion of O/W continuous phase gives rise to the formation of W/O and O/W microemulsions, respectively. However, the continuous phase is decided by the solubility of the emulsifier in the same phase, i.e., more water-soluble emulsifiers produce O/W microemulsion systems and vice versa.

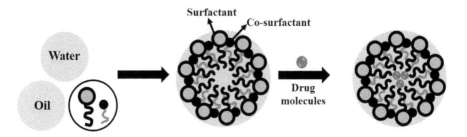

FIGURE 9.7   Schematic presentation of lipophilic drug encapsulation in O/W microemulsion systems.

As evident from Figure 9.7, in O/W microemulsion systems, oil is in the inner core (discontinuous phase) of the system and water is in the outer part (continuous phase) of the system. The reverse is true for the W/O microemulsion systems. However, O/W microemulsions attain much importance in lipophilic DDSs, i.e., the drug molecule resides within the hydrophobic core.

9.4.1.3 Solid-Lipid Nanoparticles (SLNs)

SLNs (10–1000 nm) are submicron colloidal carriers made of lipid core (solid at room temperature and body temperature) that is stabilized by surfactants (Figure 9.8) [19, 20]. Lipids such as stearate and lecithin are most suitable for the purpose and are obtained from natural sources like meat, poultry, fish, etc., and soy, sunflower, egg, respectively. The surfactant molecules help to maintain a balance between the inner lipid phase and the external aqueous phase. SLNs serve as a beneficial drug carrier due to their green synthesis, low toxicity, better biocompatibility, larger surface area, ability to deliver hydrophobic drugs, safe administration ability, etc. Additionally, customizing the surface of SLNs by targeting agents triggers the targeted delivery of drugs from the system.

9.4.2 Self-Assembled DDSs

9.4.2.1 Polymersome

Polymersomes (20–500 nm) are reservoir-like vesicles prepared from synthetic amphiphilic block copolymers. Like liposomes, polymersomes can also encapsulate hydrophilic as well as hydrophobic drugs into their aqueous core and amphiphilic polymer bilayer, respectively (Figure 9.9) [21]. However, they show prolonged circulation time, better storage stability, improved functional properties, compared to conventional liposomal systems. Commonly, PEG constructs the hydrophilic block of the copolymer and contributes to the prolonged circulating phenomenon of polymersomes. PEGylation additionally shows the EPR effect that helps to accumulate polymersomes at desired sites [22]. Apart from PEG, hydrophilic polymers such as polyacrylamide, polyurethanes, etc., are also used for constructing hydrophilic segments. Designing stimuli-responsive systems (pH, light, overexpressed enzymes, temperature, redox, viscosity, etc.) by introducing stimuli-active segments is much required for targeted

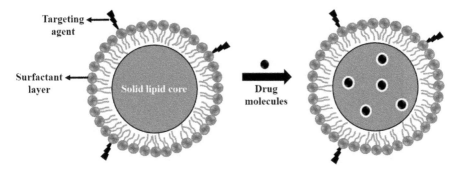

FIGURE 9.8 Schematic presentation of hydrophobic drug encapsulation in SLN systems.

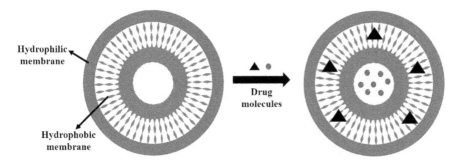

FIGURE 9.9 Schematic presentation of hydrophobic (▲) and hydrophilic (●) drug encapsulation in polymersomes.

delivery of a drug to the affected tissues. Polymersomes are an excellent choice of many researchers for this purpose. The presence of cross-linking via cross-linkable groups further enhances the stability and controlled release nature of polymersomes.

9.4.2.2 Macrocycle

Macrocycles act as 'host' molecules to drugs (guest molecules) and thus can accommodate hydrophobic drug molecules partially or entirely by forming a host-guest complex [23]. The forces of attraction involved in these types of host-guest complexes are hydrophobic and van der Waals interactions. Generally, the complexation enhances the solubility and bioavailability of drug molecules. As evidenced from Figure 9.10, host macrocycles enclose the guest molecule inside their cage-like structure and thus protect the adverse environment (heat, oxygen, radiation, etc.). The key feature of these 'host' molecules is their biodegradable nature. The formation of host-guest complexes not only augment the water solubility of drug molecules but also help to mask their bad smell or bitter taste and lower their toxicity [9]. Inclusion complex additionally causes the easy formation of solid dosage for drugs that are liquid at room temperature. Few examples of conventional macrocycles are cyclodextrins (α, β, and γ), cucurbit[n]urils, pillar[n]arene, calix[n]arene, etc. (Table 9.2).

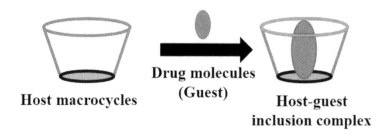

Host macrocycles

Drug molecules
(Guest)

Host-guest
inclusion complex

FIGURE 9.10   Schematic presentation of host-guest complex formation.

These components are often found in different natural sources like corn, potato (α, β, and γ cyclodextrins), cucumber (cucurbit[n]urils). However, chemical modifications on conventional macrocycles are possible and are known to enhance their functional properties.

9.4.2.3 Micelle

Micelles (5–500 nm) are another important class of organic nanocarriers that are highly utilized in DDSs and are designed by self-assembly of amphiphilic (polar heads and nonpolar tails) block copolymers [24]. The parameter that decides the formation of proper micellar systems is 'critical micellar concentration' (CMC) and is defined as the concentration above which the amphiphilic molecules aggregate to form micelles. The CMC value of any aggregate can be determined experimentally using different probe molecules such as pyrene, nile red, etc., using fluorimetry [25]. Polymersomes form vesicles, while micelles self-assemble to produce a compact core-shell structure (Figure 9.11) and encapsulate hydrophobic drug molecules within their hydrophobic core. Like polymersomes, micelles can also accumulate at desired sites due to the EPR effect [24]. In this case also, stimuli-responsive micellar systems have a potential application in the targeted delivery of drug molecules.

9.4.2.4 Dendrimer

Dendrimers (1–100 nm) are branched macromolecules having plenty of advantages such as low polydispersity, biocompatibility, multi-drug-loading capacity, smaller size, etc. An initiator core, branching units, and surface groups (or terminal end groups) are the main three structural units of any dendrimer [26, 27]. Dendrimers maintain a monotonous layered structural pattern where each new layer generates by doubling the number of end groups. As can be evidenced from Figure 9.12, drug molecules entrapped in the interior of a dendrimer produce a host-guest interaction and such drugs are called 'endoreceptors'. However, drug molecules can also interact with the periphery of dendrimers and such interacting drugs are called 'exoreceptors'.

9.4.2.5 Nanogel

Nanogels having characteristics of both hydrogels and nanoparticles are cross-linked (physically or chemically) three-dimensional hydrogel particles with sizes of 10–100 nm (Figure 9.13) [28]. Several biopolymers like chitosan, agar, alginates

TABLE 9.2

Classification of some of the macrocycles based on their structure and size

Macrocycles	Structure	Size	Drugs used for host-guest complex
Cyclodextrins (CD)	α-CD: n = 1, β-CD: n = 2, γ-CD: n = 3	~0.57 nm (α-CD), ~0.78 nm (β-CD), ~0.95 nm (γ-CD)	Thymol, cyclosporine, quercetin, morin, doxorubicin
Cucurbit[n]uril (CB[n])	CB[n]: n = 5, 6, 7, 8, 10	0.44–1.24 nm	Doxorubicin, albendazole, carbendazim, indomethacin

(Continued)

TABLE 9.2
(Continued)

Macrocycles	Structure	Size	Drugs used for host-guest complex
Pillar[n]arene	 n = 5, 6, 7, 8, 9, 10	~0.47 nm ($n=5$), ~0.67 nm ($n=6$), ~0.8 nm ($n=7$)	Doxorubicin, chlorambucil
Calix[n]arene	 n = 4, 5, 6, 8	NA	Paclitaxel, fluorouracil, docetaxel

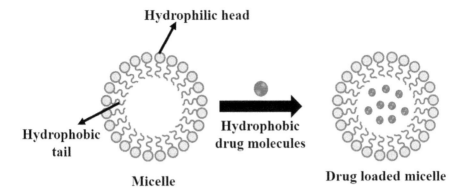

FIGURE 9.11   Schematic presentation of hydrophobic drug encapsulation within micelles.

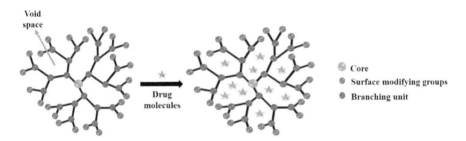

FIGURE 9.12   Schematic presentation of dendrimer and dendrimer-drug complex.

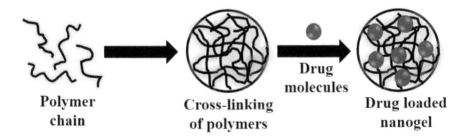

FIGURE 9.13   Schematic presentation of cross-linked nanogel and drug loading within nanogel.

obtained from natural sources have been used to prepare nanogels. Unlike other nanocarriers, nanogels have their own advantages such as good drug-loading capacity, prolonged circulation time in the serum, biodegradability, biocompatibility, high stability, and good water/fluid absorbing capacity. The swelling property of nanogels is attributed to the presence of cross-linked polymer networks. The drug release mechanism in nanogels involves both active and passive pathways. Release based on size, swelling, surface charge, and other physicochemical conditions follows an active pathway mechanism whereas release based on stimuli-responsive behaviour follows a passive pathway mechanism [29].

9.4.3 Polymeric Nanocomposites

9.4.3.1 Carbon Nanotube (CNT)

CNT is one of the recently developed DDSs having single-walled or multi-walled graphene sheets, a nano-needle shape, and a hollow structure (Figure 9.14). Such materials are mostly obtained from the high-temperature treatment of biomass resources. The effectiveness of CNTs as promising DDSs is due to their high surface area, light weight, needle-like shape, etc. The ultrahigh surface area of single-walled CNTs makes them more useful in drug delivery applications than multi-walled CNTs. The outer layers of nanotubes can be functionalized to enhance their efficacy as DDS and also attachment of drug molecules can occur via functional groups. This is generally termed 'wrapping' like binding. Functionalization of the outer surface with PEG is known to enhance water solubility and other functional properties of CNTs. However, drug molecules can also be loaded within CNTs, which is known as "filling" modes of binding. Interestingly, the movement of CNTs within cells can be tracked by labelling them with different fluorescent probes [30].

9.4.3.2 Layered Silicate

Layered silicates (thickness $\sim 1\,\text{nm}$), particularly the modified ones, are extensively exploited as drug delivery tools due to their unique structure, large surface area, reduced toxicity, easy availability, and, most importantly, good adsorption efficiency. These types of silicates have a sandwich-like structure constructed from a two-dimensional layer of two fused silica tetrahedral sheets (Figure 9.15). The area between two stacked layers is known as *gallery*, occupied by cations that are exchangeable for further functionalization of the surface of layered silicates (Figure 9.15). A few examples of natural layered silicate materials include laponite, saponite, montmorillonite (MMT), rectorite, magnesium aluminium silicate, magadiite (MAG), etc. Among these, MMT and MAG are majorly utilized for drug delivery applications because of their superior properties. However, the chemical stability and ion exchange competence of MAG is better than that of MMT and thus chemical modifications in MAG are easy through ion exchange [31, 32].

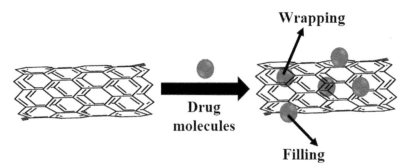

FIGURE 9.14   Schematic presentation of CNT.

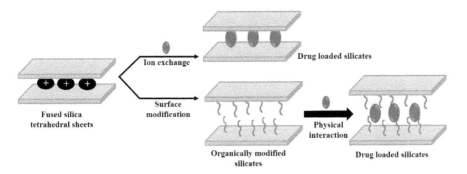

FIGURE 9.15   Schematic description of drug loading into unmodified and modified layered silicate.

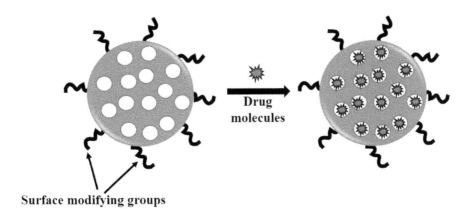

FIGURE 9.16   Schematic presentation of drug loading into mesopores of MSN.

9.4.4 Inorganic Nanomaterial-Based DDSs

9.4.4.1 Mesoporous Silica Nanoparticles (MSNs)

In recent years, MSNs (2–50 nm) seem to have become an excellent option as a targeted carrier for the delivery of numerous therapeutic agents. Some of the advantages of MSNs are ordered porous interior, large pore volume, biocompatibility, high rigidity, huge surface area, excellent stability (mechanical, thermal, and chemical), and, most importantly, safety with regard to use [33]. Due to the adequate pore volume and surface area (Figure 9.16), MSNs are capable of loading a larger quantity of drug molecules. Importantly, the surface and pore size of MSNs is very much tunable leading to enhanced drug-loading capacity [34]. Drug molecules are loaded inside the pores of MSN via the diffusion/adsorption method and drug release is triggered via internal or external stimuli.

9.4.4.2 Gold Nanoparticles (GNPs)

The key advantage of nanomaterials is their large modifiable surface and noncyto-toxicity in normal cells. GNPs (1–150 nm) are one of the most applied nanomaterials for drug delivery purposes (Figure 9.17). As already mentioned, a tunable surface triggers passive/active targeting to the affected tissues, and thus a promising thera-peutic effect can be reached. Targeting probes used here include specific antibodies, ligands, antigenic agents, etc. [35].

9.4.4.3 Quantum Dots (QDs)

QDs are another recently developed fluorescent semiconductor nanoparticle with excellent applications in the drug delivery field. QDs (2–10 nm) are composed of a core semiconductor element surrounded by another semiconductor shell (Figure 9.18) and display a large spectrum of advanced properties such as narrow emission profile, huge brightness, excellent photostability, large stokes shift, etc. The core materials of QDs such as cadmium selenide (CdSe), cadmium telluride (CdTe), indium-arsenate (In-As), etc., are toxic [36]. The shell provides stabilization to the fluorescent core and also reduces its toxicity. The shell materials for QDs vary depending upon their field of application, e.g., in DDS biocompatible QDs (carbon, graphene, and zinc oxide) are desirable. It is noteworthy that carbon-based QDs (carbon dots) are much utilized as DDSs [37]. As in the case of GNP systems, here also drug molecules are

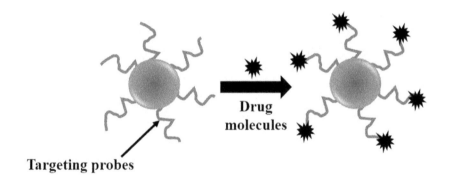

FIGURE 9.17   Schematic description of drug loading using GNP.

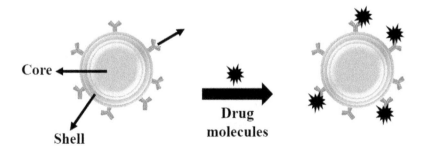

FIGURE 9.18   Schematic illustration of the structure of QDs and drug loading.

absorbed on the surface of QDs through capping agents (biocompatible materials) already present on the shell surface.

One of the major factors that needs to be considered in the case of every single DDS is its drug-loading capacity or in other words its encapsulation efficacy. The therapeutic dose of the system is considerably dependent on its encapsulation efficacy. Equation 9.1 is used to calculate the encapsulation efficacy (EE) of any system [9].

$$EE\,(\%) = \frac{m_L}{m_I} \times 100 \qquad (9.1)$$

where m_L and m_I represent the mass of drug-loaded and initial mass of drug added to the system, respectively.

Another common factor that determines the effective controlled release capacity of a DDS is the cumulative release percentage and it is calculated by performing release experiments according to the available protocols. The percentage of cumulative release is then calculated from the following equation (Equation 9.2) [38]:

$$\text{Cumulative release}\,(\%) = \frac{\text{amount of drug released}}{\text{amount of drug loaded}} \times 100\,(\%) \qquad (9.2)$$

9.5 APPLICATIONS OF DDSs

Above we have discussed various types of DDSs that have been evaluated for better therapeutic behaviour of many drugs. Now we are going to discuss some consequences of using DDSs. Most importantly, the introduction of DDSs has turned out to be one of the best strategies for cancer therapy in recent times. Many of the researches on DDSs are being focused on controlled, site-specific, prolonged delivery of anticancer drugs to get relief from the deadliest disease of our time. The anticancer drug doxorubicin is known for its cardiac myocyte toxicity [21]. As a result, usage of free doxorubicin can worsen its harmful effects on healthy human cells and this limitation has been diminished by researchers using proper DDSs. Many biodegradable and non-biodegradable amphiphilic copolymeric systems were introduced and established as efficient solutions for controlled delivery of doxorubicin in cancer treatment where the toxicity of the drug had been minimized [39–42].

Similarly, to reduce the adverse effects of subcutaneous injection of insulin (allergic reactions, pain, hyperinsulinemia, low patient compliance, etc.) on diabetic patients, researches based on the oral delivery of insulin have become very important since the last two decades. Challenges related to the presence of digestive enzymes and acidic pH conditions in the gastrointestinal (GI) tract decrease the resultant bioavailability of insulin. Against this background, pH-responsive hydrogels have been successfully applied by researchers to safely deliver insulin orally, thus escaping the adverse environment of the GI tract [43, 44].

Another commonly known disease is tuberculosis (TB) where the treatment gets hampered by patient noncompliance since it requires long-term treatment and combination drug therapy (fixed dose combination). The popular combination drugs used in the treatment of TB are isoniazid and rifampicin. However, the instability of rifampicin in the lower pH environment of the GI tract leads to its poor bioavailability. Polymeric nanocarriers such as liposomes or micelles are excellent candidates for enhancing the bioavailability and stability of such combinations and other anti-TB drugs (rifabutin, amikacin, pyrazinamide, etc.) [45, 46].

As we all know, nonsteroidal anti-inflammatory drugs (NSAIDs) are an excellent class of drugs used to treat various inflammations [47]. Ibuprofen is an NSAID used in several pain treatments; at the same time, poor water solubility restricts its bioavailability. Lipid-based DDSs for the delivery of ibuprofen have been developed by researchers that enhanced the bioavailability of the drug by increasing its water solubility and $t_{1/2}$ in plasma [48]. Another NSAID is tenoxicam (oxicam class), which is specially used to treat arthritis-related issues. Tenoxicam also suffers from low bioavailability due to its poor aqueous solubility. Besides, oral doses of the drug are limited since it causes several adverse effects like diarrhoea, vomiting, etc. [49]. However, researchers have developed alternative pathways via the transdermal route for the administration of tenoxicam in the patient's body. For this purpose, modern DDSs such as microemulsions, lipid-based carriers, nanogels, etc., have been successfully utilized by different research groups [49–51].

9.6 THEORETICAL MODELLING

After all the discussions in light of the experimental tools, at last, we must consider the theoretical aspects of DDSs. Especially, the release kinetics of drugs are considered through different theoretical models. Several kinetic models are available in the literature that describe the drug release mechanism from DDSs [52–54]. In this segment, we will discuss some of the reported models and the corresponding equations that have been widely applied by researchers. To be specific, the data obtained from in vitro release experiments (mentioned above) are fitted to the below-mentioned kinetic models to study the release kinetics.

- Zero-order model: In this model, the following equation of zero-order kinetics (Equation 9.3) is used to represent drug release.

$$Q_0 - Q_t = k_0 \times t \qquad (9.3)$$

  where Q_0 is the initial cumulative release of drug from the DDS, Q_t is the cumulative release of drug from the DDS at time t, and k_0 is the zero-order release rate constant.
- First-order model: In this model, Equation 9.4 is used to describe the drug release.

$$\ln \frac{Q_0}{Q_t} = k \times t \qquad (9.4)$$

  where k is the first-order release rate constant.

- Higuchi model: This model is based on certain hypotheses such as the initial concentration of the drug in the matrix which is higher than its solubility; constant one-dimensional drug diffusion process; constant drug diffusivity; drug particles that are much smaller than system thickness, etc. [55]. Equation 9.5 is used to describe the drug release mechanism:

$$Q_t = k_H \times t^{1/2} \tag{9.5}$$

  where k_H is the Higuchi dissolution constant.

- Korsmeyer-Peppas model: In this model, Equation 9.6 is used to describe the drug release.

$$\frac{Q_t}{Q_\infty} = k \times t^n \tag{9.6}$$

  where n is the release exponent and its numerical value provides insight into the drug release mechanism. Here, $n < 0.45$ implies Fickian diffusion mechanism; $0.45 < n < 0.89$ implies non-Fickian diffusion mechanism (both diffusion and dissolution); $n > 0.89$ suggests a dissolution mechanism, i.e., outer wall dissolution [53, 55].

- Weibull model: In this model, Equation 9.7 is utilized to describe the drug release.

$$Q = Q_0 \left[1 - e^{-\frac{(t-T)^b}{a}} \right] \tag{9.7}$$

  where Q is the amount of drug dissolved as a function of time t. Q_0 is the total amount of drug released from the DDS. T represents the lag time measured as a result of drug dissolution process. The parameter 'a' signifies a scale factor that defines the time dependence, whereas parameter 'b' represents the shape of the drug dissolution curve progression.

9.7 FINAL FATE OF DDSs

The design and development of a suitable DDS should be accompanied by detailed information of the consequences after the release of the drug to the site. To comment on that a thorough study of the in vivo fate and drug release mechanism of DDS should be carried out. Though a huge number of reports on the development of DDSs have been published in recent times, yet the resultant numbers of commercialized DDSs are very few due to the lack of complete in vivo monitoring of the stability and drug release mechanisms of the same. Thus, understanding of physicochemical properties, biostability, bioavailability, biodegradation, and pharmacokinetic behaviour of any DDS is most important to get a perfect transformation from lab to market for real-life application. Our body has a diverse chemical milieu that consists of various physicochemical parameters such as pH, enzymatic activities, reducing/oxidizing amino acids or compounds, etc. So it is expected that stimuli-responsive DDSs should undergo degradation inside the cellular environment to produce smaller segments that can be eliminated from the body or can remain inside the body as non-toxic components. For instance, macrocycles such as cyclodextrins, cucurbit[n]urils, etc., are biodegradable and hence show no toxicity towards the human body. But if degradation leads to the formation of toxic side products, it can cause serious damage to our body in the long course of time. Some of the DDSs show toxicity in the human body due to their non-biodegradable property. Specifically, dendrimers are made from non-degradable materials and are thus very difficult to remove from the human body, producing toxicity with time. Thus, a proper investigation should be carried out for a particular DDS to establish its final fate to conclude its applicability in in vivo purposes. Many new strategies have been evolved by researchers but proper clarification about "what happens to DDSs after drug release?" has not been promoted till now. Due to this, practical applications of DDSs are still a tough challenge though excellent theoretical and in vitro applicability have been established.

REFERENCES

1. He, H., Liang, Q., Shin, M.C., Lee, K., Gong, J., Ye, J., Liu, Q., Wang, J. and Yang, V., Significance and strategies in developing delivery systems for bio-macromolecular drugs. *Front. Chem. Sci. Eng.*, 7, 496, 2013.
2. Scherrmann, J.M., Drug delivery to brain via the blood–brain barrier. *Vasc. Pharmacol.*, 38, 349, 2002.
3. Martinho, N., Damgé, C. and Reis, C.P., Recent advances in drug delivery systems. *J. Biomater. Nanobiotechnol.*, 2, 510, 2011.
4. Alalaiwe, A., The clinical pharmacokinetics impact of medical nanometals on drug delivery system. *Nanomedicine*, 17, 47, 2019.
5. Yadav, H.K., Almokdad, A.A., Sumia, I.M. and Debe, M.S., Polymer-based nanomaterials for drug-delivery carriers. In *Nanocarriers for Drug Delivery*, pp. 531–556, Elsevier, 2019.
6. Nicolas, J., Mura, S., Brambilla, D., Mackiewicz, N. and Couvreur, P., Design, functionalization strategies and biomedical applications of targeted biodegradable/biocompatible polymer-based nanocarriers for drug delivery. *Chem. Soc. Rev.*, 42, 1147, 2013.
7. Patel, S., Bhirde, A.A., Rusling, J.F., Chen, X., Gutkind, J.S. and Patel, V., Nano delivers big: designing molecular missiles for cancer therapeutics. *Pharmaceutics*, 3, 34, 2011.
8. Savjani, K.T., Gajjar, A.K. and Savjani, J.K., Drug solubility: importance and enhancement techniques. *ISRN Pharm.*, 2012, 1, 2012.
9. Sengupta, P., Bose, A. and Sen, K., Liposomal encapsulation of phenolic compounds for augmentation of bio-efficacy: a review. *ChemistrySelect*, 6, 10447, 2021.
10. Zhou, J., Rao, L., Yu, G., Cook, T.R., Chen, X. and Huang, F., Supramolecular cancer nanotheranostics. *Chem. Soc. Rev.*, 50, 2839, 2021.
11. Karimi, M., Ghasemi, A., Zangabad, P.S., Rahighi, R., Basri, S.M.M., Mirshekari, H., Amiri, M., Pishabad, Z.S., Aslani, A., Bozorgomid, M. and Ghosh, D., Smart micro/nanoparticles in stimulus-responsive drug/gene delivery systems. *Chem. Soc. Rev.*, 45, 1457, 2016.

12. Rao, N.V., Dinda, H., Ganivada, M.N., Sarma, J.D. and Shunmugam, R., Efficient approach to prepare multiple chemotherapeutic agent conjugated nanocarrier. *Chem. Commun.*, 50, 13540, 2014.

13. Ganivada, M.N., Rao N, V., Dinda, H., Kumar, P., Das Sarma, J. and Shunmugam, R., Biodegradable magnetic nanocarrier for stimuli responsive drug release. *Macromolecules*, 47, 2703, 2014.

14. Santra, S., Sk, M.A., Mondal, A. and Molla, M.R., Self-immolative polyurethane-based nanoassemblies: surface charge modulation at tumor-relevant pH and redox-responsive guest release. *Langmuir*, 36, 8282, 2020.

15. Lo, Y.L., Tsai, M.F., Soorni, Y., Hsu, C., Liao, Z.X. and Wang, L.F., Dual stimuli-responsive block copolymers with adjacent redox-and photo-cleavable linkages for smart drug delivery. *Biomacromolecules*, 21, 3342, 2020.

16. Li, J., Li, X., Liu, H., Ren, T., Huang, L., Deng, Z., Yang, Y. and Zhong, S., GSH and light dual stimuli-responsive supramolecular polymer drug carriers for cancer therapy. *Polym. Degrad. Stab.*, 168, 108956, 2019.

17. Callender, S.P., Mathews, J.A., Kobernyk, K. and Wettig, S.D., Microemulsion utility in pharmaceuticals: implications for multi-drug delivery. *Int. J. Pharm.*, 526, 425, 2017.

18. Lawrence, M.J. and Rees, G.D., Microemulsion-based media as novel drug delivery systems. *Adv. Drug Deliv. Rev.*, 45, 89, 2000.

19. Mo, R., Jiang, T., Di, J., Tai, W. and Gu, Z., Emerging micro-and nanotechnology based synthetic approaches for insulin delivery. *Chem. Soc. Rev.*, 43, 3595, 2014.

20. Gastaldi, L., Battaglia, L., Peira, E., Chirio, D., Muntoni, E., Solazzi, I., Gallarate, M. and Dosio, F., Solid lipid nanoparticles as vehicles of drugs to the brain: current state of the art. *Eur. J. Pharm. Biopharm.*, 87, 433, 2014.

21. Levine, D.H., Ghoroghchian, P.P., Freudenberg, J., Zhang, G., Therien, M.J., Greene, M.I., Hammer, D.A. and Murali, R., Polymersomes: a new multi-functional tool for cancer diagnosis and therapy. *Methods*, 46, 25, 2008.

22. Nahire, R., Haldar, M.K., Paul, S., Ambre, A.H., Meghnani, V., Layek, B., Katti, K.S., Gange, K.N., Singh, J., Sarkar, K. and Mallik, S., Multifunctional polymersomes for cytosolic delivery of gemcitabine and doxorubicin to cancer cells. *Biomaterials*, 35, 6482, 2014.

23. Van De Manakker, F., Vermonden, T., Van Nostrum, C.F. and Hennink, W.E., Cyclodextrin-based polymeric materials: synthesis, properties, and pharmaceutical/biomedical applications. *Biomacromolecules*, 10, 3157, 2009.

24. Chamundeeswari, M., Jeslin, J. and Verma, M.L., Nanocarriers for drug delivery applications. *Environ. Chem. Lett.*, 17, 849, 2019.

25. Mondal, S., Saha, M., Ghosh, M., Santra, S., Khan, M.A., Saha, K.D. and Molla, M.R., Programmed supramolecular nanoassemblies: enhanced serum stability and cell specific triggered release of anti-cancer drugs. *Nanoscale Adv.*, 1, 1571, 2019.

26. Nanjwade, B.K., Bechra, H.M., Derkar, G.K., Manvi, F.V. and Nanjwade, V.K., Dendrimers: emerging polymers for drug-delivery systems. *Eur. J. Pharm. Sci.*, 38, 185, 2009.

27. Kesharwani, P., Jain, K. and Jain, N.K., Dendrimer as nanocarrier for drug delivery. *Prog. Polym. Sci.*, 39, 268, 2014.

28. Zhang, H., Zhai, Y., Wang, J. and Zhai, G., New progress and prospects: the application of nanogel in drug delivery. *Mater. Sci. Eng. C*, 60, 560, 2016.

29. Wang, H., Chen, Q. and Zhou, S., Carbon-based hybrid nanogels: a synergistic nanoplatform for combined biosensing, bioimaging, and responsive drug delivery. *Chem. Soc. Rev.*, 47, 4198, 2018.

30. Madani, S.Y., Naderi, N., Dissanayake, O., Tan, A. and Seifalian, A.M., A new era of cancer treatment: carbon nanotubes as drug delivery tools. *Int. J. Nanomed.*, 6, 2963, 2011.

31. Schmidt, D., Shah, D. and Giannelis, E.P., New advances in polymer/layered silicate nanocomposites. *Curr. Opin. Solid State Mater. Sci.*, 6, 205, 2002.

32. Ge, M., Tang, W., Du, M., Liang, G., Hu, G. and Alam, S.J., Research on 5-fluorouracil as a drug carrier materials with its in vitro release properties on organic modified magadiite. *Eur. J. Pharm. Sci.*, 130, 44, 2019.

33. Iturrioz-Rodríguez, N., Correa-Duarte, M.A. and Fanarraga, M.L., Controlled drug delivery systems for cancer based on mesoporous silica nanoparticles. *Int. J. Nanomed.*, 14, 3389, 2019.

34. Barui, S. and Cauda, V., Multimodal decorations of mesoporous silica nanoparticles for improved cancer therapy. *Pharmaceutics*, 12, 527, 2020.

35. Ajnai, G., Chiu, A., Kan, T., Cheng, C.C., Tsai, T.H. and Chang, J., Trends of gold nanoparticle-based drug delivery system in cancer therapy. *J. Exp. Clin. Med.*, 6, 172, 2014.

36. Probst, C.E., Zrazhevskiy, P., Bagalkot, V. and Gao, X., Quantum dots as a platform for nanoparticle drug delivery vehicle design. *Adv. Drug Delivery Rev.*, 65, 703, 2013.

37. Gidwani, B., Sahu, V., Shukla, S.S., Pandey, R., Joshi, V., Jain, V.K. and Vyas, A., Quantum dots: prospectives, toxicity, advances and applications. *J. Drug Delivery Sci. Technol.*, 61, 102308, 2021.

38. Cheng, C., Peng, S., Li, Z., Zou, L., Liu, W. and Liu, C., Improved bioavailability of curcumin in liposomes prepared using a pH-driven, organic solvent-free, easily scalable process. *RSC Adv.*, 7, 25978, 2017.

39. Ghoroghchian, P.P., Li, G., Levine, D.H., Davis, K.P., Bates, F.S., Hammer, D.A. and Therien, M.J., Bioresorbable vesicles formed through spontaneous self-assembly of amphiphilic poly(ethylene oxide)-block-polycaprolactone. *Macromolecules*, 39, 1673, 2006.

40. Du, J.Z., Du, X.J., Mao, C.Q. and Wang, J., Tailor-made dual pH-sensitive polymer–doxorubicin nanoparticles for efficient anticancer drug delivery. *J. Am. Chem. Soc.*, 133, 17560, 2011.

41. Mane, S.R., Rao, N.V., Chaterjee, K., Dinda, H., Nag, S., Kishore, A., Das Sarma, J. and Shunmugam, R., Amphiphilic homopolymer vesicles as unique nano-carriers for cancer therapy. *Macromolecules*, 45, 8037, 2012.

42. Chen, J., Qiu, X., Ouyang, J., Kong, J., Zhong, W. and Xing, M.M., pH and reduction dual-sensitive copolymeric micelles for intracellular doxorubicin delivery. *Biomacromolecules*, 12, 3601, 2011.

43. Lowman, A.M., Morishita, M., Kajita, M., Nagai, T. and Peppas, N.A., Oral delivery of insulin using pH-responsive complexation gels. *J. Pharm. Sci.*, 88, 933, 1999.

44. Jafari, B., Rafie, F. and Davaran, S., Preparation and characterization of a novel smart polymeric hydrogel for drug delivery of insulin. *BioImpacts: BI*, 1, 135, 2011.

45. Mane, S.R., Chatterjee, K., Dinda, H., Sarma, J.D. and Shunmugam, R., Stimuli responsive nanocarrier for an effective delivery of multi-frontline tuberculosis drugs, *Polym. Chem.*, 5, 2725, 2014.

46. Pinheiro, M., Lúcio, M., Lima, J.L. and Reis, S., Liposomes as drug delivery systems for the treatment of TB. *Nanomedicine*, 6, 1413, 2011.

47. Chakraborty, H., Banerjee, R. and Sarkar, M., Incorporation of NSAIDs in micelles: implication of structural switchover in drug–membrane interaction. *Biophys. Chem.*, 104, 315, 2003.

48. Lamprecht, A., Saumet, J.L., Roux, J. and Benoit, J.P., Lipid nanocarriers as drug delivery system for ibuprofen in pain treatment. *Int. J. Pharm.*, 278, 407, 2004.

49. Bawazeer, S., El-Telbany, D.F.A., Al-Sawahli, M.M., Zayed, G., Keed, A.A.A., Abdelaziz, A.E. and Abdel-Naby, D.H., Effect of nanostructured lipid carriers on transdermal delivery of tenoxicam in irradiated rats. *Drug Delivery*, 27, 1218, 2020.

50. Elkomy, M.H., El Menshawe, S.F., Eid, H.M. and Ali, A.M., Development of a nanogel formulation for transdermal delivery of tenoxicam: a pharmacokinetic–pharmacodynamic modeling approach for quantitative prediction of skin absorption. *Drug Dev. Ind. Pharm.*, 43, 531, 2017.

51. Goindi, S., Narula, M. and Kalra, A., Microemulsion-based topical hydrogels of tenoxicam for treatment of arthritis. *AAPS Pharmscitech*, 17, 597, 2016.

52. Liu, X., Wang, P., Zou, Y.X., Luo, Z.G. and Tamer, T.M., Co-encapsulation of vitamin C and β-Carotene in liposomes: storage stability, antioxidant activity, and in vitro gastrointestinal digestion. *Food Res. Int.*, 136, 109587, 2020.

53. Ota, A., Istenič, K., Skrt, M., Šegatin, N., Žnidaršič, N., Kogej, K. and Ulrih, N.P., Encapsulation of pantothenic acid into liposomes and into alginate or alginate–pectin microparticles loaded with liposomes. *J. Food Eng.*, 229, 21, 2018.

54. Irizarry, R., Skomski, D., Chen, A., Teller, R.S., Forster, S., Mackey, M.A. and Li, L., Theoretical modeling and mechanism of drug release from long-acting parenteral implants by microstructural image characterization. *Ind. Eng. Chem. Res.*, 57, 15329, 2018.

55. Dash, S., Murthy, P.N., Nath, L. and Chowdhury, P., Kinetic modelling on drug release from controlled drug delivery systems. *Acta Pol. Pharm.*, 67, 217, 2010.

10 Self-Healing and Self-Cleaning Nature-Inspired Surfaces

Vickramjeet Singh, Shikha Indoria, Madhu Bala, Ramesh L. Gardas and Yan-Ling Yang

10.1 INTRODUCTION

Naturally occurring materials have large structural and functional diversities, and these diversities evolved over millions of years due to "survival of the fittest" [1–3]. Mutations and selection over time help biological surfaces to adapt to various environmental conditions [1, 3]. Over these years, humans have observed, learned, analyzed, adopted, and mimicked natural materials to improve and enjoy a quality of life [3]. Biodegradable, protective, antireflective, self-healing, and self-repairing are some of the properties associated with biomaterials [3]. Thus, scientists have designed various surfaces inspired by nature, whereby the inspiration involves the earliest observation of a particular function or design which leads to the creation or development of an idea/product to have a similar function or design [4]. Mimetics are used for further advancement in the fabrication of material/surfaces inspired by nature, which can perform the specific functional task typically performed by natural materials. For instance, superhydrophobic surfaces which demonstrate low hysteresis (low droplet adhesion to the surface) have been fabricated inspired by the "lotus leaf" and the "rose petal" provides inspiration for superhydrophobic surfaces which show high adhesion (significant contact angle hysteresis [CAH]) [5–7].

Nature-inspired materials are in continuous combat with various environmental conditions such as those involving harsh conditions from a broad temperature range to humid conditions. The materials which can sustain environmental wear and tear are highly desirable for technological applications. Inspiration from nature leads to the development of not only processes but also product design and functionality [4, 8]. The lack of stability against mechanical damage or against chemical agents hinders the widespread application of nature-inspired materials/surfaces [9–11]. Again, inspired by nature, the surfaces/materials can be fabricated with the capability to self-heal and maintain the required surface wetting properties. In this chapter, we have discussed and focused on self-healing and self-cleaning surface.

DOI: 10.1201/9781032636368-10

10.2 SELF-HEALING SURFACES

Nature has provided various fascinating multifunctional engineered materials having complex surface structures and functionalities [12–13]. Biological materials not only show stability against various chemical or mechanical stresses but also demonstrate self-healing ability when damaged [9]. Inspired by these natural materials, scientists, biologists, and engineers have developed materials that can demonstrate self-repairing behavior when damaged [9]. Self-repairing means that after damage, the surfaces can repair themselves autonomously with assistance from available resources [9]. Numerous classifications and definitions are given to explain the self-repairing materials [9, 14]. Speck and co-workers have explained the terms "self-repairing", "self-healing," and "self-sealing" [9]. The focus of the chapter is on self-healing materials that, according to one of the classifications, can be divided into two: intrinsic self-healing or extrinsic self-healing [9]. The former involves healing as an inherent (intrinsic) ability of the material while for the latter the healing process is based on release of the encapsulated healing agent (rooted inside the vascular system, hollow fibers, or in microcapsules) [9, 14–17]. Further materials can be classified as (a) autonomous self-healing material which does not require any trigger (as damage can act as a trigger) to start the healing process and (b) non-autonomic self-healing material which requires an external trigger to induce healing [9, 11].

Living organisms have shown remarkable ability to autonomously self-heal and this healing process is observed from DNA to veins, tissues to organs [18]. As shown in Figure 10.1, *Delosperma cooperi* leaves demonstrate remarkable self-healing properties to mechanical damage, whereby the healing process involves tissue bending or contraction closing the wound within an hour [18]. Speck and co-workers [19] have shown the self-healing ability of *Delosperma cooperi* leaf and the healing process is attributed to a combination of driving forces including swelling and hydraulic shrinking and growth-related mechanical stresses present inside the tissues [18, 20]. During growth of the leaf elastic stress is stored within the heterogeneous structure and mechanical damage disturbs the equilibrium, releasing energy which causes viscoelastic shape conversion influencing the entire leaf to bend, thus closing the wound [18–20].

Mammals also show recovery (self-healing behavior) to mechanical damage owing to heterogeneous microstructures of cancellous bones [18, 21]. Inspired by nature various biomimic materials have been synthesized to maintain their long life span against damage while materials perform their desired functions [9, 18]. Self-healing smart materials demonstrate a healing tendency due to: (a) reconstruction of damaged bonds after material damage [9, 11, 18, 22–24]; (b) incorporation of encapsulated fluids that rupture to release the fluid for repair of damaged areas [9, 11, 18]; (c) re-mending structural damages by embedded living organisms [19]; (d) stimuli-responsive dispersed material that remotely responds toward external stimulus for initiating the healing (e.g., magnetic nanoparticles responding to an external magnetic field) [9, 18].

The self-healing behavior occurs due to chemical or physical events or a combination of both [9, 25]. When plants are damaged, biomolecules such as

FIGURE 10.1   The succulent flowering plant *Delosperma cooperi* and (b) the self-healed wound depicted by the arrow. Scale bar 5 mm [19]. Image reprinted from © [19] licensee Beilstein-Institut under the CC BY 4.0.

oligosaccharides or oligo-peptides or other biomolecules are responsible for inducing changes which sense damage and start the chemical process responsible for repair [25]. For biomimic polymeric materials the self-healing mechanism involving (a) chemical, (b) physical, and (c) physicochemical changes are shown in Scheme 10.1. Various polymeric materials or their composites which show self-healing ability have been prepared [11, 26–27]. In case of chemical process, the self-healing occurs owing to dynamic processes/bonds such as covalent bonds, supramolecular interactions, covalent bonding, or free-radical processes [9, 23, 25, 28, 29].

Polymeric composite materials such as MXene or graphene-loaded polymers have been utilized as strain sensors owing to improved conductivity [10, 30]. A continuous polymeric network can be formed by incorporation of planar 2-D material inside the polymeric matrix and the resultant composite shows high conductivity [10, 30]. The polymer under stress (in the form of elongation) is damaging to the conductive structure influencing the path of charge carriers (electrons) and may contribute toward resistance [25, 30, 31]. As shown in Figure 10.2, the self-healable crosslinked poly-siloxane elastomer has dynamic (Diels-Alder [DA]) bonds. The mechanism involves "elastomer-viscous-elastomer" transformation which is responsible for the self-healing process. The mechanism is shown in Figure 10.2, where a reversible DA reaction occurs, enabling the healing process with improvement in tensile strength after incorporation of graphene [10]. Not only dynamic DA bonds, but disulfide bonds, hydrogen bonds, reversible imine bonds, boronic ester bonds, acyl hydrazone bonds, and host-guest interactions and other weak interactions too have been employed in various polymeric materials for inducing the healing behavior once damage occurs [10, 32–39].

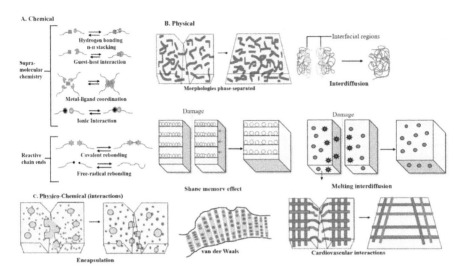

SCHEME 10.1 Various self-healing mechanisms showing the (a) chemical approaches which induce self-healing such as dynamic reversible bonds; (b) physical processes involved in self-healing, e.g., shape-memory effect; and (c) physico-chemical processes involved in achieving self-healing property such as embedded microcapsules containing the repair agents [25].

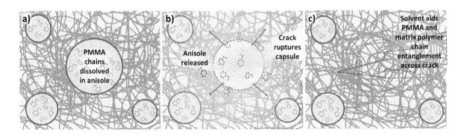

FIGURE 10.2 The self-healing mechanism based on solvent welding showing (a) the original polymeric matrix with embedded microcapsules; (b) the formation of cracks which propagate and the consequent breakage of embedded microcapsules and the release of anisole into the polymer matrix (c) solvent-aided polymeric entanglement healing the cracks [43]. Reused from [43] under CC BY 4.0.

Hydrogels and polyurethane are widely researched polymeric materials that can demonstrate self-healing property [10, 11]. The former are soft materials which are an intermediate form between a polymer solution and a solid polymer [11]. Such materials have been widely utilized for biomedical applications, e.g., lens, cancer treatment, cell or drug delivery, bioelectronics, wound healing, etc. [11]. The three-dimensionally crosslinked polymeric material can entrap water owing to its hydrophilic structure and can enable exchange of materials with the surrounding aqueous medium [11]. The mechanical damage to the 3-D structure hinders the

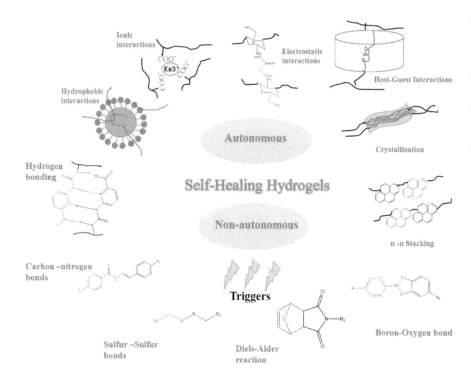

SCHEME 10.2 Various strategies employed for autonomous or non-autonomous self-healing polymeric hydrogels [11].

functions of these hydrogels. The damage can cause cracks which may propagate, further weakening the networked structure [11]. To enhance life span and mechanical durability and to maintain their function, hydrogels are being developed which can show autonomous self-repair/self-healing and this approach has been gaining attention [11]. As mentioned above, the self-healing ability can be autonomous or non-autonomous, whereby the latter requires a trigger for self-healing. Nonetheless, various interactions or dynamic bonds required for initiating the repair process are depicted in Scheme 10.2.

Polyurethane (PU) are formed by reacting polyols (R–(OH)$_n$) and isocyanates (R′–(N=C=O)$_n$) [11, 40]. The healing mechanism can be used to classify PU as intrinsic or extrinsic PU [28, 11, 41, 42]. The presence of dynamic reversible covalent bonds enables the self-healing property of an intrinsic PU providing repairing (or healing) agents pre-embedded in a microvascular network or microcapsules once a damage occurs [11]. Extrinsic polymers can show healing mechanism until healing agents are available; thus, intrinsic polymers are considered better with more durability and reliability [11].

The self-healable materials or composite materials with embedded healing agents have also shown potential for applications while maintaining their functions even under external stress [10, 26–27]. One such example is the formation of a self-healing nanocomposite material based on photocurable anisole and PMMA (polymethyl methacrylate)-filled resins [44]. Here, stereolithographic (SL) three-dimensional printing was used for creating computer-aided design with UV-curable resins having microcapsules embedded [43]. A 405-nm laser was used for crosslinking (i.e., polymerization) and the self-healing property was based on the solvent welding mechanism [43]. As depicted in Figure 10.2, light curable resins have microcapsules containing a solvent (anisole) loaded with PMMA/anisole-filled urea formaldehyde. Upon mechanical damage, cracks are formed, which moves in the structure thus rupturing the microcapsules [43], releasing the healing materials which fill the cracks. The healing liquid contains healing agents with various compositions, and curing was observed at different time intervals. The solvent welding system shows maximum 87% recovery [43]. Once the solvent is released into the cracks, it initiates polymer diffusion and entanglement around cracks formed in the polymer matrix causing cracks to heal [43].

Such crack healing is also well-known in other materials such as in liquid-infused surfaces [44–49]. Such surfaces are inspired by nectarine morphologies of the Nepenthes (pitcher) plant [44]. The plants secrete mucus from their glands not only to lure but also to capture their prey [44]. Such secreted liquid along with the porous surface structure enables the liquid to remain atop the surface, with the liquid penetrating inside the porous surface. Once such a surface is damaged externally, the liquid refills the damaged cracks, thus healing because of its slippery repellant property [44, 50].

The liquid refilling system not only enables self-healing but healing can also be done based on the shape-memory effect (SME) [44]. The heterogeneous biological morphologies can induce built-in SME, causing the material to be self-healable [44]. Microcapsules can also be loaded with living organisms such as bacterial spores for self-healing of concrete materials [19, 44, 51]. Such bacterial spores embedded inside the concrete material can remain viable for months [44–51]. Material can also initiate the self-healing property upon exposure to external stimuli. The PU-based composite containing few-layers graphene (FLG) has been shown to be mechanically stronger with repeated healing process using various external sources such as electromagnetic waves, oscillating magnetic field, electricity, and IR (infra-red) radiations [19, 52, 53]. Thus, various materials/composite materials/surfaces have been synthesized and they have been demonstrated to show mechanical durability against damage while maintaining their desired function or properties. Such materials tend to work in various environmental conditions encountering not only mechanical wear and tear but also exposure to harsh temperature, pressure, or humid conditions. Sometimes, the material needs to be cleaned to maintain it to increase the life span. Such materials have also been modified to demonstrate self-cleaning ability. In the next section, self-cleaning surfaces and their cleaning mechanism have been discussed.

10.3 SELF-CLEANING SURFACES

Inspired by nature, materials that can selectively repel water have been synthesized. The inspiration can be the lotus leaf or the rose petal [6]. The former involves superhydrophobic surface with negligible hysteresis and the latter has high adhesion power, as shown in Figure 10.3 [54–56]. The material which can sustain environmental

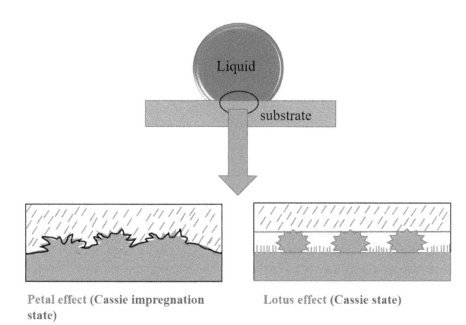

Petal effect (Cassie impregnation
state)

Lotus effect (Cassie state)

FIGURE 10.3   A liquid drop on a substrate showing the petal effect (i.e., Cassie impregnation state) and the Cassie state for lotus effect.

wear and tear is highly desirable for technological applications. The lack of stability against mechanical damage or against chemical agent hinders the widespread application of repellent surfaces. Again, inspired by nature, surfaces/materials which show the ability to self-heal along with maintaining the required surface wetting properties can be fabricated [6, 55, 56].

For fabrication and application of superhydrophobic surfaces, wettability characterization is required. How the surface interacts with a given liquid in air or in submerged conditions is influenced by surface wettability [56, 58]. The surface wettability characterization involves measuring the three-phase contact angle (CA) and also the CAH [56]. Surface wettability is the interaction of a liquid in contact with a solid substrate surrounded by other liquid or gas. These types of interactions are due to intermolecular forces (both adhesive and cohesive) and such interactions can be controlled or modulated [56–58].

Solid surfaces can be divided into two categories, i.e., low-energy surfaces and high-energy surfaces [56]. Low-energy surfaces such as Teflon and other polymers show poor wettability. In contrast, high-energy surfaces show complete wetting due to a lowering in interfacial energy. High-energy surfaces include glass, metal, and ceramic [59]. These solids are held together by strong chemical forces such as covalent, metal, and ionic bonds. When combined with surface topography, partially wetting a low-energy surface causes the surface to exhibit very high CAs (superhydrophobic, CA > 150°) [56, 60].

Surface wettability can be characterized by measuring CA. The most commonly used method for CA measurement is a sessile drop method [61–66]. CA can be estimated by optical measurements by visualizing the drop cast onto a surface and analyzing it using software. Lower CA ($\leq 90°$) defines good wettability while higher CA ($\geq 90°$) defines poor wettability. Surfaces having CA $\leq 90°$ are known as hydrophilic surfaces and surfaces having CA $\geq 90°$ are known as hydrophobic surfaces, when water is used as the measuring liquid. Complete wetting can be achieved when CA is zero while a CA of 180° shows no wetting at all.

The earliest study of geometrically determined CA was by Thomas Young in 1805 [64, 67]. If the solid-liquid interfacial energy is denoted as γ_{SL}, the solid-vapor interfacial energy as γ_{SV}, and the liquid-vapor interfacial energy as γ_{LV}, then the equilibrium CA by the Young's equation is described as

$$\cos\theta = (\gamma_{SV} - \gamma_{SL}) / \gamma_{LV} \qquad (10.1)$$

where θ is the CA between the drop and any other solid substrate. Surface roughness plays an important role in governing CA and wettability of a surface. The Cassie-Baxter and Wenzel model represent the CA on a real substrate. Any real substrate has some roughness or we can say no surface is perfectly smooth [64–67]. The Wenzel model is applicable for a rough surface but the surface should have chemical homogeneity [67]. In contrast, the Cassie-Baxter model is applicable for a flat surface but the surface should have chemical heterogeneity [68]. The Wenzel model correlates this in mathematical form as

$$\cos\theta_A = r\cos\theta_T \qquad (10.2)$$

where θ_A is the calculated CA; θ_T is the Young's CA as described in Equation 10.1; r is the surface roughness ratio between the actual and projected surface area. The Wenzel model concluded that the surface roughness describes the wettability of the real surface. This presence of surface roughness can make a hydrophilic surface superhydrophilic and a hydrophobic surface superhydrophobic [67].

The Cassie-Baxter model equation is given as

$$\cos\theta = f_1 \cos\theta_1 - f_2 \qquad (10.3)$$

where θ is the angle for the liquid-air interface; f_1 is the contact area of the surface; f_2 is the contact area of the air underneath the drop; θ_1 is the CA of the water with the surface. In the Cassie-Baxter model, higher CA indicates the larger area fraction of the drop with air trapped beneath [68].

The given solid-liquid-vapor combination yields a broad range of CAs. The increasing contact line is called advancing contact angle while the decreasing contact line is called receding contact angle [61–66]. The difference between them (advancing and receding CA) is called contact angle hysteresis (CAH), which also gives the measure of non-uniformity of the surface [54–56]. Some methods that study CAH measurement are the needle method, the tilted method, and the Wilhelmy method. Johnson and Dettre proposed a model which accurately calculates the advancing and receding CA [54–56]. In this model, the rough surface causes metastable states that trap a three-phase contact line.

Superhydrophobic surfaces can be fabricated by combining surface topology (nano-/microstructures) with suitable surface chemistry (low-energy material) [69]. The surface topology can be made using a number of techniques such as solvent evaporation [70], etching of metals/alloys/substrates [71, 72], template or replica, laser abrasion, light irradiation [73–75], plasma treatment [76], chemical/physical vapor deposition [77], sol-gel method [78], and high temperature curing [79, 80]. Surface chemistry can be performed by using fluorinated (e.g., perfluorochemicals) or non-fluorinated chemicals (e.g., inorganic nanoparticles) [79, 80]. Chemicals can be deposited on the surface using various methods such as dip-coating, electro-spun, spray-coating, spin-coating, vapor deposition, and the flame method [80]. Nonetheless, the application of superhydrophobic surfaces is limited by the lack of mechanical and chemical stability. Moreover, the superhydrophobic surface fails under the influence of a surface-active agent or against hot water [81]. Surfaces lose topology or low-surface energy material due to mechanical abrasion, chemical exposure, or under high temperature conditions [80]. Surfaces can be easily damaged by a pencil scratch and to overcome these stability-related issues, mechanochemical durable surfaces must be fabricated [80]. One of the alternatives is to fabricate superhydrophobic surfaces that can show self-healing ability against external stimuli [80–82]. Again, inspired by nature, the self-healing mechanism involves replenishment of surface low-energy molecules or repair/reorganization of surface topology [80–82]. Moreover, the surface can remain free from contamination owing to self-cleaning ability. Thus, in the next section natural-inspired materials showing autonomous cleaning behavior are highlighted.

10.4 NATURE-INSPIRED SURFACES

The self-cleaning behavior of natural surfaces inspired scientists to develop materials having self-cleaning abilities [6, 82]. In recent years, the surfaces inspired by the superhydrophobic "lotus leaf" having self-cleaning ability have attracted great attention. The leaf of the lotus plant has the tendency to remain clean even when present in muddy water. The combination of vsurface chemistry and nano-hierarchical structures provides a very high CA for the liquid drop. On such surfaces the water can roll off with ease, taking impurities along with it, as shown in Figure 10.4.

Strong adhesive strength with reusable ability of adhesive tapes is inspired by gecko [6, 82]. The ability of gecko to produce adhesive making their toes sticky and clean during the climbing motion has inspired not only adhesives but also tiny

FIGURE 10.4   The self-cleaning action on a superhydrophobic surface. The moving drop picks up dirt along the surface when rolling on it, thus cleaning the surface.

click and climb robots [6, 82]. In addition to self-cleaning ability other functionalities are also offered by smart surfaces/materials; these functions range from anti-icing to anti-fog, anti-biofouling to oil-water separation membranes/filters [6, 82]. The self-cleaning ability of a surface depends upon the surface wettability and its tuning toward various liquid drops. The surface can either become highly partial wetting (superhydrophobic/superoeleophobic CA > 150°) or complete wetting (or superhydrophilic/oeleophilic) [83]. The two extreme surface wettabilities can provide an opportunity for the materials to demonstrate self-cleaning ability. So, the surface can be made either extremely wettable or non-wettable. In the former case, the self-cleaning ability is due to fluid film flow on the surface, while the latter involves dynamic drop motion which can carry dirt particles along with it, the so-called "lotus effect" originating from the sacred leaf of purity [83].

10.4.1 Superhydrophobic Surfaces

The surfaces which demonstrate water contact angle (WCA) in air greater than 150° are termed as superhydrophobic and those surfaces which have WCA < 5° in air are called superhydrophilic [84]. However, total wetting surfaces showing WCA ≈ 0° in air are sometimes also referred to as superhydrophilic [85, 86]. On such total wetting surface CA cannot be estimated [87]. The Young's equation is used for calculation of CA on a smooth and chemically homogeneous surface, as shown in Equation 10.2. The roughness factor, r, and the chemical nature of the surface are the factors which influence the CA. For a smooth flat solid surface, the surface energy can be minimized (~6.7 mJ m^{-2}) by using fluorinated groups (CF$_3$); however, in such a case the maximum CA reached can be 120°. According to the well-known Wenzel and Cassie-Baxter models, the roughness factor, r, can influence WCA [84]. For reaching WCA greater than 150°, air trapped inside the surface roughness (grooves) along with low-energy surface groups is required. This unique roughness combination with surface chemistry can provide a superhydrophobic surface [84]. However, the surface roughness can also influence the CAH. Ultra-low to very high adhesion is possible on a superhydrophobic surface [84]. The Cassie-impregnated state gives high CAH and the air-trapped Cassie-Baxter angle provides low CAH. The CA on an air-trapped solid surface with roughness is given by the equation $\cos \theta_{CB} = f(\cos\theta + 1) - 1$, where θ_{CB} is the Cassie-Baxter CA and f is the fraction of the solid surface. Since the fraction f cannot be zero, so CA cannot be 180°. For a total wetting surface [86], the roughness factor r should $r \geq 1/\cos\theta$ [86]. Bioinspired artificial surfaces fabricated for various applications are provided in Scheme 10.3 [7].

The selected surfaces inspired by nature are discussed below.

10.4.1.1 Plant Leaves

Various natural surfaces such as plant leaves, insect legs, and wings demonstrate the so-called lotus effect [87–89]. Here a water drop with high CA (>150°) does not remain static but moves spontaneously when the surface is disturbed (trembling) [87–89]. During the drop motion, if dirt is present on the surface, it is removed [87–89]. The self-cleaning lotus leaf (*Nelumbo nucifera*), which is a symbol of sacred purity, is known for more than 2,000 years [5–7]. The combination of surface roughness and

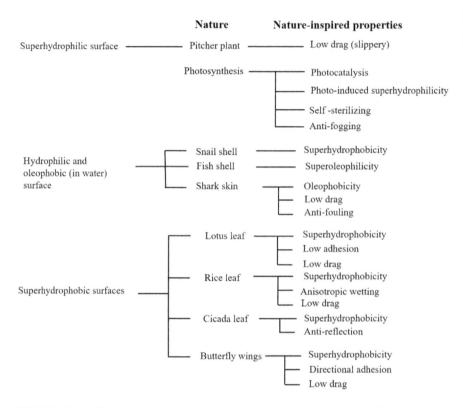

SCHEME 10.3 The schematic representation between natural materials (surfaces) and their functions/properties which inspires the fabrication of bio-mimicked functional materials [7].

hydrophobic (epicuticular) wax makes the lotus leaf superhydrophobic [5, 6, 90]. The leaf surface has roughness due to papillose epidermal cells forming microasperities or papillae [5, 90]. The papillae also have a rough surface having nanoscale asperities composed of 3-D epicuticular waxes [5, 90]. The waxes are long-chain hydrophobic hydrocarbons that can exist in various morphologies, for instance, tubules in the case of lotus leaf [5, 90]. Cyclic compounds are also present in some plant species [5, 91]. The water drops on these hierarchical structures sit on the surface such that air pockets are formed beneath. That is why the static WCA is about 163° with low CAH of 3° [5, 92, 93]. Natural wax such as beeswax is also produced by insects and is extensively used in the cosmetic industry [91].

The non-wetting properties are thus inspired by nature, i.e., leaf, stem, or fruits of various plant species, for fabrication and applications of super-repellent engineered surfaces [91, 93]. The SEM characterization of natural surfaces provided an opportunity to study and mimic the micromorphology so as to duplicate these structural features in superhydrophobic surfaces. The dual-scale surface topology on lotus leaves [94] provides surface roughness due to hierarchical structures which are nanostructures on top of micropapillae [94]. Such structures are mimicked to obtain engineered surfaces with superhydrophobicity [94].

FIGURE 10.5 (a) The optical photograph of taro plant (*Colocasia esculenta*) leaves; (b) plucked taro leaf; and (c) water drops on the surface of the taro leaf. Reference [93]. Reused under CCA 4.0 International License from [93].

Self-cleaning plant species (around 200 species) were studied for their surface morphological textures and wettability by Neinhuis and Barthlott [6]. Most of these plant species are superhydrophobic with 3-D cuticular wax morphologies on the surface, thus exhibiting Cassie-Baxter structure for high water repellency [6, 92, 94, 95]. The water repellent behavior of taro plant leaves (*Colocasia esculenta*) was also recently reported [93]. The average WCA of 150.5° was reported with a CAH of about 9°. Taro leaves and drop standing on taro leaves is shown in Figure 10.5.

The SEM images of the taro leaf with different level of zoom are shown in Figure 10.6 [93]. The leaf depicts a combination of a pentagon and a hexagon forming a honeycomb structure. These structures on further zoom can be seen to have nanoscales on top of the pentagon and hexagon [93]. These images show that a two-tier structure along with a primary honeycomb and secondary nanoscale flake-like structures are present. Inspired by these, superhydrophobic surfaces with honeycomb type structure were fabricated onto silicon wafers by the lithographic technique [93]. The fabricated surface demonstrated non-wetting properties with significant hysteresis as water drops remain stuck even when the surface is tilted 90° or 180° [93].

The air trapped between the suspended drops on a superhydrophobic surface inspired by the lotus leaf was also fabricated using suspended single-walled carbon

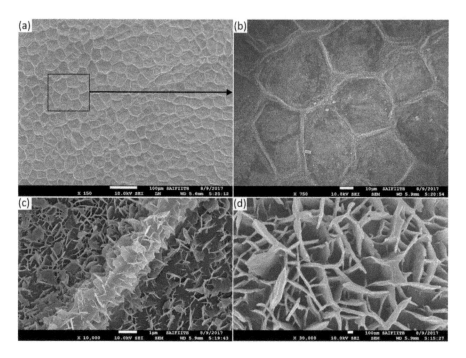

FIGURE 10.6   The scanning electron microscopic images of taro leaves magnified at (a) 150×, (b) 750×, (c) 10,000×, and (d) 30,000× [93]. Reused under CCA 4.0 International License.

nanotubes (SWCNTs) networks onto a 3-D patterned flexible micro-patterned poly-meric substrate (polydimethylsiloxane, PDMS) [96]. Wang et al. [97] have also fabri-cated a superhydrophobic self-cleaning surface on an Al substrate. The surface was made superhydrophobic by combination of surface topology with chemistry involv-ing treatment with 1H,1H,2H,2H-perfluorooctadecyltrichlorosilane (FDTS) [97]. The Al surface was chemically etched in HCl to form surface facets followed by the creation of a porous alumina layer (Al$_2$O$_3$). Finally, after surface topology creation, the surface was dip-coated with FDTS in ethanol [97].

A year-aged sample depicts superhydrophobicity indicating that the surface can maintain its non-wetting behavior even when kept in air for a very long period of time [97]. There was a tendency for the water drop to slide at a low tilted angle of 1° and the surface was evaluated for self-cleaning behavior using sand as dirt [97]. As shown in Figure 10.7, the water drops placed on the surface easily clean the sand particles leaving the surface clean. The surface was coated with sand particles as shown in Figure 10.7a, and water drops are placed on the surface to remove sand (swallowed) showing the self-cleaning properties in Figure 10.7b–d [97]. The Al surface depicted mechanical stability against sand and fabric abrasion, the impact of water, and also corrosion [97].

Leaf-inspired surfaces were fabricated on a copper substrate having dual-scale surface structures, i.e., micro-/nanohierarchical structures [94]. The fabrication steps involve cleaning copper to remove native oxides and electroless deposition using

FIGURE 10.7   Optical images depicting the self-cleaning behavior of a fabricated superhydrophobic surface. The surface was contaminated by sand particles and cleaned by sliding drops of water [97]. Reused under CCA 4.0 international license.

salts (NiSO$_4$·6H$_2$O, Na$_3$C$_6$H$_5$O$_7$·2H$_2$O; NaH$_2$PO$_2$·H$_2$O) as main/reducing/complexing agents and polyethylene glycol as a crystal modifier. Such electrodeposition formed a micropapillary layer onto the copper surface. Further, electrodeposition was performed to obtain nano-nickel structures. Various cone-shaped to hemispherical papillae with nanoscale branch-like structures were formed and demonstrated superhydrophobicity. This nanoscale branch-like nickel topology resembles lotus leaf surface structures, as shown by SEM image analysis [94].

Liu et al. [98] have fabricated a transparent self-cleaning superhydrophobic surface by the sol-gel method using fluoroalkylsilane. The solution-based coating used ethanol as a solvent, ammonia as a catalyst, and water for hydrolysis. The methoxy groups of (heptadecafluoro-1,1,2,2-tetrahydrodecyl) trimethoxysilane undergo hydrolysis and polycondensation reaction forming a networked structure with non-hydrolyzable groups (fluoro-carbons) [98]. The surface covering of low-energy material was controlled to obtained a combination of surface chemistry and roughness. Such surface demonstrated CA of about 169° with low sliding angle of less than 5° when the coating was deposited in a time span of 300 s (termed as T3 coating). This superhydrophobic state occurs owing to the Cassie-Baxter state, i.e., low surface free energy material and a rough surface with air pockets. The surface interstices were able to entrap air beneath the liquid water drop showing very high CA. The lotus effect is easily evident in water drops by running a stream of water jet on the surface. Self-cleaning ability was demonstrated when ten water drops of volume 10 μL each were able to clean an area of $2 \times 2\,cm^2$ having dirt particles (Oil Red O) while water drops rolled on the surface [98].

For low surface energy hydrophobic materials, expensive and non-degradable fluorinated compounds have been used [99]. However, these fluorinated molecules are toxic and have shown bioaccumulation and birth defects (due to perfluoro sulfonate and perfluorooctanoic acid) [100, 101]. Thus, the future need is to replace toxic fluorinated coatings with biocompatible, simple, substrate independent, completely fluoride-free anti-wetting coatings [99]. In literature, fluoride-free coatings were made using long-chain alkanes, inorganic nanoparticles (TiO_2, SiO_2), waxes, PDMS, or carbon nanomaterials, metal alkyl carboxylates, and montmorillonites with organosiloxane [99, 101, 102–106].

Recently, a composite system based on nano-TiO_2 and an amorphous SiO_2 microsphere was fabricated by the wet grinding method. The surface was modified with PDMS and spray cast onto the surface. The resultant surface demonstrated tuneable adhesion and wettability triggered by UV irradiation and calcination [107]. The changes in surface wettability due to exposure to UV radiation are shown in Figure 10.8 [107]. As can be seen in Figure 10.8, the surface shows tunable wettability changes by modulating the coating from a "lotus-like" to "rose-like" surface upon treatment with UV irradiation and calcination [107]. Such surface also demonstrated the self-cleaning behavior showing the lotus effect, as demonstrated in Figure 10.8d.

The dust-containing surface was irradiated with UV light and the surface was cleaned with water falling on the tilted dirty surface. The surface turned out to be clean at the end of the experiment, demonstrating self-cleaning ability. Leaves of other plants such as taro (*Colocasia esculenta*)) leaves, rice (*Oryza sativa*) leaves, India canna leaves, water ferns, Mimosaceae, etc. [6, 7], also exhibit water repelling and self-cleaning. Among these, rice leaves interestingly demonstrate surface anisotropic wettability [7, 108]. Bixler and Bhushan [108] have for the first time reported that rice leaves and the butterfly wing combine the shark skin and lotus effect, thereby providing a surface showing water repellency, anisotropic flow, self-cleaning ability, low adhesion properties, which were believed to be promoting self-cleaning, low drag, directed fluidic control, and antifouling properties [26, 108]. The water drops on rice leaves roll off along the veins and demonstrate low sliding angle in that direction [26]. Scientists [27, 109–112] have adopted various methodologies to obtain rice-mimicking surfaces with anisotropic CAs and one such report involves template synthesis using the natural green rice leaf surface and the chemicals used for two-step transcription were PDMS and heptadecafluoro-1,1,2, 2-tetradecyl trimethoxysilane [113].

SEM image analysis revealed that the surface contains arrays of sub-millimeter-scaled macro grooves, with an approximate width of 50 μm and height of approximately 200 μm. The papilla structures were approximately of 3–5 and 5–8 μm heights and widths, respectively, [113]. The zoomed image depicts various nanostructures on the papilla structures [113]. The micro-/nanostructures on the surface make the surface rough and superhydrophobic. The anisotropic wetting property arises due to

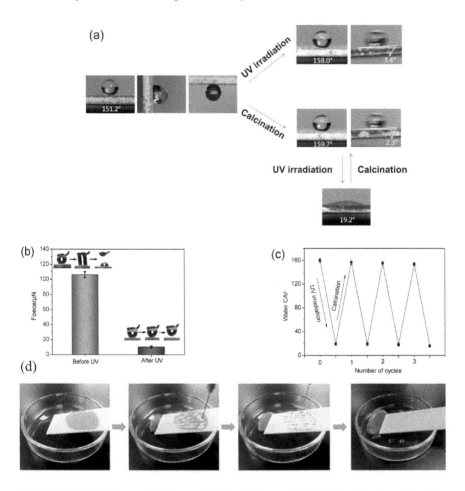

FIGURE 10.8 (a) The wettability modulation before and after UV irradiation and calcination; (b) force of adhesion measured before and after treatment with the stimuli (UV); (c) the WCA before and after UV treatment and calcination [107]; (d) the self-cleaning experiment to demonstrate the "lotus effect" [107]. Reused under CCA International license.

micro-grooves extending in one direction parallel to the vein direction, with arrays of papilla ordered in the same direction [113]. The hierarchical structure makes water drops move along the grooves, demonstrating very low sliding angle in a particular direction [113]. Natural rice leaves and an artificial surface demonstrated two sliding angles along two orthogonal directions. For rice leaves the sliding angles were 9° and 3° along the perpendicular and parallel directions, respectively. For the artificial mimicked surface the sliding angles were 40° and 25°, respectively [113].

10.4.1.2 Insect Wings

Insect wings have self-cleaning behavior owing to the topological superhydrophobic surface of the wings [114]. The ability to prevent contamination (dust, pollen fragments, soil, pathogenic bacteria) and self-cleaning of wings is because of non-wetting surface [91, 115]. Hydrodynamics, aerodynamics, self-cleaning surfaces have been inspired by insect wings [91, 115, 116]. In humid conditions, the self-cleaning action is achieved by the jumping condensate. The smaller drops coalescence forming larger drops, and these larger drops themselves get ejected from the surface picking contaminants before leaving the surface. The surface energy released due to coalescence causes autonomous drop ejection [91, 117]. Such surfaces made of lipids, proteins, tissues are superhydrophobic and lack silicone or fluorochemicals. These biological surfaces provide an inspiration to synthesize/fabricate fluorine-free repellent surfaces [91, 117].

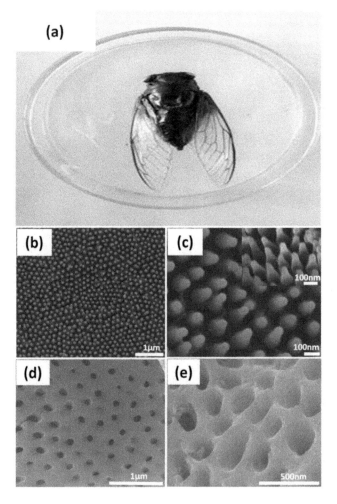

FIGURE 10.9 (a) Optical image of the insect black cicada wing; (b) SEM image at low resolution showing nano-nipple; (c) magnified image of nano-nipple arrays image; (d) SEM image of the negative TiO$_2$ replica; and (e) magnified view of biomorphic TiO$_2$ (replica) [118]. Reused under CCA 4.0 International license.

Insect-inspired surfaces have been fabricated for various applications including photocatalytic activity and solar cell applications (antireflective coatings) [7, 118]. The nano surface topology also enhances photocatalytic activity. The black cicada wing, as shown in Figure 10.9, has nanopillar (nano-nipple) arrays on the wing, as observed by SEM analysis, with separation between nano-nipples, as shown in Figure 10.9c. The cicada wings have hexagonally packed nano-nipples (nanopillars) with pitch values between 110 and 140 nm [7, 119–120]. The nanosurface hierarchical structure is responsible for superhydrophobic, antireflection, and self-cleaning properties [7, 118]. The cicada wing-inspired negative replica structures of biomorphic TiO_2 nanoholes are depicted in Figure 10.9d and e [118].

Here, Atomic Force Microscope (AFM) and Scanning Electron Microscope (SEM) analysis reveals that the self-cleaning mechanism is similar to that observed for the lotus leaf [7, 118]. The self-cleaning properties maintain the antireflecting behavior as contaminants are removed from the surface and the inspired coatings which remain clean and transparent are used for solar cell applications [7, 118]. Like cicada wings, butterfly wings which are superhydrophobic (CA 152° ± 2°) have also inspired scientists to fabricate smart surfaces [7]. The wings also demonstrated directional adhesion behavior (anisotropic grooves providing a low drag) [7, 108]. Nature has always inspired not only to mimic but also to learn so that technological advancement and development can be sustainable for the environment as a whole.

10.5 CONCLUSION

Natural materials such as proteins, saccharides, biopolymers have shown potential to be utilized in diverse biomedical applications. Wound healing and sealing are functions commonly observed in nature [4, 9, 92]. Not only function but structural features of biomaterials have also evolved over a period of time [4, 9, 92]. Industries and academia have been fascinated by the functions and properties (self-cleaning and self-healing) of biomaterials. Thus, inspired by nature various surfaces/interfaces/materials have been fabricated to mimic not only their structural features but also their functions [9, 19]. The largest interface on earth is the interface provided by biological surfaces (plant surfaces), which evolved according to change over a billion of evolution years. The concept of "living-like" has thus grown and synthetic capabilities have improved by continuous development of new fabrication methodologies [9, 19, 25]. Technological development and application of nature-inspired material becomes feasible by systematic transmission of understandings of natural materials and their functions.

Mimicking the biological world to have products readily available for human use has advantages as well as disadvantages. Smart materials that can self-heal have recently been employed for soft robotics [121]. The adaptation to environmental conditions requires not only durability against harsh temperature/humid conditions but also stability against mechanical wear and tear [25, 121]. Polymeric materials have been used in fields from biomedical to soft computing, whereby the ecological and economic remedy is to have a material which shows its robustness against external stimuli (i.e., can self-heal when damaged or self-clean when contaminated). Commercial realization of nature-inspired materials is still a challenge which must consider affordable and scalable development [4, 25]. Biocompatibility, large-scale production, biodegradability, and price are key challenges associated with fabrications and applications of nature-inspired materials [4, 25]. Thus, there is a need for sustainable and reliable solutions to overcome these problems. Living-like materials will be constructed using material that can show adaptive and autonomous properties [25].

ACKNOWLEDGMENT

MB and VS would like to acknowledge the Science and Engineering Research Board (SERB), India, for their financial support (SRG/2019/000846) and Dr. B. R. Ambedkar NIT Jalandhar, India, for the necessary infrastructure facilities. The author S. Indoria would like to acknowledge LPU, Jalandhar. RLG acknowledges the financial support from IIT Madras through the Institute Research & Development Award (IRDA): CHY/20–21/069/RFIR/008452.

REFERENCES

1. Koch, K., Barthlott, W., Superhydrophobic and superhydrophilic plant surfaces: an inspiration for biomimetic materials. *Phil. Trans. R. Soc. A*, 367, 1893, 2009.
2. Youngblood, J.P., Sottos, R.N., Bioinspired materials for self-cleaning and self-healing. *MRS Bull.*, 33, 732–741, 2009.
3. Makhlouf, A.S.H., Rodriguez, R., Chapter 15 - Bioinspired smart coatings and engineering materials for industrial and biomedical applications. In *Advances in Smart Coatings and Thin Films for Future Industrial and Biomedical Engineering Applications*, Abdel Salam Hamdy Makhlouf, Nedal Y. Abu-Thabit (eds.), Elsevier, 407–427, 2020.
4. Katiyar, K.N., Goel, G., Hawi, S., Goel, S., Nature-inspired materials: emerging trends and prospects. *NPG Asia Mater.*, 13, 56, 2021.
5. Bhushan, B., Jung, Y.C., Natural and biomimetic artificial surfaces for superhydrophobicity, self-cleaning, low adhesion and drag reduction. *Prog. Mater Sci.*, 56, 1, 2011.
6. Barthlott, W., Neinhuis, B., Purity of scared lotus or escape from contamination in biological surfaces. *Planta*, 202, 1, 1997.
7. Nishimoto, S., Bhushan, B., Bio-inspired self-cleaning surfaces with superhydrophobicity superoleophobicity and superhydrophilicity. *RSC Adv.*, 3, 3, 2013.
8. Troy, E., Tilbury, M.A., Power, A.M., Wall, J.G., Nature based biomaterials and their applications in biomedicine. *Polymers*, 13, 3321, 2021.
9. Speck, O., Speck, T., An overview of bioinspired and biomimetic self-repairing materials. *Biomimetics*, 4, 26, 2019.
10. Ma, Z., Li, H., Jing, X., Liu, Y., Mi, H.Y., Recent advancement in self-healing composite elastomer for flexible strain sensors: materials, healing systems, and features. *Sens. Actuators A*, 329, 112800, 2021.
11. Zuo, X.L., Wang, S.F., Le, X.X., Lu, W., Chen, T., Self-healing polymeric hydrogels: Toward multifunctional soft smart materials. *Chinese J. Polym. Sci.*, 39, 1262–1280, 2021.
12. Vijayan, P.P., Puglia, D., Biomimetic multifunctional materials: a review. *Emergent Mater.*, 2, 1, 2019.
13. Kim, J.R., Netravali, A.N., Self-healing green polymers and composites. In *Advanced Green Composites*, Anil Netravali (ed.), New York: Wiley, 135–186, 2018.

14. Döhler, D., Michael, P., Binder, W., Principles of self-healing polymers. In *Self-Healing Polymers: From Principles to Applications*, W. Binder (ed.). Weinheim: Wiley-VCH, 34, 203–220, 2013.
15. Diesendruck, C.E., Sottos, N.R., Moore, J.S., White, S.R., Biomimetic self-healing. *Angew. Chem. Int. Ed.*, 54, 36, 2015.
16. Bekas, D.G., Tsirka, K., Baltzis, D., Paipetis, A.S., Self-healing materials: a review of advances in materials, evaluation, characterization and monitoring techniques. *Compos. Part B Eng.*, 87, 92, 2016.
17. Hager, M.D., Greil, P., Leyens, C., van der Zwaag, S., Schubert, U.S., Self-healing materials. *Adv. Mater.*, 22, 47, 2010.
18. Yang, Y., Davydovich, D., Hornat, C.C., Liu, X.M.W., Urban, M.W., Leaf inspired self-healing surface. *Nat. Rev. Mater.*, 4, 1928–1936, 2018.
19. Speck, O., Schlechtendahl, M., Borm, F., Kampowski, T., Speck, T., Humidity-dependent wound sealing in succulent leaves of Delosperma cooperi-an adaption to seasonal drought stress. *Beilstein J. Nanotechnol.*, 9, 175, 2018.
20. Quan, Y.Y., Zhang, L.Z., Qi, R.H., Cai, R.R., Self-cleaning of surfaces: the role of surface wettability and dust types. *Sci. Rep.*, 6, 38239, 2016.
21. Torres, A.M., Matheny, J.B., Keaveny, T.M., Taylor, D., Rimnac, C.M., Hernandez, C.J., Material heterogeneity in cancellous bone promotes deformation recovery after mechanical failure. *Proc. Natl. Acad. Sci. USA*, 113, 11, 2016.
22. Chen, S., Li, X., Li, Y., Sun, J., Intumescent flame-retardant and self-healing superhydrophobic coatings on cotton fabric. *ACS Nano*, 9, 4, 2015.
23. Ghosh, B., Urban, M.W., Self-repairing oxetane-substituted chitosan polyurethane networks. *Science*, 323, 5920, 2009.
24. Imato, K., Nishihara, M., Kanehara, T., Amamoto, Y., Takahara, A., Otsuka, H., Self-healing of chemical gels cross-linked by diarylbibenzofuranone-based trigger-free dynamic covalent bonds at room temperature. *Angew. Chem. Int. Ed.*, 51, 5, 2012.
25. Wang, S., Urban, M.W., Self-healing polymers. *Nat. Rev. Mater.*, 6, 20933, 2015.
26. Yao, Y., Xiao, M., Liu, W., A short review on self-healing thermoplastic polyurethanes. *Macromol. Chem. Phys.*, 222, 8, 2021.
27. Zhang, F., Zhang, L., Yaseen, M., Huang, K., A review on the self-cleaning ability of epoxy polymers. *J Appl. Polym. Sci.*, 138, 16, 2021.
28. Chen, Y., Kushner, A.M., Williams, G.A., Guan, Z., Multiphase design of autonomic self-healing thermoplastic elastomers. *Nat. Chem.*, 4, 6, 2012.
29. Yang, Y., Urban, M.W., Self-repairable polyurethane networks by atmospheric carbon dioxide and water. *Angew. Chem. Int. Ed.*, 53, 45, 2014.
30. Liu, S., Lin, Y., Wei, S., Chen, S., Zhu, J., Liu, L., A high performance self-healing strain sensor with synergetic networks of poly (-caprolactone) microspheres, graphene and silver nanowires. *Compos. Sci. Technol.*, 146, 110–118, 2017.
31. Kavitha, A.A., Singh, N.K., Smart "all acrylate" ABA triblock copolymer bearing reactive functionality via atom transfer radical polymerization (ATRP): demonstration of a "click reaction" in thermoreversible property. *Macromolecules*, 43, 7, 2010.
32. Yarmohammadi, M., Shahidzadeh, M., Ramezanzadeh, B., Designing anelastomeric polyurethane coating with enhanced mechanical and self-healing properties: the influence of disulfide chain extender. *Prog. Org.Coat.*, 121, 2018.
33. Wu, X., Wang, J., Huang, J., Yang, S., Interfaces, robust, stretchable, and self-healable supramolecular elastomers synergistically cross-linked by hydrogen bonds and coordination bond. *ACS Appl. Mater. Interfaces*, 11, 7, 2019.
34. Hu, J., Mo, R., Jiang, X., Sheng, X., Zhang, X., Towards mechanical robust yet self-healing polyurethane elastomers via combination of dynamic mainchain and dangling quadruple hydrogen bonds. *Polymer*, 183, 121912, 2019.

35. Ling, L., Liu, F., Li, J., Zhang, G., Sun, R., Wong, C.P., Self-healable and mechanically reinforced multidimensional-carbon/polyurethane di-electric nano-composite incorporates various functionalities for capacitive strain sensor applications. *Macromol. Chem. Phys.*, 219, 1800369, 2018.

36. Xiao, L., Shi, J., Wu, K., Lu, M., Self-healing supramolecular waterborne polyurethane based on host–guest interactions and multiple hydrogen bonds. *React. Funct. Polym.*, 104482, 2020.

37. Schiff, H., Mittheilungen aus dem Universitätslaboratorium in Pisa: eineneue reihe organischer Basen. *Archiv Der Pharmazie*, 131, 1, 1864.

38. Zhao, W., Qu, X., Xu, Q., Lu, Y., Dong, X., Ultrastretchable, self-healable, and wearable epidermal sensors based on ultralong Ag nanowires composited binary-networked hydrogels. *Adv. Electron. Mater.*, 6, 7, 2020.

39. Chen, Y., Tang, Z., Zhang, X., Liu, Y., Wu, S., Guo, B., Covalently cross-linked elastomers with self-healing and malleable abilities enabled by boronic ester bonds. *ACS Appl. Mater. Interfaces*, 10, 28, 2018.

40. Wu, T., Chen, B., Facile fabrication of porous conductive thermoplastic polyurethane nanocomposite films via solution casting. *Sci. Rep.*, 7, 17470, 2017.

41. Hia, I.L., Vahedi, V., Pasbakhsh, P., Self-healing polymer composites: prospects, challenges, and applications. *Polym. Rev.*, 56, 2, 2016.

42. Tittelboom, K.V., Belie, N.D., Self-healing in cementitious materials—a review. *Materials*, 6, 6, 2013.

43. Sanders, P., Young, A.J., Qin, Y., Fancy, K.S., Reithofer, M.R., Nicolas, R.G., Kleitz, F., Pamme, N., Chin, J.M., Stereolithographic 3D printing of extrinsically self-healing composites. *Sci. Rep.*, 9, 388, 2019.

44. Sun, L., Wang, Y., Zhang, X., Bian, F., Shang, L., Zhao, Y., Sun, W., Bio-inspired self-replenishing and self-reporting slippery surfaces from colloidal co-assembly templates. *Chem. Eng. J.*, 426, 131641, 2021.

45. Kan, Y., Zheng, F., Li, B., Zhang, R., Wei, Y., Yu, Y., Zhang, Y., Ouyang, Y., Qiu, R., Self-healing dual biomimetic liquid-infused slippery surface in a partition matrix: fabrication and anti-corrosion capability for magnesium alloy. *Colloids Surf. A*, 630, 663–700, 2021.

46. Xiang, T., Liu, J., Liu, Q., Wei, F., Lv, Z., Yang, Y., Shi, L; Li, C.; Chen, D.; Xu, G., Self-healing solid slippery surface with porous structure and enhanced corrosion resistance. *Chem. Eng. J.*, 417, 128083, 2021

47. Scheiger, J.M., Kuzina, M.A., Eigenbrod, M., Wu, Y., Wang, F., Heißler S., Hardt, S., Nestler, B., Levkin, P.A., Liquid wells as self-healing, functional analogues to solid vessels. *Adv. Mater.*, 33, 23, 2021.

48. Sakuraba, K., Kitano, S., Kowalski, D., Aoki, Y., Habazaki, H., Slippery liquid-infused porous surfaces on aluminium for corrosion protection with improved self-healing ability. *ACS Appl. Mater. Interfaces*, 13, 37, 2021.

49. Wang, X., Long, Y., Mu, P., Li, J., Silicone oil infused slippery candle soot surface for corrosion inhibition with anti-fouling and self-healing properties. *J. Adhes. Sci. Technol.*, 35, 10, 2020.

50. Singh, V., Sheng, Y.-J., Tsao, H.-K., Self-healing atypical liquid infused surfaces: superhydrophobicity and superoleophobicity in submerged conditions. *J. Taiwan Inst. Chem. Eng.*, 97, 96–104, 2019.

51. Wang, J.Y., Soens, H., Verstraete, W., De Belie, N., Self-healing concrete by use of microencapsulated bacterial spores. *Cem. Concr. Res.*, 56, 139–152, 2014.

52. Huang, L., Yi, N., Wu, Y., Zhang, Y., Zhang, Q., Huang, Y., Ma, Y., Chen, Y., Multichannel and repeatable self-healing of mechanical enhanced graphene-thermoplastic polyurethane composites. *Adv. Mater.*, 25, 15, 2013.

53. Corten, C.C., Urban, M.W., Repairing polymers using oscillating magnetic field. *Adv. Mater.*, 21, 48, 2009.
54. Starov, V., Static contact angle hysteresis on smooth, homogeneous solid substrates. *Colloid Polym. Sci.*, 291, 2, 2013.
55. Nystrom, D., Lindqvist, J., Ostmark, E., Antoni, P., Carlmark, A., Hult, A., Malmstrom, E., Superhydrophobic and self-cleaning bio-fiber surfaces via ATRP and subsequent post functionalization. *ACS Appl. Mater. Interfaces*, 1, 816, 2009.
56. McHale, G., Shirtcliff, N.J., Newton, M.T., Contact angle hysteresis on super-hydrophobic surfaces. *Langmuir*, 20, 23, 2004.
57. Parvate, S., Dixit, P., Chattopadhyay, S., Superhydrophobic surfaces: insights from theory and experiment. *J. Phys. Chem. B*, 124, 8, 2020.
58. Bhushan, B., Nosonovsky, M., Jung, Y.C., Towards optimization of patterned superhydrophobic surfaces. *J. R. Soc. Interface*, 4, 15, 2007.
59. Manoharan, K., Bhattacharya, S., Superhydrophobic surfaces review: functional application, fabrication techniques and limitations. *J. Micromanuf.*, 2, 1, 2019.
60. Hozumi, A., Jiang, L., Lee, H., Shimoomura, M.S., *Stimuli-responsive Dewetting/wetting Smart Surfaces and Interfaces*, Cham: Springer, 1–456, 2018
61. Mundo, R.D., Palumbo, F., Comments regarding 'an essay on contact angle measurements'. *Plasma Process Polym.*, 8, 1, 2011.
62. Müller, M., Oehr, C., Comments on "an essay on contact angle measurements" by Strobel and Lyons. *Plasma Process Polym.*, 8, 1, 2011.
63. Strobel, M., Lyons, C.S., An essay on contact angle measurements. *Plasma Process Polym.*, 8, 1, 2011.
64. Bourges-Monnier, C., Shanahan, M.E.R., Influence of evaporation on contact angle. *Langmuir*, 11, 7, 1995.
65. Ruiz-Cabello, F.J.M., Rodriguez-Valverde, M.A., Marmur, A., Cabrerizo-Vilchez, M.A., Comparison of sessile drop and captive bubble methods on rough homogeneous surfaces: a numerical study. *Langmuir*, 27, 15, 2011.
66. Rio, O.I., Neumann, A.W., Axisymmetric drop shape analysis: computational methods for the measurement of interfacial properties from the shape and dimensions of pendant and sessile drops. *J. Colloid Interface Sci.*, 196, 2, 1997.
67. Wenzel, R.N., Resistance of solid surfaces to wetting by water. *Ind. Eng. Chem.*, 28, 8, 1936.
68 Cassie, A.B.D., Baxter, S., Wettability of porous surfaces. *Trans Faraday Soc.*, 40, 546–551, 1944.
69. Quéré, D., Reyssat, M., Non-adhesive lotus and hydrophobic materials. *T. R. Soc. A*, 366, 1539, 2008.
70. Quéré, D., Wetting and roughness. *Ann. Rev. Mater. Res.*, 38, 71, 2008.
71. Erbil, H.Y., Demirel, A.L., Avci, Y., Mert, O., Transformation of a simple plastic into a superhydrophobic surface. *Science*, 299, 1377, 2003.
72. Wang, S., Feng, L., Jiang, L., One-step solution-immersion process for the fabrication of stable bionic superhydrophobic surfaces. *Adv. Mater.*, 18, 767, 2006.
73. Xi, J., Feng, L., Jiang, L., A general approach for fabrication of superhydrophobic and superamphiphobic surfaces. *Appl. Phys. Lett.*, 92, 053102, 2008.
74. Ichimura, K., Oh, S., Nakagawa, M., Light-driven motion of liquids on a photoresponsive surface. *Science*, 288, 1624, 2000.
75. Feng, C., Zhang, Y., Jin, J., Song, Y., Xie, L., Qu, G., Jiang, L., Zhu, D., Bioinspired surfaces with special wettability. *Langmuir*, 17, 4593, 2021.
76. Chen, W., Fadeev, A.Y., Hsieh, M.C., Öner, D., Youngblood, J., McCarthy, T.J., Ultrahydrophobic and ultralyophobic surfaces: some comments and examples. *Langmuir*, 15, 10, 1999.

77. Li, S., Li., H., Wang, X.B., Song, Y., Liu, Y., Jiang, L., Zhu, D., Super-hydrophobicity of large-area honeycomb-like aligned carbon nanotubes. *J. Phys. Chem. B*, 106, 36, 2002.
78. Tadanaga, K., Morinaga, J., Matsuda, A., Minami, T., Superhydrophobic–superhydrophilic micropatterning on flowerlike alumina coating film by the sol–gel method. *Chem. Mater.*, 12, 3, 2000.
79. Luo, Z.Z., Zhang, Z.Z., Hu, L.T., Liu, W.M., Guo, Z.G., Zhang, H.J., Wang, W.J., Stable bionic superhydrophobic coating surface fabricated by a conventional curing process. *Adv. Mater.*, 20, 5, 2008.
80. Zhang, C., Laing, F., Zhang, W., Liu, H., Ge, M., Zhang, Y., Dai, J., Wang, H., Xing, G., Lai, Y., Tang, Y., Constructing mechanochemical durable and self-healing superhydrophobic surfaces. *ACS Omega*, 5, 2, 2019.
81. Singh, V., Nguyen, T.P., Sheng, Y.-J., Tsao, H.-K., Stress-driven separation of surfactant-stabilized emulsions and gel emulsions by superhydrophobic/superoleophilic meshes. *J. Phys. Chem. C*, 122, 43, 2018.
82. Zhang, M., Feng, S., Wang, L., Zheng, Y., Lotus effect in wetting and self-cleaning. *Biotribology*, 5, 31–43, 2016.
83. Blossey, R., Self-cleaning surfaces-virtual realities. *Nat. Mater.*, 2, 5, 2003.
84. Gennes, P.G.D., Wyart, F.B., Quere, D., *Capillarity and Wetting Phenomena, Drops, Bubbles, Pearls.* Springer-Verlag, 1–291, 2004.
85. Liu, M., Wang, S., Jiang, L., Nature-inspired superwettability surfaces. *Nat. Rev. Mater.*, 2, 7, 2017.
86. Majhy, B., Iqbal, R., Sen, A.K., Facile fabrication and mechanistic understanding of a transparent reversible superhydrophobic-superhydrophillic surfaces. *Sci. Rep.*, 8, 1, 2018.
87. Singh, V., Wu, C.J., Sheng, Y.-J., Tsao, H.K., Self-propulsion and self-restoration of aqueous drop on sulfobetaine silane surfaces. *Langmuir*, 33, 24, 2017.
88. Baker, E.A., Chemistry and morphology of plant epicuticular waxes. In *The Plant Cuticle*, D.F. Cutler, K.L. Alvin, C.E. Price (eds.). London: Academic Press, 1982, pp. 139–166.
89. Jetter, R., Kunst, L., Samuels, A.L., Composition of plant cuticular waxes. In *Biology of the Plant Cuticle*, M. Riederer, C. Muller (eds.). Oxford: Blackwell Publishing, 145–181, 2006.
90. Bhushan, B., Jung, Y.C., Koch, K., Micro, nano and hierarchical structures for super-hydrophobicity, self-cleaning and low adhesion. *Phil. Trans. R. Soc. A*, 367, 1894, 2009.
91. Bayer, I.S., Superhydrophobic coatings from eco-friendly materials and processes: a review. *Adv. Mater. Interfaces*, 7, 13, 2020.
92. Koch, K., Bhushan, B., Jung, Y.C., Barthlott, W., Fabrication of artificial lotus leaves and significance of hierarchical structure for superhydrophobicity and low adhesion. *Soft Matter*, 5, 7, 2009.
93. Kumar, M., Bhardwaj, R., Wetting characteristics of Colocasia esculenta (Taro) leaf and a bioinspired surface thereof. *Sci. Rep.*, 10, 935, 2020.
94. Lu, Y., Superior lubrication properties of biomimetic surfaces with hierarchical structure. *Tribol. Int.*, 119, 131–142, 2018.
95. Koch, K., Bhushan, B., Barthlott, W., Multifunctional surface structures of plants: an inspiration for biomimetics. *Prog. Mater Sci.*, 54, 2, 2009.
96. Li, B., Wang, X., Jung, H.Y., Kim, Y.L., Robinsons, J.T., Zalalutdinov, M., Hong, S., Hao, J., Ajayan, P.M., Wan, K.T., Jung, Y.J., Printing highly controlled suspended carbon nanotube network on micro-patterned superhydrophobic flexible surface. *Sci. Rep.*, 5, 15908, 2015.

97. Wang, G., Liu, S., Wei, S., Liu, Y., Lian, J., Jiang, Q., Robust superhydrophobic surface on Al substrate with durability, corrosion resistance and ice-phobicity. *Sci. Rep.*, 6, 20933, 2015.
98. Liu, S., Liu, X., Latthe, S.S., Gao, L., An, S., Yoon, S.S., Liu, B., Xing, R., Self-cleaning transparent superhydrophobic coatings through simple sol-gel processing of fluoroalkylsilane. *Appl. Surf. Sci.*, 351, 897–903, 2015.
99. Schlaich, C., Yu, L., Cuellar Camacho, L., Wei, Q., Haag, R., Fluorine-free superwetting systems: construction of environmentally friendly superhydrophilic, superhydrophobic, and slippery surfaces on various substrates. *Polym. Chem.*, 7, 48, 2016.
100. Martin, J.W., Mabury, S.A., Solomon, K.R., Muir, D.C.G., Bioconcentration and tissue distribution of perfluorinated acids in rainbow trout (*Oncorhynchus mykiss*). *Environ. Toxicol. Chem.*, 22, 1, 2003.
101. Wu, W., Wang, X., Liu, X., Zhou, F., Spray-coated fluorine-free superhydrophobic coatings with easy repairability and applicability. *Appl. Mater. Interfaces*, 1, 8, 2009.
102. Kong, X., Zhu, C., Lv, J., Zhang, J., Feng, J., Robust fluorine-free superhydrophobic coatings on polyester fabrics by spraying commercial adhesive and hydrophobic fumed SiO_2 nanoparticles. *Prog. Org. Coat.*, 138, 105342, 2020.
103. Qu, M., Xue, M., Yuan, M., He, J., Abbas, A., Zhao, Y., Wang, J., Liu, X., He, J., Fabrication of fluorine-free superhydrophobic coatings from montmorillonite with mechanical durability and chemical stability. *J. Coat. Technol. Res.*, 16, 1043–1053, 2019.
104. Zhang, L., Xue, C.H., Cao, M., Zhang, M.M., Li, M., Ma, J.Z., Highly transparent fluorine-free superhydrophobic silica nanotube coatings. *Chem. Eng. J.*, 320, 244–252, 2017.
105. Cai, Y., Zhao, Q., Quan, X., Feng, W., Wang, Q., Fluorine-free and hydrophobic hexadecyltrimethoxysilane-TiO_2 coated mesh for gravity-driven oil/water separation. *Colloids Surf. A*, 586, 124189, 2019.
106. Wang, Q., Sun, G., Tong, Q., Yang, W., Hao, W., Fluorine-free superhydrophobic coatings from polydimethylsiloxane for sustainable chemical engineering: preparation methods and applications. *Chem. Eng. J.*, 426, 130829, 2021.
107. Wang, X., Ao, W., Sun, S., Zhang, H., Zhou, R., Li, Y., Wang, J., Ding, H., Tunable adhesive self-cleaning coating with superhydrophobicity and photocatalytic activity. *Nanomaterials*, 11, 1486, 2021.
108. Bixler, G.D., Bhushan, B., Bioinspired rice leaf and butterfly wing surface structures combining shark skin and lotus effects. *Soft Matter*, 8, 11271, 2012.
109. Wongsamut, C., Demleitner, M., Suwanpreedee, R., Alttadt, V., Manuspiya, H.J., Copolymerization approach of soft segment towards the adhesion improvement of polycarbonate-based thermoplastic polyurethane. *Adhesion*, 97, 15, 2020.
110. Imato, K., Nakajima, H., Yamanaka, R., Takeda, N., Self-healing polyurethane elastomer based on charged transfer interactions for bio-medical applications. *Polym. J.*, 53, 355, 2021.
111. Li, Y.H., Guo, W.J., Guo, W.J., Li, X. Liu., Zhu, H., Zhang, J.P., Liu, X.J., Wei, L.H., Sun, A.L., Effect of mechanical properties on the self-healing behavior of waterborne polyurethane coatings. *Chem. Eng. J.*, 393, 124583, 2020.
112. Cho, S., Anderson, H., White, S., Sottos, N., Braun, P., Polydimethylsiloxane based self-healing materials. *Adv. Mater.*, 18, 8, 2006.
113. Yao, J., Wang, J., Yu, Y., Yang, H.H., Xu, Y., Biomimetic fabrication and characterization of an artificial rice leaf surface with anisotropic wetting. *Chin. Science Bull.*, 57, 2631, 2012.
114. Kim, S., Wu, Z., Esmailia, E., Dombroskie, J.J., Jung, S., How a raindrop gets shattered on biological surfaces. *PNAS*, 117, 25, 2020.

115. Nguyen, S.H., Webb, H.K., Mahon, P.J., Crawford, R.J., Ivanova, E.P., Natural insect and plant micro-/nanostructured surfaces: an excellent selection of valuable templates with superhydrophobic and self-cleaning properties. *Molecules*, 19, 9, 2014.

116. Schmuser, L., Zhang, W., Marx, M.T., Encinas, N., Vollmer, D., Gorb, S., Baio, J.E., Rader, H.J., Weidner, T., Role of surface chemistry in the superhydrophobicity of the springtail orchesella cincta (Insecta: Collembola). *ACS Appl. Mater. Interfaces*, 12, 12294–12304, 2020.

117. Wisdom, K.M., Watson, J. A., Qu, X., Liu, F., Watson, G.S., Chen, C.H., Self-cleaning of superhydrophobic surfaces by self-propelled jumping condensate. *Proc. Natl. Acad. Sci. USA*, 110, 20, 2013.

118. Imran, Z., Wang, Z., Zheng, W., Zhu, Y., Zhang, Z., Zhang, J., Imtiaz, M., Abbas, W., Zhang, D., The highly efficient photocatalytic and light harvesting property of Ag-TiO$_2$ with negative nano-holes structure inspired from cicada wings. *Sci. Rep.*, 7, 1, 2017.

119. Lee, W., Jin, M.K., Yoo, W.C., Lee, J.K., Nanostructuring of a polymeric substrate with well-defined nanometer-scale topography and tailored surface wettability. *Langmuir*, 20, 18, 2004.

120. Watson, G.S., Myhra, S., Cribb, B.W., Watson, J.A., Putative functions and functional efficiency of ordered cuticular nanoarrays on insect wings. *Biophys. J.*, 94, 8, 2008.

121. Terryn, S., Langenbach, J., Roels, E., Brancart, J., Camille, B.-H., Poutrel, Q.-A., Georgopoulou, A., Thuruthel, T.G., Safaei, A., Ferrentino, P., Sebastan, T., Norvez, S., Lida, F., Bosman, W.A., Tournilhac, F., Clemens, F., Assche, G.V., Vanderborght, B., A review on self-healing polymers for soft robotics. *Mater. Today*, 47, 187–205, 2021.

11 Natural Cosmetics

Dharmesh Varade

11.1 BACKGROUND

The cosmetics in today's world are becoming more "ecofriendly". The market demand for natural and organic cosmetics has increased immensely in the last few decades because of the growing importance of personal and beauty care in various parts of the world. The growing demand for natural and organic cosmetics indicates consumer's awareness about safeguarding the environment and health by consuming sustainable products [1–3].

Cosmetics have been in use for thousands of years and they have a rich and long history. The foremost primeval make-up was developed from nature's flowers, trees, water or soil. With the development of social structures, cultural beliefs and traditional practices played a substantial role in the application of cosmetics all over the world. Currently, cosmetics science not only is based on natural science but it also contemplates psychological aspects because the emotional and mental value of cosmetics products is more important than their scientific functionality.

The demand for natural cosmetics has increased tremendously because consumers worldwide are more focused on their health, well-being and appearance. Consumers are becoming aware that many products that are marketed as natural are not necessarily natural. And, hence, terms such as natural, organic, no artificial preservatives and no animal testing are gaining arduous attention. The long list of ingredients written on cosmetics products available in the market often confuses consumers, making it difficult for them to determine whether an ingredient is chemically synthesized or natural.

Petroleum-based ingredients such as phthalates, oxybenzene or parabens are present in high amount in conventional cosmetics. Though petroleum-based materials are considered as safe for use in cosmetics, these are unreliable and are not natural in the real sense. Dermatitis reactions caused by parabens in some individuals on cutaneous exposure are reported. Formaldehyde, Quartenium-15, Euxyl K 400, etc., are synthetic preservatives commonly found in cosmetics products. These preservatives are reported to cause dermatitis. In Japan and Sweden the use of formaldehyde is prohibited due to safety issues. Methyl di-bromoglutaronitrile present in Euxyl K 400 is the main ingredient responsible for dermatitis [1, 4].

The composition of natural and organic cosmetics includes minerals, plant extracts, products from microorganisms, or constituents extracted from these sources. The natural properties of these complex mixtures of plant extracts and oils protect, hydrate and nourish the skin. Several biological functions and toxicological assessment are performed for many natural products obtained from plants and herbs that are used in cosmetics products. The authenticity, originality and quality

DOI: 10.1201/9781032636368-11

of products are ensured by the presence of organic, biodegradable and sustainable ingredients.

Natural plus organic cosmetics not simply take our well-being into consideration by providing natural protection and care to our skin but also represent nature through their natural fragrances and textures. Natural essential oil like rose, mint and lavender are used in aromatherapy to relax, activate and soothe our bodies.

11.2 COMPOSITION OF COSMETICS

The properties of the functional materials used in cosmetics formulations and the user experience form the major criteria for the development of cosmetics products. Hence, planning the formulation qualitatively and quantitatively is very much needed. Cosmetic ingredients can be classified into three essential categories based on formulation, e.g., aqueous or hydrophilic base, oil or hydrophobic base and amphiphilic substances, as depicted in the Figure 11.1. An extensive range of cosmetics formulations are produced by mixing these three types.

Aqueous or hydrophilic base ingredients: The most common ingredient used in cosmetics because of its capability to dissolve many other substances is water. Properties of water like low refractive index and volatility make it the best solvent for chemical reactions. Ethanol and polyols are other hydrophilic base materials that can be used as a substitute to water in formulating cosmetics materials. However, the replacement changes the solvent and moisture-retaining capacity.

Oil or hydrophobic base ingredients: Oleaginous base ingredients consist of oils and fat that are used as structuring materials in many cosmetics. These bases are anhydrous, greasy and insoluble in water which includes hydrocarbons, esters, ether, silicones and fluoride compounds. Because of the comprehensive diversity in the

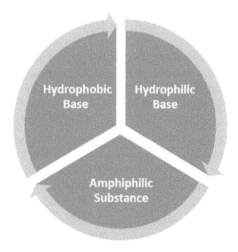

FIGURE 11.1   Cosmetics ingredients that play essential roles in creating the assembly of formulations.

structures of these compounds, oleaginous base materials exhibit various properties like polarity, melting point or compatibility. This variety of properties lends them with unique applications or specific functions like UV light absorption increase in percutaneous absorption.

Amphiphilic substances: Amphiphiles have a hydrophilic head and a hydrophobic tail in the same structure, thereby making them excellent candidates as emulsifiers and solubilizers or to disperse hydrophobic or hydrophilic bases to the other. Amphiphiles are used as dispersion promoters in various cosmetics formulations. A higher fatty acid may offer surfactant or oleaginous properties depending on pH. Apart from being structuring materials in cosmetics, certain amphiphiles are used for their functional applications like moisturizing to adjust the feel of touch.

Materials that add functions and effects: Certain ingredients apart from oleaginous bases, water-soluble bases and amphiphiles are added to cosmetics to impart functions and effects. These ingredients add value to the cosmetics products, thereby appealing to consumers. These ingredients can be classified as follows:

- *Materials added to increase functional value*: The functional value of the products are associated with their main features and cosmetics enhancement aims to improve these functional values. The physiological and physical functions of cosmetics are improved by adding functional values to the cosmetics. Cosmetics products utilized in skin care, hair care and certain make-up products impart physiological effects. Various substances like amino acids, vitamins, and botanical extracts are utilized as active ingredients in cosmetics products to develop physiological effects. Many new experiments have been undertaken in dermatology since the 1980s for new evaluation methods on healthy skin such as experiments designed using culture cells. Also, several studies on whitening and antiaging have been performed in dermatology to enhance physiological effects.

  The physical functions include cleaning, moisturizing, occlusion and UV protection, which add value to cosmetics. These physical functions of cosmetics are determined by the evaluation of the physical property and human use testing. Over the years, there have been various developments in order to add and improve the physical functions of the cosmetics by adding specific ingredients. For example, structuring oil-based materials are replaced by substances that perform unique physical functions like water retention, occlusion, UV absorption or increase percutaneous absorption to add their functions and effects to the formulation. However, maximizing the functions and effects in cosmetics is not much recommended and optimum effects are preferable. For example, robust cleaning products that take away too much of oil are not preferable; rather, cleansers with optimum cleansing activity are recommended while designing the product [4].

- *Materials added to increase emotional value*: There are certain cosmetic products where emotional value dominates functional values for the consumer. Sight, smell and touch are the three main senses that create an emotional value and preference for a cosmetics product. Various pigments, coloring agents and aromatic substances are added to cosmetics to improve their emotional value as many consumers can identify the slight differences in color and smell, which influences their preference for the product. The emotional value of cosmetics products like perfumes and even antiaging creams are more vital than their functionality because of preferential customer choice for certain products. Antiaging creams which render certain biological properties are not just measured for percutaneous absorption but also to enjoy the feel and scent of the creams. A variety of cosmetics products are designed to enhance their feel, the sense of touch being the utmost imperative sense. The formulation assembly of cosmetic products determines the touch and feel of cosmetics products. Very minute specifications of ingredients like polyols, powders and polymers, etc., are considered while formulating cosmetics products to enhance the feeling of touch on application and user experience. These preferences constitute the emotional values that form very prominent criteria in designing cosmetic products [4].

• *Materials used for quality control*: An extended shelf-life of at least three years is expected in most cosmetics products. There are many similar materials that are added to consumer goods like foods, pharmaceutical products and detergents. Materials used in cosmetics products are regulated, used for different purposes and have different quality targets. Majorly, quality control is performed for raw materials, intermediates and finished products in different stages. While formulating cosmetics products, shelf-life plays an important role, and care is taken to maintain the formulation's quality in terms of its condition, color and smell. Many materials are formulated to increase the shelf-life and stabilize the cosmetics product to provide consumers with a safe and sound product. Various methodologies for formulating and cleaning of raw materials are used in cosmetics products. For example, the emulsification method is used for many cosmetics formulations, but emulsions are thermodynamically unstable and hence require stabilizing agents like higher fatty alcohols and polymers like carbomers and xantham gum. Putrefaction is the process that leads to a change in the composition of products causing health issues and lowering the appeal of the products. Hence, to prevent bacteria from growing and wax from oxidizing in cosmetics products, preservatives are used. Several antimicrobial agents act on the cell membrane of the microbes and can disturb the stability of the emulsion. Hence, formulators take a lot of care while adding or increasing the preservative content in the cosmetics formulation. Several other materials such as antioxidants, pH buffering agents, chelating reagents and discoloration-preventing reagents are formulated for the purpose of stabilization [1, 4].

11.3 CHOOSING COSMETICS INGREDIENTS

A lot of care is to be taken while selecting the ingredients for cosmetics formulation. Since the quality targets of the materials used in cosmetics are different from other consumer goods like foods and pharmaceutical products, precautions are taken while selecting ingredients for cosmetics products. The long shelf-life of cosmetics products is achieved by using ingredients with very high stability. Moreover, it is very difficult to control the frequency and amount of the cosmetics product used. Hence, safety is the highest priority during the cosmetics formulation.

There are certain regulations in most countries under which the cosmetics and pharmaceutical products are produced and sold and their usage is managed. Even though these regulations are similar for pharmaceutical and cosmetics products in most countries, the essential guidelines of cosmetics vary as to the pharmaceutical products. These regulations likewise vary contingent upon the region. Like, in Europe, animal testing for the evaluation of cosmetics products is strictly prohibited. Therefore, choosing ingredients for cosmetics products is very challenging due to varying regulations.

Personal principles of consumers based on their opinion, belief and tastes influence the cosmetics market largely. For example, many consumers do not prefer the usage of materials like ethanol and methyl parahydroxybenzoate that have been used as generic materials for a long time. Even though there is no scientific evidence for these claims, these substances are avoided in cosmetics products to fulfill the criteria of consumer's personal beliefs. Also, manufacturers refrain from choosing such materials in cosmetics formulations. Cosmetics formulators often consider such cultural differences of consumers against certain materials in the stage of development and also suggest and give their ideas when they explain their approach. For example, the choice of ingredients like alkyl benzoate in a cosmetics product is largely dependent upon region. It is an acceptable and extensively used ingredient in Europe; on the other hand, it is not preferable in Japan because of the formulator's preferential choice of certain ingredients influenced by consumer's beliefs.

The material regulations for cosmetics products in each country also play a vital role in choosing cosmetics ingredients. Today, the development, production and sales of cosmetics products are not limited to specific regions. Therefore, it is essential to understand the material regulations of each country as the advanced ethics of cosmetics differ depending upon the region. The stability standards of products also influence the development style of the cosmetics formulation.

11.4 ORGANIC VS. NATURAL

Organic and natural cosmetics are qualitatively similar but quantitatively altered. They do not contain any ingredients that are synthetic or semi-synthetic in their formulation with very few exceptions. All the raw materials are naturally derived and from allowed processes with organic origin. The major difference in organic and natural products is the composition of the raw materials. There are several agencies that issue certificates and levy criteria which the manufacturing industry must fulfill to confirm with the quality of the finished product. A few major agencies that certify natural and organic products are listed below [5]:

• Bundesverband Deutscher Industrie und Handelsunternehmen (BDIH) in Germany

- National Association for Sustainable Agriculture, Australia (NASAA), in Australia
- Soil Association Organic Standard in the United Kingdom
- Instituto Biodinâmico de Certificações (IBD) in Brazil
- ECOCERT in France
- Istituto per la Certificazione Etica e Ambientale (ICEA) in Italy
- Quality Assurance International (QAI) in the United States of America
- Oregon Tilth in the United States of America
- Cosmetics Organic Standard (Cosmos) in the European Union (EU)

As per IBD, organic and natural cosmetics can be categorized as follows:

Organics: About 95% of the components in the formulation are organic raw materials with the certificate of extraction. Stringent criteria must be followed for the extraction, production, purification and handling of raw materials. The rest 5% composition of formulation can be other agricultural natural raw materials or extracts of allowed non-certified organic formulations or water. These raw materials can be derived from crops that are certified. The chemical characteristics of these products are preserved as these raw materials are biodegradable. Raw materials that are organic are always natural.

Natural: The composition of natural cosmetics includes raw materials where certificate of extraction is not mandatory. The production of these raw materials does not adhere to the principles of organic production and these materials are usually produced in a conventional way from mineral or vegetable products. Raw materials that are naturally occurring are not necessarily organic.

The International Trend of Organic and Natural Cosmetics: The consumer market for natural and organic cosmetics is increasing rapidly because of the growing awareness related to the safety of cosmetics and concern for the environment by following an ecofriendly lifestyle among consumers. Organic and natural cosmetics with botanical ingredients are being marketed globally. The global nongovernmental organization – the International Federation of Organic Agriculture Movements – has established the global standard for organic food. Even though there are many certification authorities as mentioned earlier that certify as per their own standards but a globally recognized unified certification agency for organic and natural cosmetics is yet to be recognized. The initial natural cosmetics standard was set by BDIH, a Germany-based certificate agency, and then followed by several endorsement authorities that started their own certification. In 2010, a certification authority Cosmetics Organic Standard (COSMOS) formed by the union of several European certification authorities established a global standard for natural and organic cosmetics. According to COSMOS, the use of synthetic surfactants and preservatives is approved but this is questionable from the viewpoint of consumers.

11.5 FORMULATION CHALLENGES OF NATURAL COSMETICS

Cosmetics formulations are often degraded by microbial agents like yeast, fungi and bacteria, causing infections. Customer usage and storage also play a crucial role in the preservation of cosmetics products. Microbial degradation, expiry of shelf-life of cosmetics and storage at extremely high temperature are the reasons for the instability of cosmetics. Hence, using preservatives in cosmetics for storage is a prerequisite.

The biggest challenge for formulators is using natural ingredients as preservatives without causing any sensitizing effect. These preservatives must be as effective as synthetic preservatives. Natural preservatives are classified as plant-derived, animal-derived, certain microbes and certain antimicrobials. On the basis of their function, they are categorized as antimicrobial and antioxidant agents. The antimicrobials inhibit the growth of bacteria, yeast, molds or fungi, and antioxidants inhibit the oxidation process in the product, thereby increasing the shelf-life of cosmetics.

Even though many cosmetics are labeled as "natural" or "organic", there is no guarantee for the originality of the products. Checking the ingredients mentioned on the product may provide some information about the ingredients used in the product but the accuracy of these claims of being organic or natural is uncertain as mentioning the origin of the ingredient is not a necessary requirement by law. To resolve such issues, NATRUE (Brussels-based international non-profit association committed to promoting and protecting Natural and Organic Cosmetics worldwide), provides verifiable criteria for the assurance of consumers. The ingredients, formulations and processes for organic or natural cosmetics are strictly met under the NATRUE label. Most of the conventional cosmetics like petroleum-based products are obtained from mining, which causes soil depletion and is a serious threat to wildlife habitat. Natural and organic cosmetics formulations are mainly derived from the extracts of plants and flowers. Sustainable production of various plants and flowers and preservation of biodiversity by protecting the wildlife habitat is achieved by using organic and natural raw materials.

11.6 NATURAL ACTIVE INGREDIENTS USED IN COSMETICS

The following is a list of some of the natural active ingredients currently employed in cosmetics products:

Papain: Papain is an active ingredient found in the dried latex of papaya fruit as shown in Figure 11.2(a). The proteolytic enzyme like papain has antibacterial, antifungal and antiviral properties. It reduces inflammation on the skin and, therefore, is widely used in exfoliating products to remove damaged keratin on the skin that forms small bumps. This helps in the reduction of acne. Papaya contains vitamins A, B and C. A mixture of antioxidant agents such as lycopene and vitamin C reduces wrinkles, acting as an antiaging agent. Papaya is also used as a skin lightning agent due to the presence of beta carotene and phytochemicals [6].

Aloe vera: Aloe vera gel is obtained from the leaves of aloe vera and is widely used in cosmetics for its soothing effect (Figure 11.2(b)). It consists of minerals, enzymes, sugars, several phenolic compounds, amino acids, lignin, saponins, etc. Aloe vera gel blocks UV radiation and helps in retaining natural skin moisture. Thus, it is used as a moisturizing and revitalizing agent. Sunburns are prevented and the immune system is stimulated by the bradykinase enzyme present in aloe vera. It also exhibits many biological properties like antiviral, antibacterial and laxative effects. It also activates macrophages and helps in wound healing [7–9].

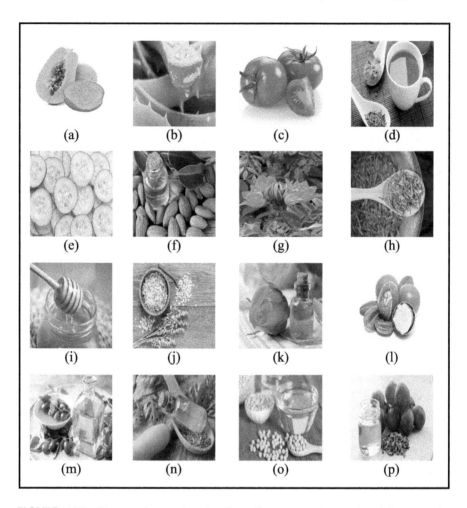

FIGURE 11.2 Photos of natural active ingredients currently employed in cosmetics products.

Tomato: Tomatoes are very rich in antioxidants like lycopene and are widely studied in the cosmetics and pharmaceutical field (Figure 11.2(c)). The scavenging activity of free radicals, reduction in lipid peroxidation and prevention against erythema caused by UV radiation are major functions exhibited by lycopene. The natural application of tomatoes is to reduce skin irritation, adjust skin pH, reduce oily skin and tighten skin by reducing skin pores, thereby preventing acne naturally [10, 11].

Green tea: Green tea contains polyphenols, flavonoids, caffeine, alkaloids and phenolic acids (Figure 11.2(d)). Polyphenols are major chemopreventive agents. The extract of topical green tea reduces UV-damaged skin [12–14].

Cucumber: Cucumber extract has deep moisturizing ability and mild astringent ability (Figure 11.2(e)). By removing dead cells, it helps to tighten the skin. In addition to containing vitamin C, caffeic acid and fiber, cucumbers are rich in water and many beneficial minerals to reduce skin irritation, help in water retention and offer soothing feel to the skin. Hence, cucumbers are topically applied on swollen eyes, burns or inflamed and tired skin [15–17].

Indian beech tree: A very good absorbance of UV light is exhibited by the extracts of leaves of the Indian beach tree. Hence, this extract is highly effective in sunscreen formulations [18, 19].

Almond: Almond seeds are extremely rich in polyphenolic compounds like flavonoids and phenolic acids (Figure 11.2(f)). These compounds are very good antioxidant agents. Application of almond cream results in anti-photo aging. Almond oil is used as a moisturizer. It has photo-protective effects and prevents the damage caused by sun exposure. It is also used to reduce stretch marks. It contains vitamin E that removes skin dullness and makes the skin lighter in complexion [20].

African tulip tree: The stem bark of the African tulip tree has antioxidant, antimalarial and anti-hyperglycemic and wound healing properties (Figure 11.2(g)). The extracts of this plant protect against UV light and are a very good substitute for chemical sunscreen creams [21, 22].

Saffron: Saffron contains antioxidants like crocin, crocetin, safranal and kaempferol (Figure 11.2(h)). It also contains vitamin C and has anti-inflammatory and antibacterial properties. Several other minerals and carotenoids present in saffron help in damage repair and ensure cellular health leading to skin recovery from environmental and oxidative stress. Saffron is also a natural UV absorbing agent and protects skin from UV radiations [23, 24].

Ceramide: Ceramide is a major component found in the epidermis of the skin and is naturally secreted in the body during the biosynthesis of complex sphingolipids. Ceramide has great potential in cosmetics such as in hair and skin care products. Water retention is the important function performed by ceramides. It is the best skin barrier agent to prevent permeability of the skin, reducing dryness and skin irritation and keeping it plump and firm. Ceramide is a highly valued component and is an active cosmetic at very low concentrations. Ceramide is used in shampoos, conditioners and hair serums to repair hair damage by limiting protein release and strengthening cuticular cohesion [1, 25, 26].

Honey: Organic honey is used as an emollient and humectant in cosmetics formulation (Figure 11.2(i)). It keeps the skin juvenile, regulates pH and protects the skin from pathogenic infections. Many cosmetics products like lip ointments, hydrating creams, sunscreen lotions, etc., include honey as an active ingredient. Organic honey has antimicrobial and anti-inflammatory properties and helps in healing skin wounds. Cosmetics products usually contain honey in 1%–10% concentration but nearly 70% can be reached by adding additives like oils, gel and emulsifiers. Several skin-related problems like acne, psoriasis and eczema can be treated with honey [27, 28].

Oat: Skin care products often contain oats to soothe itchy skin (Figure 11.2(j)). Oat extract and colloidal oat are mostly found in skin care products. Due to its anti-histamine and anti-inflammatory properties, it is used to soothe allergic reactions [29, 30].

Rose water: Pure rose water is used as a skin cleansing agent (Figure 11.2(k)). It is used to remove excess oil from the skin thereby preventing blackheads and whiteheads, acne and pimples on the skin caused duo to clogging of pores by oxidation. Rose water is a natural skin toner and rose aroma is used to uplift mood. Due to its anti-inflammatory properties, rose water is used to reduce puffiness under eyes. It helps in hydrating and moisturizing skin [31, 32].

11.6.1 NATURAL SUNSCREEN OILS

Shea butter: Shea is obtained from the shea nut (Figure 11.2(l)). It melts at body temperature and penetrates rapidly into the skin without any oily feeling. The abundance of triglycerides in shea butter makes it an excellent emollient and helps in preventing moisture loss from the skin. It also contains antioxidants and vitamins like A and E, which promote cell regeneration in the skin thereby leading to improved blood circulation below the skin surface. It also exhibits anti-inflammatory properties and acts like ceramides and is often used in skin care products [33, 34].

Jojoba oil: Jojoba oil is effective for treating psoriasis, eczema and dry skin (Figure 11.2(m)). It is an excellent moisturizer. A long-chain triglyceride like myristic acid present in the oil gives protection to the skin [35, 36].

Carrot seed oil: It is an essential oil with many antioxidant, antifungal, antiseptic properties and has a very good fragrance (Figure 11.2(n)). It is extremely rich in vitamin A. When diluted carrot seed oil is applied directly on the skin as a carrier oil, it provides natural protection from the sun. A natural sun protection factor (SPF) of 38 and 40 is offered by this oil [37, 38].

Soybean oil: A considerable amount of essential fatty acids, iron, protein, and lecithin are found in soybean oil (Figure 11.2(o)). It also contains natural antioxidants, thereby protecting the skin from free radicals. It contains vitamin E, thereby making it easily infusable in cosmetics formulations. This helps cosmetics products like antiaging or skin brightening creams to maximize their effect. When compared to other oils, it is a cost-effective moisturizer and offers natural SPF of 10 [39, 40].

Grapeseed oil: Grapeseed oil has been growing in popularity over the past few decades (Figure 11.2(p)). It's often promoted as healthy due to its high amounts of polyunsaturated fat and vitamin E. Marketers claim it has all sorts of health benefits, including lowering your blood cholesterol levels and reducing your risk of heart disease.

11.6.2 OTHER NATURAL OILS USED IN COSMETICS

Evening primrose oil: It is rich in γ-linolenic acid and promotes skin repair and healthy skin. It is used to reduce many skin problems like psoriasis, eczema, acne or any type of dermatitis. It provides skin with a hydrating and soothing effect. It is effective in reducing dry skin by moisturizing it and preventing skin from premature aging. It enhances the texture and elasticity of the skin, providing a clear and healthy skin [41, 42].

Coconut oil: This is rich in fatty acids and has antibacterial properties. It helps to boost collagen production making skin healthy, and essential fatty lipids present in it can enhance the barrier function of the skin improving the moisture content of the skin. Coconut oil can easily penetrate into the skin when compared with other oils; hence, it prevents water loss in the top layer of the skin and reduces cracks. It is found in many cosmetics products like natural lip glosses, hair masks, moisturizers, etc. [43].

Olive oil: It is rich in antioxidants like vitamin E and is often used in antiaging products. It has anti-inflammatory properties and offers effective repair to skin from free radical damage caused by exposure to sun. The topical application of olive oil helps to reduce oxidative stress of the skin and hair [44, 45].

Avocado and avocado oil: It is extracted from the pulp and increases the production of collagen in the skin. It contains waxes, minerals, proteins and vitamins, making it an excellent antioxidant for the skin [46, 47].

Sunflower seed oil: This oil is rich in linoleic acid, oleic acid and vitamin E and helps in lipid synthesis and hydration of the skin without causing irritation. It also enhances barrier function of the skin. It is used as an emollient in cosmetics products. It is also non-comedogenic.

11.7 NATURAL COSMETICS PRODUCTS

In this context some of the natural cosmetic products and the ingredients present in them are discussed.

   i. *Moisturizers/emollients*: Moisturizing agents are mainly classified as occlusives, emollients and humectants.
   - Occlusives prevent transepidermal water loss and offer physical barrier protection to skin. Examples of natural occlusives are coconut oil, candelilla wax, beeswax, jojoba oil and coconut oil.
   - Emollients are ingredients that soften skin and make it smooth and supple by reducing dry and rough skin. These may be fluid or thick in texture on application. Butters like shea, cocoa, mango and different oils like almond, argan, avocado, olive, evening primrose, sunflower, etc., act as natural emollients to the skin.
   - Humectants are natural hydration-boosting agents. Humectants pull water from the dermis toward the stratum corneum (outer layer of epidermis) and bind water vapor from the atmosphere. Hyaluronic acid, glycerin, honey, glycerol are natural humectants. Lactic acid is also used in cosmetics for hygienic and aesthetic products due to its moisturizing effect and antimicrobial and skin rejuvenating properties. Lactate esters are used in cosmetics due to their hygroscopic and emulsifying properties.

   ii. *Detergents/shampoos*: When a shampoo is made up of organic and plant-based ingredients and sulfate-free components, it is considered as natural. Natural shampoos are typically made with essential oils, fruit extracts, botanicals or organic ingredients instead of synthetic ones. Ingredients that are plant-based are generally considered mild, including seed oils and fruit extracts. They're less likely to disrupt the hair and scalp's natural pH and oil balance. Plant oils like argan, coconut, jojoba, geranium oil are used in natural shampoos. Other natural ingredients used in shampoos are organic honey, root extracts like burdock root and fruit extract of star anise and certain essential oils like lemon, peppermint, lavender and bergamot.

iii. *Sunscreen*: Aloe vera, tomato, green tea, cucumber, Indian beech tree, almond, African tulip tree, saffron are several natural ingredients that provide protection against UV rays and hence can be used in sunscreens [48].

iv. *Antiwrinkle and antiaging agents*: These are categorized as below [49]:

- *Barrier repair agents*: Natural barrier repair agents include ceramide and cholesterol. Certain natural essential fatty acids like omega-3 present in flaxseed, chia and walnut oil and omega-6 in grapeseed oil are excellent skin barrier repair agents possessing anti-inflammatory and anti-irritancy properties. These are excellent building blocks for healthy cell membranes.

- *Antioxidants*: These are essential ingredients that prevent the damaging effect of free radicals. Different kinds of vitamins and polyphenols act as antioxidant agents. Vitamins enhance collagen production in the body making the skin plump and reducing wrinkles and fine lines. Vitamin A is important for the production of new healthy cells in the skin minimizing natural aging, scars and stretch marks on the skin. Vitamin E boosts the free radical scavenging activity leading to conditioning and softening effect on the skin. Vitamin B3 is essential to maintain oil balance, retaining moisture and treating solar keratosis [50].

- *Hydroxy acids*: These acids are most commonly found in fruits and are excellent exfoliating agents. These acids rejuvenate the skin by reducing signs of aging like wrinkles, spots, dryness or discolored skin. Citrus fruits, sugarcane, fermented fruits are major source of alpha-hydroxy acids [51].

- *Resveratrol*: It has very good antiaging property and is found in red wine. It is used in skin care formulations for its antioxidant property. Biochanin A and isoflavanoid derived from peanuts are also used in the prevention of wrinkles and aging.

- *Anti-inflammatory ingredients*: Turmeric, oats, licorice root, lavender and chamomile are natural ingredients possessing anti-inflammatory properties. Argan oil, almond oil, shea butter and several other treenut oils are used to reduce inflammation by decreasing free radicals, stimulating enzymes for detoxification and antioxidant systems.

v. *Acne cosmetics*: Tea tree oil with anti-inflammatory and antimicrobial properties reduces irritation and redness caused by acne-causing bacteria and helps to unclog skin pores. Natural clay like kaolin, bentonite clay and charcoal is used since ancient times as a natural remedy for acne. These detoxify skin pores by absorbing excess oil. These also possess antibacterial properties. Fruit acids like citric acid, glycolic acid and salicylic acid have astringent and exfoliating properties. These remove skin irritation like acne and help in unclogging pores [52].

vi. *Hair growth agent*: The Indian gooseberry also commonly known as "aamla" is a seed used to promote hair growth. Its herbal hair oil or polyherbal ointment is very effective in traditional medicine for hair growth. Herbal formulations containing extracts of several plants like *Tridax procumbens* (jayantiveda), *Hibiscus rosa-sinensis* (Chinese hibiscus), *Trigonella foenum-graecum* (fengreek seeds) and *Emblica officinalis* (aamla) have synergistic effects leading to increase in hair growth. "Brahmi" also promotes hair growth because of alkaloids present in it which enhance protein kinase activity.

11.8 NATURAL POLYMERS IN COSMETICS

Polymers are an essential class of cosmetics formulation for high-performance products. Polymers are used for a variety of functions like emulsifiers, fixatives, conditioning agents, rheology tuners, foam stabilizers or destabilizers and film formers. These wide varieties of functions performed by polymers are attributed to the structural diversity of polymers. Natural polymers like starch, polysaccharides, collagen, cellulose, carrageenan, gelatin and hyaluronic acid are widely used in natural cosmetics products [53–55]. These are biocompatible, biodegradable, ecofriendly and safe to be used in cosmetics in hair and skin care products. Natural polymers most commonly used in cosmetics application are depicted in Table 11.1.

Starch: It is a natural polymer derived from potatoes, cassava root, corn and pinion. Starch obtained from these sources offers different properties to cosmetics formulations due to the varying content of amylose present in them. Different types of starch used in cosmetics are soluble and granular starch. Soluble starch is used to provide smooth skin and hair whereas granular starch is used for soft, moisturized and non-greasy skin and hair. Natural starch is often used in combination with other natural polymers like chitosan in cosmetics formulations for improved properties like antioxidative release.

Chitosan: Chitosan is a deacetylated derivative of chitin, a compound naturally derived from marine resources like fungi (Figure 11.3(a)). The physico-chemical, unique biological and technological properties of chitosan make it an excellent ingredient in cosmetics products. Chitosan is used in a variety of hair care products like

---

TABLE 11.1
Natural polymers used in cosmetics

Polymers	Function
Starch	Emulsifier, offers high viscosity, film formation, moisturizer
Chitosan	Moisturizer, cleansing agent, skin protecting agent from UV rays, antimicrobial and antioxidant
Cellulose	Spreadability of creams, adhesion of fragrance, antistatic hair
Collagen	Moisturizer, film formation, antioxidants
Hyaluronic acid	Humectant, protection from UV, antiaging agent
Ulvan	Emulsifier, gelling, moisturizer, perfuming agent

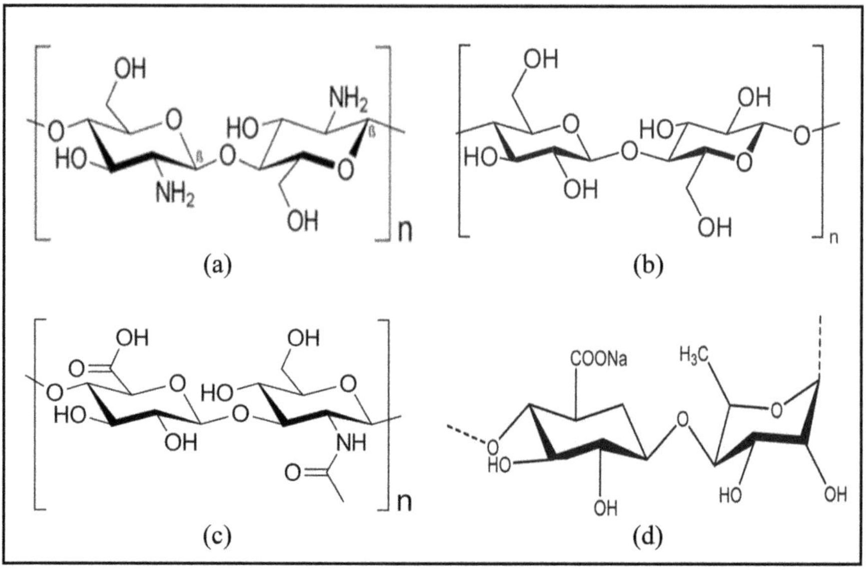

FIGURE 11.3   Structural formulae of various natural polymers.

shampoos, conditioners, hair colorants and hair lotions, various skin care products, nail lacquers and deodorants. Chitosan is used as an emollient, moisturizer, humectant, cleansing agent, skin protecting agent from UV rays and an antimicrobial and antioxidant agent in natural cosmetics products. In addition to improving fragrance adhesion, chitosan is also used to stabilize cosmetics formulations. Cosmetics products use chitosan for its high molecular weight, positive electrical charge, filming properties and aesthetic appeal.

Cellulose: Cellulose is a fibrous polymer of plant origin (Figure 11.3(b)). It cannot be directly used in cosmetics because of its insolubility in aqueous media. It is often used in cosmetics applications in combination with several other natural polymers like carrageenan for its thickening property. It is used as a rheology modifier and colloidal stabilizer for increasing adhesion and tuning viscosity of the formulations. Cellulose fibrils offer very good spreadability of the creams, thin film forming of nail polish and anti-caking agent in foundation cream. Polyquaternium-10, a hydroxyethyl cellulose polymer, is used in hair conditioning products for its antistatic, smooth and shiny properties [55–57].

Hyaluronic acid: Hyaluronic acid is a polymer of disaccharides naturally present in the skin (Figure 11.3(c)). Its main function is to keep skin moist and all tissues lubricated with water of hydration. Cosmetics formulations usually contain 0.025% to 0.050% hyaluronic acid to give a smooth and viscous feel. It helps to reduce problems like sagging skin, fine lines or wrinkles on skin and also reduces extreme dryness [58].

Collagen: Collagen used in cosmetics is mainly derived from marine resources like sponges, squids and special species of fishes. Collagen forms the structural protein of the skin and is used in many cosmetics products because of its biocompatibility and water retention ability. It is used as a natural moisturizer and humectant for the skin. Antiaging and anti-wrinkling products mainly contain collagen. The film-forming ability of collagen can be boosted by other biopolymers like chitosan, hyaluronic acid, silk fibrion, etc. Hydrolyzed collagen exhibits excellent antioxidant property.

Ulvan: It is a water-soluble sulfated polysaccharide of algal origin (Figure 11.3(d)). It is used in cosmetics products owing to its amphiphilic and edible nature. It is used as an emulsifying and stabilizing agent in colloidal cosmetics solutions. It exhibits gelling properties in a pH range of 7.5 to 8.0 with various divalent cations like Ca^{2+}, Cu^{2+} and Zn^{2+}. It is extremely stable at high temperatures like 180°C. The rheological and self-aggregation behavior of ulvan make them unique raw materials for cosmetics formulation. They are used as perfuming agents in body cream milk products.

11.9 NATURAL SURFACTANTS USED IN COSMETICS

Natural surfactants are derived from natural products and are biodegradable, biocompatible, ecofriendly and safe to be used. Natural surfactants are gaining attention in cosmetics due to growing demand for animal-friendly products and concern for environment conservation. Natural surfactants are categorized as follows:

Biosurfactants: Surfactants that are produced from microorganisms are known as biosurfactants. They are widely used due to their biodegradability and ecological acceptability and are obtained from renewable resources. Based upon the microorganisms, the substrates employed in the bioprocessing and the fermentation conditions, there is a difference in the structure of biosurfactants. The characteristics of biosurfactants are largely influenced by surface properties, emulsification and foaming ability.

Based on chemical structures various kinds of biosurfactants are glycolipids, lipopeptides, phospholipids, fatty acids and polymeric compounds [59, 60]. Glycolipids are most widely used in cosmetics due to their physico-chemical properties and biological activity. They are used as multi-functional agents in cosmetics. Glycolipids are classified as:

- *Sophorolipids*: These are originally obtained from yeast Candida species and commercially produced from *Starmerella bombicola*. They are composed of the glucose disaccharide sophorose and hydroxylated fatty acid. Sophorolipids are used as emulsifiers, wetting agents, solubilizers and foaming agents in cosmetics products. These have excellent moisturizing ability and antibacterial properties and are used in the treatment of acne, body odors and dandruff. Sophorolipids enhance collagen neosynthesis, scavenge free radicals, increase elasticity of the skin enhancing antiaging property and are used in skin lightening cosmetics products. Sophorolipids are used as humectants, lip creams, eye shadow and in compressed powder cosmetics.
- *Rhamnolipids*: *Psedomonas aeruginosa* is a pathogenic Gram-negative bacteria that produces rhamnolipids. Their structure is composed of a rhamonose moiety linked to a beta-hydroxy fatty acid chain. These are effective surfactants with excellent antimicrobial activities against fungi. Because of biological properties and surface activity rhamnolipids are used in several

health-care and cosmetics formulations. Rhamnolipids are a class of chemical compounds that are found in antiwrinkle and antiaging cosmetics products. Certain commercial skin care products based on these compounds have been launched. Good skin compatibility and non-irritancy of rhamnolipids make them excellent cosmetics ingredients.

- *Mannosylerythritol lipids*: These surfactants are primarily produced from Pseudozyma species. The structural composition contains 4-O-β-D-mannopyranosyl-meso-erythritol as the hydrophilic and fatty acids as the hydrophobic moieties. These biosurfactants are used in cosmetics because of their excellent surface activity. They also possess antimicrobial activity and are non-irritants to the skin. They are widely used in cosmetics skin care products especially in antiwrinkle products to prevent roughness of the skin.

11.10 COSMETICS MARKET

The cosmetics market has grown largely over the years globally. The cosmetics market was valued at $380.2 billion in 2019 and is expected to grow to $463.5 billion by 2027 with a 5.3% compound annual growth rate (CAGR). Currently, cosmetics have become an essential part in the daily lifestyle of individuals. Apart from women, men too are using cosmetics in their daily routine, leading to an increase in the global cosmetics market.

Manufacturers all over the world have changed their marketing strategy for cosmetics products understanding the growing awareness among consumers for eco-friendly cosmetics products. There is special emphasis on the marketing strategy of natural cosmetics wherein an explanation of the ingredients is provided while launching a new product in the market. Also, appealing packaging of cosmetics products also impacts the cosmetics market as many consumers prefer to carry easy to handle packing. Moreover, synthetic cosmetics products like shampoos, conditioners and serum containing several chemicals have severe side effects such as hair fall, roughness. Therefore, replacing these products with natural cosmetics has also led to an increase in the demand for natural cosmetics.

The market across Europe, North America, Asia Pacific and The Latin America, Middle East and Africa (LAMEA) is studied on the basis of category, gender and distribution channel. Figure 11.4 represents the market on the basis of category for mainly skin care products, hair care products, deodorants and fragrance and finally make-up and color cosmetics. In 2019, skin and sun care products constituted the major category; however, the deodorants & fragrances category is expected to grow highly in the forecast period.

Figure 11.5 represents the cosmetics market by gender divided into men, women and unisex. In 2019, cosmetics for women formed the major category of the market and the market is expected to grow robustly in the forecast period.

Figure 11.6 represents the cosmetics market on the basis of distribution channel, which is categorized into supermarkets, specialty stores, pharmacies, online sales channels and others. In 2019, supermarkets formed the major category in the cosmetics market and it is projected to be the dominant category even during the forecast period.

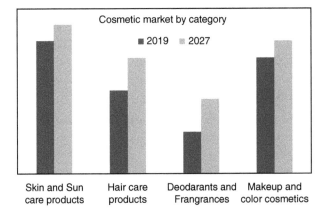

FIGURE 11.4   The skin and sun care products segment dominates the global market and is expected to retain its dominance throughout the forecast period.

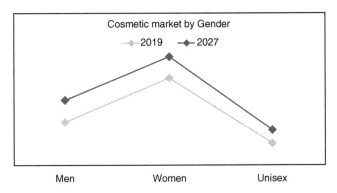

FIGURE 11.5   The unisex segment is expected to grow at the highest CAGR of 6.6% during the forecast period.

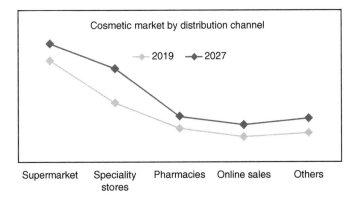

FIGURE 11.6   The online sales channel segment is expected to grow at the highest CAGR of 7.6% during the forecast period.

Some of the major key players in the cosmetics market are Avon Products Inc., Kao Corporation, L'Oreal S.A., Oriflame Cosmetics S.A., Revlon, Inc., Shiseido Company Limited, Skin Food Co., Ltd., The Estee Lauder Companies Inc., and The Procter & Gamble Company and Unilever Plc.

11.11 CONCLUSION

Cosmetics sales of natural and organic products are growing at the fastest rate in the industry. Consumers are increasingly looking for environmentally friendly products. Today, cosmetics are more "ecofriendly". A process of selecting raw materials for testing and evaluating the effectiveness and toxicity of the test creates a greater impact on cosmetics products, which makes consumers search for this type of product. Natural and organic products must meet the guiding principle and ethics as recognized by the regulatory agencies, but there is no synchronization between them. In addition to establishing packaging processes, these standards propose a sustainable extraction process and allowable extracting methods. It is a challenge to formulate natural or organic cosmetics to ensure stability, safety and effectiveness. There are many natural products that are used in cosmetics products that perform a biological function besides being toxicologically evaluated. In general, the use of plants and herbs results in cosmetics products with a greater level of sustainability, which allows companies to gain a greater share of the market. It should be noted that many cosmetics products contain natural products that perform specific biological functions, but their efficacy and toxicological properties should be evaluated. Formulated products must be biodegradable for their environmental profile to be considered sustainable and preferred by the environment.

REFERENCES

1. Aziz, A. A., Taher, Z. M., Muda, R., & Aziz, R., Cosmeceuticals and natural cosmetics, in: *Recent trends in research into Malaysian medicinal plants research*, pp. 126–175, Penerbit UTM Press, 2017. Edited by Rosnani Hasham
2. Soni, M. G., Taylor S. L., Greenberg N. A., & Burdock G. A., Evaluation of the health aspects of methylparaben: a review of the published literature. *Food and Chemical Toxicology*, 40(10), 1335–1373, 2002.
3. Elder, R. L., Final report on the safety assessment of methylparaben, ethylparaben, propylparaben and butylparaben. *Journal of the American College of Toxicology*, 3(5), 147–209, 1984.
4. Sakamoto, K., Lochhead, R., Maibach, H., & Yamashita, Y. (Eds.), *Cosmetic science and technology: theoretical principles and applications*, Elsevier, 2017.
5. Fonseca-Santos, B., Corrêa, M. A., & Chorilli, M., Sustainability, natural and organic cosmetics: consumer, products, efficacy, toxicological and regulatory considerations. *Brazilian Journal of Pharmaceutical Sciences*, 51(1), 17–26, 2015.
6. Starley, I. F., Mohammed, P., Schneider, G., & Bickler, S. W., The treatment of paediatric burns sing topical papaya. *Burn*, 25(7), 636–639, 1999.
7. Eshun, K., & He, Q., Aloe vera: a valuable ingredient for the food, pharmaceutical and cosmetic industries - a review. *Critical Reviews in Food Science and Nutrition*, 44(2), 91–96, 2004.

8. Chithra, P., Sajithlal, G. B., & Chandrakasan, G., Influence of aloe vera on the glycosaminoglycans in the matrix of healing dermal wounds in rats. *Journal of Ethnopharmacology*, 59(3), 179–186, 1998.

9. Maan, A. A., Nazir, A., Khan, M. K. I., Ahmad, T., Zia, R., Murid, M., & Abrar, M., The therapeutic properties and applications of aloe vera: a review. *Journal of Herbal Medicine*, 12, 1–10, 2018.

10. Tan, S., Ke, Z., Chai, D., Miao, Y., Luo, K., & Li, W., Lycopene, polyphenols and antioxidant activities of three characteristic tomato cultivars subjected to two drying methods. *Food Chemistry*, 338, 128062, 2021.

11. Zhao, B., Liu, H., Wang, J., Liu, P., Tan, X., Ren, B., & Liu, X., Lycopene supplementation attenuates oxidative stress, neuroinflammation, and cognitive impairment in aged CD-1 mice. *Journal of Agricultural and Food Chemistry*, 66(12), 3127–3136, 2018.

12. Prasanth, M. I., Sivamaruthi, B. S., Chaiyasut, C., & Tencomnao, T., A review of the role of green tea (*Camellia sinensis*) in antiphotoaging, stress resistance, neuroprotection, and autophagy. *Nutrients*, 11(2), 474, 2019.

13. Roh, E., Kim, J. E., Kwon, J. Y., Park, J. S., Bode, A. M., Dong, Z., & Lee, K. W., Molecular mechanisms of green tea polyphenols with protective effects against skin photoaging. *Critical Reviews in Food Science and Nutrition*, 57(8), 1631–1637, 2017.

14. Saeed, M., Naveed, M., Arif, M., Kakar, M. U., Manzoor, R., Abd El-Hack, M. E., & Sun, C., Green tea (Camellia sinensis) and l-theanine: medicinal values and beneficial applications in humans—a comprehensive review. *Biomedicine & Pharmacotherapy*, 95, 1260–1275, 2017.

15. Sharma, V., Sharma, L., & Sandhu, K. S, Cucumber (*Cucumis sativus* L.), in: *antioxidants in vegetables and nuts - properties and health benefits*, pp. 333–340, Springer, 2020. Edited by Gulzar Ahmad Nayik & Amir Gull.

16. Ugwu, C., & Suru, S, Cosmetic, culinary and therapeutic uses of cucumber (Cucumis sativus L.), in: *Cucumber economic values and its cultivation and breeding*, pp. 1–9, IntechOpen, 2021. Edited by Haiping Wang.

17. Mukherjee, P. K., Nema, N. K., Maity, N., & Sarkar, B. K., Phytochemical and therapeutic potential of cucumber. *Fitoterapia*, 84, 227–236, 2013.

18. Mansuri, R., Diwan, A., Kumar, H., Dangwal, K., & Yadav, D., Potential of natural compounds as sunscreen agents. *Pharmacognosy Reviews*, 15(29), 47, 2021.

19. Sharma, S., Meher, S., Tiwari, S. P., & Bhadran, S., In vitro antimicrobial activity and phytochemical analysis of leaves, seeds, bark, flowers of *Pongamia pinnata* (Linn. Pierre) against human pathogens. *Systematic Reviews in Pharmacy*, 12(4), 234–240, 2021.

20. Li, J. N., Henning, S. M., Thames, G., Bari, O., Tran, P. T., Tseng, C. H., & Li, Z., Almond consumption increased UVB resistance in healthy Asian women. *Journal of Cosmetic Dermatology*, 20, 2975–2980, 2021.

21. Wagh, A. S., & Butle, S. R., Preliminary phytochemical analysis and in vitro anticancer activity of *Spathodea campanulata* P. Beauv. *Asian Journal of Pharmacy and Pharmacology*, 5, 37–41, 2019.

22. Padhy, G. K., Spathodea campanulata P. Beauv.—a review of its ethnomedicinal, phytochemical, and pharmacological profile. *Journal of Applied Pharmaceutical Science*, 11(12), 017–044, 2021.

23. Mzabri, I., Addi, M., & Berrichi, A., Traditional and modern uses of saffron (Crocus sativus). *Cosmetics*, 6(4), 63, 2019.

24. Rahaman, A., Kumari, A., Farooq, M. A., Zeng, X. A., Hassan, S., Khalifa, I., & Wajid, M. A., Novel extraction techniques: an effective way to retrieve the bioactive compounds from saffron (Crocus Sativus). *Food Reviews International*, 39(4), 1–29, 2021.

25. Kahraman, E., Kaykın, M., Şahin Bektay, H., & Güngör, S., Recent advances on topical application of ceramides to restore barrier function of skin. *Cosmetics*, 6(3), 52, 2019.

26. Förster, M., Bolzinger, M. A., Fessi, H., & Briançon, S., Topical delivery of cosmetics and drugs: molecular aspects of percutaneous absorption and delivery. *European Journal of Dermatology*, 19(4), 309–323, 2009.

27. Burlando, B., & Cornara, L., Honey in dermatology and skin care: a review. *Journal of Cosmetic Dermatology*, 12(4), 306–313, 2013.

28. Juliano, C., & Magrini, G. A., Methylglyoxal, the major antibacterial factor in manuka honey: an alternative to preserve natural cosmetics? *Cosmetics*, 6(1), 1–8, 2019.

29. Pootongkam, S., & Nedorost, S., Oat and wheat as contact allergens in personal care products. *Dermatitis*, 24(6), 291–295, 2013.

30. Peterson, D. M., Oat lipids. *Lipid Technology*, 14, 56–59, 2002.

31. Dimitrova, V., Kaneva, M., & Gallucci, T., Customer knowledge management in the natural cosmetics industry. *Industrial Management & Data Systems*, 109, 1155–1165 2009.

32. Verma, R. S., Padalia, R. C., Chauhan, A., Singh, A., & Yadav, A. K., Volatile constituents of essential oil and rose water of damask rose (*Rosa damascena* Mill.) cultivars from North Indian hills. *Natural Product Research*, 25(17), 1577–1584, 2011.

33. Abdul-Mumeen, I., Beauty, D., & Adam, A., Shea butter extraction technologies: current status and future perspective. *African Journal of Biochemistry Research*, 13(2), 9–22, 2019.

34. Treesh, S. A., Saadawi, S. S., Alennabi, K. A., Aburawi, S. M., Lotfi, K., & Musa, A. S. B., Experimental study comparing burn healing effects of raw South African Shea butter and the samples from a Libyan market. *Open Veterinary Journal*, 10(4), 431–437, 2020.

35. Miwa, T. K., Structural determination and uses of jojoba oil. *Journal of the American Oil Chemists' Society*, 61(2), 407–410, 1984.

36. Sánchez, M., Avhad, M. R., Marchetti, J. M., Martínez, M., & Aracil, J., Jojoba oil: a state of the art review and future prospects. *Energy Conversion and Management*, 129, 293–304, 2016.

37. Singh, S., Lohani, A., Mishra, A. K., & Verma, A., Formulation and evaluation of carrot seed oil-based cosmetic emulsions. *Journal of Cosmetic and Laser Therapy*, 21(2), 99–107, 2019.

38. Gediya, S. K., Mistry, R. B., Patel, U. K., Blessy, M., & Jain, H. N., Herbal plants: used as a cosmetics. *Journal of Natural Product and Plant Resources*, 1(1), 24–32, 2011.

39. Athar, M., & Nasir, S. M., Taxonomic perspective of plant species yielding vegetable oils used in cosmetics and skin care products. *African Journal of Biotechnology*, 4(1), 36–44, 2005.

40. Cao, Y., Zhao, L., Ying, Y., Kong, X., Hua, Y., & Chen, Y., The characterization of soybean oil body integral oleosin isoforms and the effects of alkaline pH on them. *Food Chemistry*, 177, 288–294, 2015.

41. Koo, J. H., Lee, I., Yun, S. K., Kim, H. U., Park, B. H., & Park, J. W., Saponified evening primrose oil reduces melanogenesis in B16 melanoma cells and reduces UV-induced skin pigmentation in humans. *Lipids*, 45(5), 401–407, 2010.

42. Muggli, R., Systemic evening primrose oil improves the biophysical skin parameters of healthy adults. *International Journal of Cosmetic Science*, 27(4), 243–249, 2005.

43. Burnett, C. L., Bergfeld, W. F., Belsito, D. V., Klaassen, C. D., Marks, J. G., Shank, R. C., & Andersen, F. A., Final report on the safety assessment of *Cocos nucifera* (coconut) oil and related ingredients. *International Journal of Toxicology*, 30, 5–16, 2011.

44. Gorini, I., Iorio, S., Ciliberti, R., Licata, M., & Armocida, G., Olive oil in pharmacological and cosmetic traditions. *Journal of Cosmetic Dermatology*, 18(5), 1575–1579, 2019.
45. Akgün, N. A., Separation of squalene from olive oil deodorizer distillate using supercritical fluids. *European Journal of Lipid Science and Technology*, 113(12), 1558–1565, 2011.
46. Woolf, A., Wong, M., Eyres, L., McGhie, T., Lund, C., Olsson, S., & Requejo-Jackman, C., Avocado oil, in: *Gourmet and health-promoting specialty oils*, pp. 73–125, AOCS Press, 2009. Edited by Robert A. Moreau and Afaf Kamal-Eldin
47. Cervantes-Paz, B., & Yahia, E. M., Avocado oil: production and market demand, bioactive components, implications in health, and tendencies and potential uses. *Comprehensive Reviews in Food Science and Food Safety*, 20, 4120–4158, 2021.
48. Cirlini, M., Caligiani, A., Palla, G., De Ascentiis, A., & Tortini, P., Stability studies of ozonized sunflower oil and enriched cosmetics with a dedicated peroxide value determination. *Ozone: Science & Engineering*, 34(4), 293–299, 2012.
49. Ahmed, I. A., Mikail, M. A., Zamakshshari, N., & Abdullah, A. S. H., Natural anti-aging skincare: role and potential. *Biogerontology*, 21(3), 293–310, 2020.
50. Lupo, M. P., Antioxidants and vitamins in cosmetics. *Clinics in Dermatology*, 19(4), 467–473, 2001.
51. Kornhauser, A., Coelho, S. G., & Hearing, V. J., Applications of hydroxy acids: classification, mechanisms, and photoactivity. *Clinical, Cosmetic and Investigational Dermatology: CCID*, 3, 135, 2010.
52. Espinosa-Leal, C. A., & Garcia-Lara, S., Current methods for the discovery of new active ingredients from natural products for cosmeceutical applications. *Planta médica*, 85(07), 535–551, 2019.
53. Patil, A., & Ferritto, M. S., Polymers for personal care and cosmetics: overview. *Polymers for Personal Care and Cosmetics*, 3–11, 2013.
54. Olatunji, O. (Ed.), *Natural polymers: industry techniques and applications*, Springer, 2015.
55. Hasan, N., Application of bacterial cellulose (BC) in natural facial scrub. *International Journal on Advanced Science, Engineering and Information Technology*, 2, 272–275, 2012.
56. Czaja, W. et al., Microbial cellulose—the natural power to heal wounds. *Biomaterials*, 27, 145–151, 2006.
57. Amnuaikit, T. et al., Effects of a cellulose mask synthesized by a bacterium on facial skin characteristics and user satisfaction. *Medical Devices: Evidence and Research*, 4, 77–81 2011.
58. Neudecker, B. A., H. I. Maibach, & Stern, R., Hyaluronan: the natural skin moisturizer, in: *Cosmeceuticals and active cosmetics: drugs versus cosmetics, cosmetics science and technology series*, P. Elsner and H. I. Maibach (Eds.), Taylor & Francis, 2005.
59. De, S., Malik, S., Ghosh, A., Saha, R., & Saha, B., A review on natural surfactants. *RSC Advances*, 5(81), 65757–65767, 2015.
60. Lourith, N., & Kanlayavattanakul, M., Natural surfactants used in cosmetics: glycolipids. *International Journal of Cosmetic Science*, 31(4), 255–261, 2009.

12 Role of Upcycled Foods for Global Food Security

Misha Patel

12.1 INTRODUCTION

Food upcycling is basically defined as making food products by using the ingredients that otherwise would not have been consumed by humans. Upcycled foods are sourced from the waste streams of valid food industries and manufactured at certifiable sites and such products have great economic and environmental advantages [1]. The upcycled food products are made from ingredients that would otherwise have been wasted. Converting leftovers and surplus foods and recycling them into new ingredients and food products are some other examples. An alternative form of food upcycling is to modify the unused or leftover food to energy by either chemical catalysis or enzymatic process. Scientists have invented methods to break down or modify the food particles and convert them into reusable biofuel by using the pressure cooking method. Through this process, electricity and heat are produced from methane obtained from the pressure cooking.

Food wastage is a global concern, and it occurs at all the stages of the food distribution network such as procurement, manufacturing, processing, storage, distribution, etc. About one-third of the food produced is discarded each year and ends up in landfills [2]. It leads to food insecurity, financial loss, and other adverse environmental impacts. At a global scale, around 1.3 billion tons of edible food gets wasted each year. Therefore, if we strive to decrease food wastage, it brings a golden opportunity to reduce food insecurity and feed needy people. In addition to this, the economic valuation of food wastage is estimated at $750 billion; food upcycling or food waste management can lead to financial benefits as well. Moreover, around 4.4 billion tons of $CO_{2\,eq}$ per annum are produced due to the huge amount of food waste and this contributes to global warming [3]. Thus, food waste management and food upcycling have benefits along with increased food security and economic benefits.

In upcycling a food product, a new product with an added value or a sustainable high-quality product is produced by using waste or used materials or an existing product is used in a new way to minimize the utilization of resources [4].

The process of upcycling food has a long history based on the philosophy of using all what you have. The concept behind it is to do more with less and elevate all the food produced to its highest and best use. The term 'upcycled food' was introduced in 2020 for its use in policy and research. Before the term 'upcycled food', there were terms like 'value added surplus products' or 'waste-to-value food products' used.

DOI: 10.1201/9781032636368-12

Upcycled food is nutritious, sustainable, and high-quality food prepared from the nutrients that slip through the inefficiencies of our food supply chain.

Different studies suggest that because of its connection with the term 'recycling' and its environmental benefits, consumers will pay for upcycled foods. With the right message and marketing, upcycled food products can command a higher price than conventional products in the market. It will eventually help food companies to reduce food waste and achieve greater value from products and be profitable. Consumer acceptability for upcycled foods can be compromised if the ingredients or nutrients are viewed as waste rather than resources.

12.2 LITERATURE REVIEW

12.2.1 DEFINITION

Upcycling is the process in which used materials or materials that were meant to be wasted are used to produce something of higher quality and value in their second life. The other name for upcycling is 'creative reuse'. The process of upcycling is said to be the opposite of downcycling. It represents a variety of processes by which an old product is modified and processed to get a new life or second life. In other words, upcycling refers to the materials or products that get to be repurposed in a creative manner, and whose lifespan is therefore extended [4].

12.2.2 DIFFERENCE BETWEEN UPCYCLING AND RECYCLING

The difference between upcycling and recycling is that the latter takes old or used materials or products, breaks it down, and produces a new product of inferior quality. However, upcycling takes the material and modifies it to a sustainable and high-quality new product. Upcycling basically doesn't involve breaking down; instead, it is more about sorting and reusing old products, often in innovative and cost-effective ways.

12.2.3 APPLICATION OF UPCYCLING

Upcycling has shown significant growth in various applications. It has been applied in the field of arts, music, industrial processes such as plastic, paint, electronic fabrication, etc., textile, and food. An important application of upcycling is being done in the field of art. Reusing found objects and converting them to folk art has a long history [5]. The process of upcycling has also been applied in different industries such as plastic and paint. Recent and future technologies are implemented for upcycling and reducing waste and conserving resources. Upcycling has also been applied in the textile industry where designers have started using industrial textile waste and used clothes to create new and high-quality clothes. Now, the field has its vast application in the food industry as well. Leftover or wasted materials are used to produce new, sustainable, and high-quality upcycled food products.

12.2.4 FOOD LOSS AND FOOD WASTE

Food loss comprises the largest category and includes any eatable foods that wouldn't have made it to human consumption. It can happen at any stage of food processing

TABLE 12.1

A tabular representation showing the continental bifurcation of total food loss per year alongside the wastage channels' food loss per year [7]

Food loss and waste per continent per year	Total	At the production and retail stages	By consumers
Europe	280 kg (617 lb)	190 kg (419 lb)	90 kg (198 lb)
North America and Oceania	295 kg (650 lb)	185 kg (408 lb)	110 kg (243 lb)
Industrialized Asia	240 kg (529 lb)	160 kg (353 lb)	80 kg (176 lb)
Sub-Saharan Africa	160 kg (353 lb)	155 kg (342 lb)	5 kg (11 lb)
North Africa, West and Central Asia	215 kg (474 lb)	180 kg (397 lb)	35 kg (77 lb)
South and Southeast Asia	125 kg (276 lb)	110 kg (243 lb)	15 kg (33 lb)
Latin America	225 kg (496 lb)	200 kg (441 lb)	25 kg (55 lb)

such as manufacturing, processing, packaging, storage, or distribution [6]. It also includes food that remains uneaten at home or stores. In addition to this, it also incorporates crops that are left in the fields, food products that are spoiled or lost during transit, and other food products that fail to reach the stores. Therefore, some amount of food loss occurs at almost every step of the food supply chain.

On the other side, food waste comprises a specific portion of food loss. When food is discarded and rejected by retailers due to inappropriate shape or color and is tossed as plate waste by consumers, then it is categorized as food waste. The food that is left uneaten at a restaurant also falls under the header of food waste along with the food that gets spoiled at home such as sour milk that is poured down the drain. Around 43% of food is wasted from households, 40% from restaurants, grocery stores, and food service companies, 16% is farm waste, and about 2% is a waste stream from the manufacturing plants.

At the global level, around 1,540 kg of food loss occurs across all the continents with North America being the highest of all, as shown in Table 12.1. Therefore, food is lost or wasted for a huge variety of reasons such as climate change, adverse weather, processing problems, inappropriate transportation conditions, overproduction, and unstable markets. Other than this, overbuying, poor management, unhealthy and unmonitored cooking practices, and confusion over nutritional labels can cause food wastage at domestic level [6].

12.2.5 Upcycled Food

The term 'upcycled food' has now been properly defined for policy and research purposes. Upcycled foods are produced from the materials that would have otherwise gone waste. For example, using the pulp of fruits or vegetables, after they are squeezed to make juice, to snacks like veggie chips. Moreover, a lot of fruits and vegetables are wasted by calling them 'ugly' because of their wrong shape, color, or size. Different companies collect these 'ugly' vegetables and convert them into chutneys, sauces, ketchup, and relishes [4]. Another astonishing example of upcycled food is chewing gum. A chewing gum, also named as bubblegum, is formulated to be thrown away after the flavor has disappeared. Chewing gums can be modified into a sustainable compound for the rubber and plastic industry where it is transformed into mugs, rulers, and even shoes.

12.2.6 BENEFITS OF FOOD UPCYCLING

12.2.6.1 Economic Benefits

According to the USDA, around 30%–40% of all the food grown is not used or consumed in the USA. This equals an estimated loss of $161 billion. Food wastage is a huge concern and a complex problem [8]. However, engineers and food processors are working to resolve this problem through food upcycling. Around $1 trillion is the estimated annual loss because of food wastage. However, upcycling the food that isn't used can help the global economy to flourish. In fact, upcycling is already gaining momentum. In 2019, the upcycled food industry was calculated to be worth more than $46 billion, with an anticipated 5% growth rate [8]. As per the reports of the USDA, through upcycling and reducing food waste, households can save about $370 per person every year. The US EPA (Environmental Protection Agency) plays an integral role in setting up the Food Recovery Hierarchy for reducing food loss and food waste, as shown in Figure 12.1.

FIGURE 12.1 Food recovery hierarchy by the US Environmental Protection Agency [9].

12.2.6.2 Environmental Benefits

Many waste food products have been claimed by food industries and are modified and reused where once upon a time they ended up in a landfill [10]. Therefore, food upcycling has reduced the amount of what goes into a landfill and involves minimalistic use of resources from natural sources. Upcycling existing resources means minimal use of any raw material and natural resources in the production process. Moreover, food wastage contributes to global warming and by reducing waste, we can reduce the stress on the environment [10]. Eventually, it will reduce the pollution related to air, water, land, and greenhouse gas emissions and in natural resource conservation.

12.2.6.3 Increase in Food Security

Food security is known as having accessibility to reliable food in a good quantity that is both affordable and nutritious [11]. According to the United Nations, about 690 million people face hunger every year. Food wastage is severely detrimental to a world where many people become a victim of hunger every year. Among those suffering, children are the ones most affected with an estimated 14 million children under five years of age affected by hunger by year 2019 [12]. In most of the developing and low-income countries, there is a huge quantity of food harvested, so much of it is even lost before it reaches the consumers [13]. Upcycling leftovers to produce new food products will reduce food insecurity in such countries by helping feed people.

12.2.7 Consumer Characteristics and Beliefs toward Sustainability

Consumers' sociodemographic characters such as age, gender, income, education, socioeconomic status, and attitude toward sustainability and food waste management can have a major effect on the acceptability of upcycled foods. For instance, women are relatively more conscious about health than men and do not take risks when it comes to the overall well-being of the family. Therefore, they may not be in favor of upcycled food as they are more likely to perceive these foods as unhealthy options [14]. Moreover, income and education level also affect the acceptability of upcycled foods. Those with higher education and sustainable income are more likely to choose upcycled food products than those with lower socioeconomic status.

In addition to this, the likelihood to accept upcycled foods is also influenced by the beliefs and attitudes of the consumers. For instance, the public's preference for organically grown foods which are also environmentally friendly can direct them to buy upcycled foods [14]. Moreover, people who believe that the production of upcycled foods is a way of food waste management and are sensitive to food waste issues are likely to have greater acceptability toward upcycled food products.

12.2.8 Quality Attributes of Upcycled Foods

Food quality of any food product affects the consumer acceptability toward that product. Food quality is determined by attributes like taste, flavor, freshness, color, texture, appearance, shape, price, brand, etc. Right labeling, certification, and product brand affect the consumer acceptability toward upcycled products. Price is one of the important food quality cues that has an integral role to play in consumer choices. A higher price for upcycled foods might decrease the purchasing power of the consumers, but, on the other hand, high prices indicate their high quality as well. However, despite this, high prices of such eco-friendly foods affect their affordability for consumers, especially those who belong to a low socioeconomic group [14]. However, the right message and advertisement can help increase people's willingness to pay for these foods.

Food quality attributes also have a great influence on the acceptability of upcycled foods. For example, taste is one of the major attributes that impacts public attitude toward a product. The organoleptic properties like taste, texture, appearance, etc., of an upcycled food may be different than the conventional product in the market, so these upcycled products may be more acceptable or unacceptable than their conventional counterparts. Therefore, people are more likely to choose those with better food quality attributes [14].

12.3 DISCUSSION

12.3.1 GLOBAL ASSESSMENT

One-third of worldwide food production is wasted, and 'food waste' occurs when perishable food spoils while 'food losses' occur due to improper handling, storage, or distribution of foods. Leafy greens and tropical fruits report an annual loss of 40% and it costs the $200–$300 billion annually. The WTO and FAO aim to reduce food waste and food insecurity on a global level. Food waste contributes 8% of GHG emissions and 25% of landfill volume to the world. Water, power, and land are wasted by food. For the bioeconomy to be sustainable, by-products and side streams from food and agriculture must be valorized or upcycled. The approach encourages resource efficiency and waste minimization through sustainability-focused innovation. Despite the literature's focus on the technical side, there is a small study on the sociological and economic repercussions from the consumer's perspective [15].

12.3.2 METHODS FOR FOOD UPCYCLING

Chemical recycling and upcycling refer to the process of transforming polymers or larger units into their component monomers, small units, or chemical precursors for use in the production of new, higher-value products. These methods are the most efficient in recouping costs associated with post-consumer polymer goods, although they are often more costly than conventional recycling and disposal practices. Catalysts can speed up and improve the selectivity of the chemical recycling and upcycling of polymers [16].

The Supplant Company launches sugars from fiber. Straw, stalks, and cobs of corn, wheat, and rice are utilized to make substitute sugars using a unique technique. In baked pastries and other delights, substitute sugar mimics cane sugar's texture, caramelization, and flavor. Because it's made of fiber, the substance possesses prebiotic, low-calorie, and low-glycemic properties. The Supplant Company is launching and developing supplant sugars from fiber in the US market. Restaurants, chefs, and brands are commercial partners [17].

RIND, a creator of recycled dry fruit snacks, raised $6.1 million in Series A funding (VSV). RIND debuted a chewy 'skin-on super fruit' snack in 2018. The company expects its "once sleepy category" to grow at 8.4% CAGR through 2025. The fruit snack market is primed for change, says RIND CEO Matt Weiss. VSV shares our desire to awaken this sector, allowing customers to chew better and achieve more through the 'power of the peel'. This contribution will help us grow quickly, reducing food waste and child poverty. After rescuing 120,000 pounds of edible peels from landfills in 2020, RIND plans to rescue more than a million pounds by 2025 [17].

The Dohler Group, a global food ingredient organization, manufactures technology-driven natural ingredients and ingredient systems from both fresh food and waste streams and is headquartered in Darmstadt, Germany. The Dohler Group, being a decade away from its second century, developed a unique ingredient for vegan consumers called 'Powdered Aquafaba' from chickpea water, which is usually discarded post soaking and blanching. Chickpea water, also called aquafaba water, is spray dried to make the dry version and serves as an excellent egg replacer and vegan alternative for the baking, confectionery industry, dairy, and plant-based industries. Dohler is a global leader of fruit concentrates and purees, and its orange juice concentrate development results in a waste stream of indigestible citrus fruit, which is called fruit fiber, which when dried gives a powdered ingredient 'Citrus Fiber' which has applications in plant meat and dairy, baking, beverage industries as a texturing agent.

The growing population needs new protein sources. This chapter also examines mushroom mycelia in industrial side streams. Shake flasks containing agro-industrial streams grew submerged mushrooms and analyzed crude protein, ash, lipid, fatty acid, and amino acid profiles. UV-B irradiation converts biomass ergosterol to vitamin D2. By measuring ergosterol, mycelium in total biomass can be quantified. Basidiomycetes grew quickly on apple pomace in agro-industrial side streams, and it also recycled apple pomace's nutrients. Mushroom-substrate biomass had 21% dry protein after four days. Ergosterol irradiated with UV-B produced $115\,g\,L^{-1}$ vitamin D2 and it detected fungus. Also studied were lipids, ash, and carbs [18].

Environmentalists usually criticize animal food, but cattle can improve food security by converting food waste and processing waste into nutrient-rich animal-source meals. There is no study till date that states the amount of LCF (low concentrate mixed forage) to be fed to different animals (cattle, pigs, etc.) to maximize their nutrition. LCF is given to livestock systems with the most protein which comprises pigs, hens, broilers, dairy cattle, and beef cattle and it is popular in the European region where there are three production stages required (low, mid, and high) for different nutrients. LCF availability is affected by food production, processing, waste, and grassland output. Best EU LCF conversion suggests $31\,g$ of animal protein per day. Ideal conversion requires many livestock systems and outputs. Egg-laying chickens and milk-producing cows can use low-quality LCF due to high conversion efficiency. Animal protein supply fell to $26\,g\,cap^{-1}\,d^{-1}$ while using just traditional, highly productive animals, a 16% decrease. The quality, amount, and efficiency with which cattle utilized LCF, particularly grass resources, determined the protein supply from animals exclusively fed LCF [19].

A study examined the use of almond skins, an almond milk manufacturing by-product, which is nutritionally superior and can be used for making functional cookies. Almond skins make the texture of the biscuits lighter, decrease $L*$ and $b*$, and raise $a*$ [20].

12.3.3 LIMITATIONS

Consumer's approval of waste-to-value food products depends on the consumer's perception, their surroundings, and the conventional food product they are used to eating. No study has identified a detrimental effect from food neophobia or food technophobia, and studies have produced conflicting results on whether gender and age affect the acceptability of upcycled foods. Consumers are more inclined to accept items when they are better educated, and food companies are able to market the upcycled foods for its environmental care and motives and food waste awareness.

12.3.4 RECOMMENDATIONS

More real-world study methods, a broader range of cultural contexts, and a greater emphasis on ideas from consumer behavior and environmental psychology are all things that can improve the quality of future research within the space of food waste and food upcycling. The shift toward intervention designs will also encourage consumer-focused research on how to develop the circular (bio) economy.

REFERENCES

[1] Rondeau, S., Stricker, S. M., Kozachenko, C., & Parizeau, K., Understanding motivations for volunteering in food insecurity and food upcycling projects. *Social Sciences*, 9(3), 27, 2020.

[2] Capone, R., Water footprint in the Mediterranean food chain: Implications of food consumption patterns and food wastage. *International Journal of Nutrition and Food Sciences*, 3(2), 26, 2014.

[3] Chaya, M., Tao, X., Green, A., & BaoGen, G., Impact of climate change on pests of rice and cassava. *CAB Reviews: Perspectives in Agriculture, Veterinary Science, Nutrition and Natural Resources*, 16(050), 1–10, 2021.

[4] Spratt, O., Suri, R., & Deutsch, J., Defining upcycled food products. *Journal of Culinary Science & Technology*, 19(6), 485–496, 2021.

[5] Hjemdahl, K. M., Innovation and reinventing by artistic practice: Upcycling, performing, and curating. *Mjesto izvedbe i stvaranje grada*, 95, 2016.

[6] Thyberg, K. L., & Tonjes, D. J., Drivers of food waste and their implications for sustainable policy development. *Resources, Conservation and Recycling*, 106, 110–123, 2015.

[7] Gustavson, J., Cederberg, C., Sonesson, U., van Otterdijk, R., & Meybeck, A., *Global Food Losses and Good Waste – Extent, Causes and Prevention*, FAO, 2011.

[8] Johnson, L. K., Dunning, R. D., Gunter, C. C., Bloom, J. D., Boyette, M. D., & Creamer, N. G., Field measurement in vegetable crops indicates need for reevaluation of on-farm food loss estimates in North America. *Agricultural Systems*, 167, 136–142, 2018.

[9] Food Recovery Hierarchy, Digital Image, United States Environmental Protection Agency. https://www.epa.gov/sustainable-management-food/food-recovery-hierarchy#about, accessed 10th Jan 2023.

[10] Bruine de Bruin, W., Rabinovich, L., Weber, K., Babboni, M., Dean, M., & Ignon, L, Public understanding of climate change terminology. *Climatic Change*, 167, 3–4, 2021.

[11] Prosekov, A. Y., & Ivanova, S. A., Food security: The challenge of the present. *Geoforum*, 91, 73–77, 2018.

[12] Wight, V., & Thampi, K., *Who are America's poor children? Examining food insecurity among children in the United States*, 2010. https://www.nccp.org/publication/who-are-americas-poor-children-examining-food-insecurity-among-children-in-the-united-states/

[13] Lee, B., & Lippert, A. Food insecurity among homeless and precariously housed children in the United States: Lessons from the past. *Demographic Research*, 45, 1115–1148, 2021.

[14] Bhatt, S., Ye, H., Deutsch, J., Ayaz, H., & Suri, R., Consumers' willingness to pay for upcycled foods. *Food Quality and Preference*, 86, 104035, 2020.

[15] Ali, N. S., Khairuddin, N. F., & Zainal Abidin, S., *Upcycling: Re-use and recreate functional interior space using waste materials*, 2013. Doi: 10.13140/2.1.2643.3603

[16] Kosloski-Oh, S. C., Wood, Z. A., Manjarrez, Y., de Los Rios, J. P., & Fieser, M. E., Catalytic methods for chemical recycling or upcycling of commercial polymers. *Materials Horizons*, 8(4), 1084–1129, 2021.

[17] Mirabella, N., Castellani, V., & Sala, S. Current options for the valorization of food manufacturing waste: A review. *Journal of Cleaner Production*, 65, 28–41, 2014.

[18] Van Hal, O., De Boer, I. J. M., Muller, A., De Vries, S., Erb, K. H., Schader, C.,... & Van Zanten, H. H. E., Upcycling food leftovers and grass resources through livestock: Impact of livestock system and productivity. *Journal of Cleaner Production*, 219, 485–496, 2019.

[19] Ahlborn, J., Stephan, A., Meckel, T., Maheshwari, G., Rühl, M., & Zorn, H., Upcycling of food industry side streams by basidiomycetes for production of a vegan protein source. *International Journal of Recycling of Organic Waste in Agriculture*, 8(1), 447–455, 2019.

[20] Pasqualone, A., Laddomada, B., Boukid, F., Angelis, D. D., & Summo, C., Use of almond skins to improve nutritional and functional properties of biscuits: An example of upcycling. *Foods*, 9(11), 1705, 2020.

Index

For Product Safety Concerns and Information please contact our EU
representative GPSR@taylorandfrancis.com
Taylor & Francis Verlag GmbH, Kaufingerstraße 24, 80331 München, Germany

www.ingramcontent.com/pod-product-compliance
Ingram Content Group UK Ltd.
Pitfield, Milton Keynes, MK11 3LW, UK
UKHW021117180425

457613UK00005B/132

9 781032 538761